AF411458

Anti-Aging Nutrients with Health Beneficial Effects

Anti-Aging Nutrients with Health Beneficial Effects

Editor

Yoshinori Katakura

MDPI • Basel • Beijing • Wuhan • Barcelona • Belgrade • Manchester • Tokyo • Cluj • Tianjin

Editor
Yoshinori Katakura
Kyushu University
Japan

Editorial Office
MDPI
St. Alban-Anlage 66
4052 Basel, Switzerland

This is a reprint of articles from the Special Issue published online in the open access journal *Nutrients* (ISSN 2072-6643) (available at: https://www.mdpi.com/journal/nutrients/special_issues/aging_nutrients_health).

For citation purposes, cite each article independently as indicated on the article page online and as indicated below:

LastName, A.A.; LastName, B.B.; LastName, C.C. Article Title. *Journal Name* **Year**, *Volume Number*, Page Range.

ISBN 978-3-0365-5069-5 (Hbk)
ISBN 978-3-0365-5070-1 (PDF)

Cover image courtesy of Yoshinori Katakura

© 2022 by the authors. Articles in this book are Open Access and distributed under the Creative Commons Attribution (CC BY) license, which allows users to download, copy and build upon published articles, as long as the author and publisher are properly credited, which ensures maximum dissemination and a wider impact of our publications.

The book as a whole is distributed by MDPI under the terms and conditions of the Creative Commons license CC BY-NC-ND.

Contents

Preface to "Anti-Aging Nutrients with Health Beneficial Effects"

In this Special Issue, I attempted to clarify the possibility of anti-aging via foods and food components by clarifying the functionality of foods and food components that have health-promoting effects. We believe that the results of the paper published in this Special Issue may indicate one possibility for confronting aging societies in the future. We would like to express our gratitude to the authors who contributed to this Special Issue, and we hope that the information obtained here will be fully utilized for the future development of preventive medicine.

Yoshinori Katakura
Editor

 *nutrients*

Article

A Mechanistic Study of the Antiaging Effect of Raw-Milk Cheese Extracts

Guillaume Cardin [1,*], Cyril Poupet [1], Muriel Bonnet [1], Philippe Veisseire [1], Isabelle Ripoche [2], Pierre Chalard [2], Anne Chauder [3], Etienne Saunier [3], Julien Priam [3], Stéphanie Bornes [1] and Laurent Rios [1]

[1] Université Clermont Auvergne, INRAE, VetAgro Sup, UMRF, F-15000 Aurillac, France; cyril.poupet@uca.fr (C.P.); muriel.bonnet@uca.fr (M.B.); philippe.veisseire@uca.fr (P.V.); stephanie.bornes@uca.fr (S.B.); laurent.rios@vetagro-sup.fr (L.R.)

[2] Université Clermont Auvergne, CNRS, Clermont Auvergne INP, ICCF, F-63000 Clermont-Ferrand, France; isabelle.ripoche@sigma-clermont.fr (I.R.); pierre.chalard@sigma-clermont.fr (P.C.)

[3] Dômes Pharma, ZAC de Champ Lamet, 3 Rue Andrée Citroën, F-63284 Pont-du-Château, France; a.chauder@domespharma.com (A.C.); e.saunier@domespharma.com (E.S.); j.priam@domespharma.com (J.P.)

[*] Correspondence: guillaume.cardin@vetagro-sup.fr; Tel.: +33-4-73-98-70-35

Abstract: Many studies have highlighted the relationship between food and health status, with the aim of improving both disease prevention and life expectancy. Among the different food groups, fermented foods a have huge microbial biodiversity, making them an interesting source of metabolites that could exhibit health benefits. Our previous study highlighted the capacity of raw goat milk cheese, and some of the extracts recovered by the means of chemical fractionation, to increase the longevity of the nematode *Caenorhabditis elegans*. In this article, we pursued the investigation with a view toward understanding the biological mechanisms involved in this phenomenon. Using mutant nematode strains, we evaluated the implication of the insulin-like DAF-2/DAF-16 and the p38 MAPK pathways in the phenomenon of increased longevity and oxidative-stress resistance mechanisms. Our results demonstrated that freeze-dried raw goat milk cheese, and its extracts, induced the activation of the DAF-2/DAF-16 pathway, increasing longevity. Concerning oxidative-stress resistance, all the extracts increased the survival of the worms, but no evidence of the implication of both of the pathways was highlighted, except for the cheese-lipid extract that did seem to require both pathways to improve the survival rate. Simultaneously, the cheese-lipid extract and the dried extract W70, obtained with water, were able to reduce the reactive oxygen species (ROS) production in human leukocytes. This result is in good correlation with the results obtained with the nematode.

Keywords: raw-milk cheese; *Caenorhabditis elegans*; longevity; oxidative stress; DAF-16; p38 MAPK

Citation: Cardin, G.; Poupet, C.; Bonnet, M.; Veisseire, P.; Ripoche, I.; Chalard, P.; Chauder, A.; Saunier, E.; Priam, J.; Bornes, S.; et al. A Mechanistic Study of the Antiaging Effect of Raw-Milk Cheese Extracts. *Nutrients* **2021**, *13*, 897. https://doi.org/10.3390/nu13030897

Academic Editor: Yoshinori Katakura

Received: 4 February 2021
Accepted: 7 March 2021
Published: 10 March 2021

Publisher's Note: MDPI stays neutral with regard to jurisdictional claims in published maps and institutional affiliations.

Copyright: © 2021 by the authors. Licensee MDPI, Basel, Switzerland. This article is an open access article distributed under the terms and conditions of the Creative Commons Attribution (CC BY) license (https://creativecommons.org/licenses/by/4.0/).

1. Introduction

In the last few years, many studies have highlighted the relationship between diet and health status with the aim of improving both disease prevention and life expectancy. Among the different food groups, fermented foods, which represent an important part of our diet, have a huge microbial biodiversity that makes them an interesting source of metabolites that could exhibit health benefits. Recent studies have demonstrated that fermented foods may exhibit various beneficial effects on health, such as a cardiovascular protective effect [1,2] or an antiproliferative activity in the field of cancer prevention [3]. In a previous study [4], we focused our interest on one particular fermented food, raw goat milk cheese. We highlighted the development of a new methodology allowing us to fractionate cheese, using chemical fractionation, and to highlight the effects of the whole cheese, as well as the resulting extracts, on the longevity of the nematode *Caenorhabditis elegans*. We demonstrated a pro-longevity effect of the freeze-dried cheese and of some of the extracts (a lipophilic extract, named cheese-lipid extract, and three different hydrophilic extracts, named W40, WF and W70) on an in vivo model, using the wild-type *C. elegans* N2 strain.

The freeze-dried cheese presented the ability to increase the maximum lifespan by 63%. The cheese-lipid extract increased longevity up to 37%. The three hydrophilic extracts also increased the maximum lifespan, between 13% and 73%, depending on the concentration. Another notable particularity revealed by this assay was the percentage of the population remaining alive on the extracts after all the nematodes in the control group had died: freeze-dried cheese (13%), cheese-lipid extract (up to 5%) and hydrophilic extracts (between 4% and 16%).

In the present study, we pursued the investigation of the health benefits of raw goat milk cheese, and its metabolites, using the methodology developed in our previous study [4]. In order to better understand the mechanisms of the cheese extracts in increasing longevity, an exploration of the signaling pathways involved was performed using an in vivo *C. elegans* model. This model has shown to be efficient in many studies that evaluated the health impact of some plant extracts and microorganisms [5,6]. *C. elegans* was chosen for our studies because of its similarities with humans concerning the physiology of the intestinal cells [7] and the homology of many signaling pathways [8], that make it a relevant model for mechanistic studies. Many mutant strains are available, which allow us to characterize the implication of these signaling pathways in the effects of the extracts. In *C. elegans*, the aging process is modulated by highly conserved signaling pathways, such as the DAF-2/DAF-16 [8,9]. The transcription factor DAF-16 has been demonstrated to regulate downstream genes that influence longevity [10,11] and may be involved in the mechanism of action of the extracts in increasing life expectancy. The implication of this pathway will be investigated by using a mutant strain that did not express the transcription factor DAF-16.

With aging, age-related affections become ever more prevalent. Many processes are involved in these affections, such as the oxidative process in which the reactive oxygen species (ROS) are implicated. These compounds cause damage to lipids, proteins and DNA, which results in the death of the cell [9] and, in the end, of the organism. Moreover, during aging, the defense mechanisms of the worms are weakened [9], leaving the nematode more sensitive to the oxidative stress, which can be combined with an excessive production of ROS [12]. Consequently, the capacity of the extracts to improve the nematode survival rate on an oxidative medium was investigated in parallel with the longevity test. In keeping with the exploration of the effects of the extracts on oxidative stress, an in vitro study was conducted to measure the ability of the extracts to reduce the ROS production in human leukocytes and to correlate the results obtained with those in the nematode.

As cheese is a fermented food prepared from milk, a final investigation was performed in order to estimate the impact of the milk fermentation on the production of bioactive metabolites. To do so, the same fractionation as described in our previous work [4] was performed on the nonfermented milk in order to obtain raw goat milk extracts that were then evaluated with a longevity and survival assay on the oxidative medium.

2. Materials and Methods

2.1. Milk and Cheese Samples

Raw goat milk, freshly collected, and raw goat cheese (ripened for 20 days) were taken from a local producer (Chèvrerie des Oliviers, Saint-Georges sur Allier, France). The milk was concentrated under vacuum and freeze-dried (FreeZone Triad Freeze Dryer, Labconco Corporation, Kansas City, MS, USA). The resulting solid was crushed with mortar and pestle and the freeze-dried milk (FDM) was kept in a waterproof container at 4 °C. The cheese was cut into small slices, freeze-dried and crushed with mortar and pestle. The freeze-dried cheese (FDC) was kept in a waterproof container at 4 °C [4].

2.2. Reagents and Solvents

Five-fluoro-2′-deoxyuridine (FUdR), KH_2PO_4, amphotericin B (250 µg/mL), NaOH, agarose, cholesterol, $CaCl_2$, NaCl, EDTA, RPMI, Na_2HPO_4, $MgSO_4$, potassium phosphate buffer, NH_4Cl, $NaHCO_3$, phorbol myristate acetate (PMA), fetal bovine serum (FBS), gentamicin, glutamine and resazurin were bought from Sigma Aldrich (Saint Louis, MO, USA). lysogeny broth (LB, Miller's Modification), peptone and agar were obtained from Conda (Madrid, Spain). Yeast extract was obtained from Fisher Scientific (Hampton, VA, USA). Dihydrorhodamine 123 (DHR 123) was purchased from Cayman Chemical Company (Ann Arbor, MI, USA). TRIzol was acquired from Ambion by life technologies (Carlsbad, CA, USA). The High-Capacity cDNA Archive kit was obtained from Applied Biosystems (Foster City, CA, USA). Rotor-Gene SYBR Green Mix was acquired from Qiagen GmbH (Hilden, Germany) and primers from Eurogentec (Seraing, Belgium).

2.3. Obtaining of Milk Extracts

An extraction procedure was performed on the freeze-dried milk (FDM) to recover the different extracts, as mentioned in our previous article [4]. The apolar extract of the milk metabolites was recovered by adding distilled cyclohexane (ratio $1/10$ (w/w)) to FDM powder and mechanically agitated for 4 h. The solution was then filtered with Büchner and evaporated under vacuum. The resulting solid was dissolved in cyclohexane (ratio $1/10$ (w/v)) before filtering again to eliminate the residue of the milk and evaporated under vacuum to obtain the dry extract. The milk matrix was exhausted by repeating the same procedure three times, under the same conditions at different times (4 h, 2 h and 1 h, respectively). The resulting dry extracts were combined to constitute the final extract, known as denominated milk lipid extract. The residual solid, which was retained by the filtration, was dried under vacuum and named lipid-free milk (LFM) (Figure 1).

The LFM was extracted using a chemical fractionation to recover most of the compounds, which resulted in three successive solid/liquid extractions, with an increase of the polarity of the solvent at each step: dichloromethane, ethyl acetate and absolute ethanol (Figure 1). For each solvent, the protocol used was the same as with cyclohexane, but with a $1/10$ (w/v) ratio. The final dried extracts were denominated as extracts MA, MB and MC, respectively.

The residual solid from the absolute ethanol extraction was dried under vacuum and named residual solid milk (RSM). Each dried milk extract was kept in a waterproof container at $-25\,°C$, under argon.

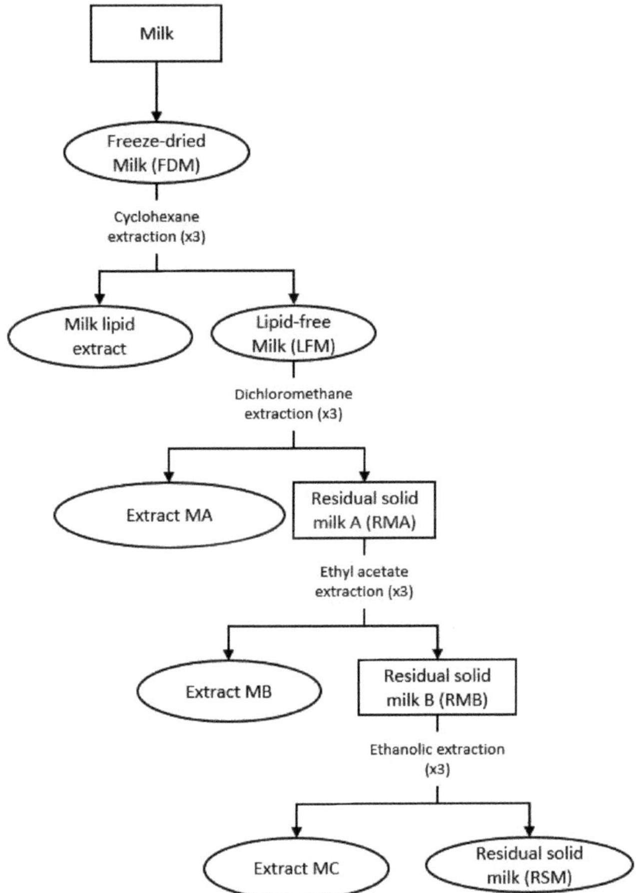

Figure 1. Preparation of the milk extracts. Ovals designate the milk extracts used for our biological studies (FDM (freeze-dried milk), milk lipid extract, LFM (lipid-free milk), and extracts MA (obtained with dichloromethane), MB (obtained with ethyl acetate), MC (obtained with absolute ethanol) and RSM (residual solid milk)).

2.4. Obtaining of Cheese Extracts

The same cheese extracts were used in this study as in our previous work [4]. The cheese extracts were obtained from the goat cheese by successive chemical extractions as described in the previous study. Briefly, the same extraction procedure as for the milk was applied to the freeze-dried cheese (FDC) to obtain a cheese-lipid extract, which was extracted with cyclohexane, as well as lipid-free cheese. From this residue, extract A and residue RA were then obtained with dichloromethane. Next, extract B and residue RB were obtained from the residue RA with ethyl acetate. To finish, extract C and residue RC were obtained from the residue RB with absolute ethanol. A fifth extraction was performed on the residual solid RC recovered from the ethanolic extraction to obtain yet another extract. HPLC-grade water was added to this solid (ratio of $1/10$ (w/v)) and the mixture was mechanically agitated for 1 h at 40 °C. The mixture was then centrifuged (8000 rpm, 15 min; Avanti J26S XPI, Beckman Coulter, Brea, CA, USA) and the supernatant was concentrated under vacuum, filtered to eliminate the residue of cheese and evaporated. Finally, the resulting solid was dried under vacuum. The cheese matrix was exhausted by repeating the same procedure three times, under the same conditions. The resulting dry extracts were combined to constitute the final extract, ground with mortar and pestle and denominated extract WF.

Two additional extracts were obtained by performing a water extraction on the lipid-free cheese, at two different temperatures (40 °C and 70 °C). The extraction procedure was the same as that for the extract WF described above, with the exception of the temperature. The resulting extracts were named W40 (extracted at 40 °C) and W70 (extracted at 70 °C).

Each dried cheese extract was kept in a waterproof container at −25 °C, under argon. The following experiments were performed using the FDC, the cheese-lipid extract and the extracts W40, W70 and WF, all of which presented a beneficial effect on the worms' lifespan in our previous study.

2.5. Microbial Strains, Growth Conditions and Heat-Killed Preparation

The *Escherichia coli* strain OP50 was provided by the *Caenorhabditis* Genetics Center (Minneapolis, MN, USA) and used as a food source during the worms' maintenance. *E. coli* OP50 was grown in lysogeny broth medium at 37 °C overnight. The microbial suspension was centrifuged (15 min, 4000 rpm; Rotofix 32A, Hettich Zentrifugen, Tuttlingen, Germany) and washed with M9 buffer (per L: 3 g of KH_2PO_4; 6 g of Na_2HPO_4; 5 g of NaCl; 1 mL of 1 M $MgSO_4$). The microbial suspension was adjusted to obtain a 100 mg/mL final concentration. During the experiments the worms were fed with heat-killed (HK) *E. coli* OP50. *E. coli* (100 mg/mL) (from the resulting suspension) was heat-killed at 70 °C for 1 h and the solution was kept at 4 °C until use.

2.6. C. elegans Maintenance

The *Caenorhabditis elegans* N2 (wild-type), and the different mutants TJ356 (*daf-16p::daf-16a/b::GFP + rol-6(su1006)*), GR1307 (*daf-16(mgDf50)*) and IG10 (*tol-1(nr2033)*) strains were acquired from the *Caenorhabditis* Genetics Center. The nematodes were cultured at 20 °C on nematode growth medium (NGM) plates (per L: 3 g of NaCl; 2.5 g of peptone; 17 g of agar; 5 mg of cholesterol; 1 mM of $CaCl_2$; 1 mM of $MgSO_4$; 25 mL of 1 M potassium phosphate buffer at pH 6), supplemented with yeast extract (4 g/L) (NGMY) and seeded with live *E. coli* OP50 [13–15].

2.7. Synchronisation of Wild-Type C. elegans and Mutant Strains

In order to avoid any variation in the results due to the age differences in the population, a synchronization of the worms was performed. Gravid worms and eggs were collected from NGMY plates and washed off using M9 buffer before centrifuging (2 min, 1500 rpm; Rotofix 32A, Hettich Zentrifugen, Tuttlingen, Germany). Five milliliters of worm bleach (2.5 mL of M9 buffer, 1.5 mL of bleach, 1 mL of sodium hydroxide 5 M) were added to the pellet and vigorously shaken until adult worm body disruption. Forty milliliters of M9 buffer were then introduced to block the effect of the worm bleach. The egg suspension was centrifuged (2 min, 1500 rpm) and washed twice with 20 mL of M9 buffer. The isolated eggs hatched at 25 °C for 24 h in 20 mL of M9 buffer, under slow agitation. The resulting L1 larvae were then transferred onto NGMY plates, seeded with live *E. coli* OP50 as a food source and maintained at 20 °C until they reached the L4/young adult stage [13–15].

2.8. Caenorhabditis elegans Longevity Assay Incubated with Dried Milk Extracts

The effect of the dried milk extracts on the life expectancy of the worms was evaluated by performing a longevity assay using the wild-type *C. elegans* N2 strain. An agar medium (per L: 3 g of NaCl and 6 g of agarose) was prepared, stored at 40 °C and split into aliquots that were individually supplemented with the dried milk extracts at the suitable concentration, according to the physicochemical properties of the extracts (Table 1). The supplemented aliquot was then poured, at 40 °C, into a 24-well plate with 0.12 mM of FUdR. The aliquot was also supplemented with amphotericin B (final concentration of 16 µg/mL), in the case of the lipid-free milk (LFM), to prevent any significant fungal development, which is commonly observed with this extract [16]. The presence or absence of the antifungal established two control conditions: with and without amphotericin B. To densify the agar, the plates were immediately moved onto ice after pouring, and kept at 4 °C until being used. Once adult, worms were incubated on a supplemented agar medium (or agar medium for the control condition) with ~15 worms per well, with HK *E. coli* OP50 as a food source, and kept at 20 °C for the duration of the assay. To avoid starvation, food was added every 3 days in the wells (20 µL of 100 mg/mL suspension). Nematodes were observed daily and were considered dead when they did not respond to a gentle mechanical stimulation. The effect of the LFM was evaluated in comparison with the control condition with amphotericin B and food (CC2), whereas the effect of the other extracts was evaluated in comparison with the control condition with food only (CC1). This assay was performed as at least three independent experiments containing three wells per condition and conducted simultaneously with the control conditions. Complementary information was taken into account to determine the effects of the extracts: the observation of the relative position of the curves, the value of the mean and maximum lifespan and the percentage of the population that was still alive on the supplemented medium when the worms in the control condition were all dead [4].

Table 1. Concentrations of the dried milk extracts used for supplementing the medium. Concentrations are expressed in percentage of extracts relative to the volume of medium.

Extracts	Concentration (*w/v*)		
	0.25%	0.5%	1%
Freeze-dried milk (FDM)	X	X	
Milk lipid extract (ML)	X	X	X
Lipid-free milk (LFM)	X	X	X
Extract MA	X	X	
Extract MB		X	
Extract MC	X	X	
Residual solid milk (RSM)	X	X	X

2.9. Longevity Assay of DAF-16 Loss of Function Mutant (GR1307 Strain) Incubated with Dried Cheese Extracts

In order to determine the implication of DAF-16 in extending the longevity of the worms incubated with dried cheese extracts, a longevity assay was conducted with the *C. elegans* GR1307 strain (DAF-16 loss-of-function). To prepare the supplemented agar medium with the cheese extracts, the same protocol described above was applied. The dried cheese extracts were used at 0.5% and 1% concentrations (*w/v*) for supplementing the medium. The aliquots were also supplemented with amphotericin B (1.6 µg/mL) in the case of the freeze-dried cheese (FDC) and the cheese-lipid extract, to prevent any significant fungal development. The effects of the FDC and the cheese-lipid extract were evaluated in comparison with the control condition with amphotericin B and food (CC2), whereas the effects of the extracts W40, W70 and WF were evaluated in comparison with the control condition with food only (CC1). This assay was performed as at least four independent experiments containing three wells per condition and conducted simultaneously with the control conditions.

2.10. Cellular Localisation of DAF-16::GFP

In order to study the biological activity of dried cheese extracts, an experiment was performed using transgenic TJ356 worms (DAF-16::GFP). The nuclear localization of the transcription factor was determined thanks to fluorescence, as described by Poupet et al. [14]. The same agar medium as described for the longevity assay was prepared, cooled down to 40 °C and supplemented with 5 mg/L of cholesterol, 1 mM of $CaCl_2$, 1 mM of $MgSO_4$ and 1 M of potassium phosphate buffer at pH 6 (25 mL for 1 L of medium). The medium was then split into aliquot, and individually supplemented with cheese extracts at 1%

(w/v) and poured into a 24-well plate before being immediately transferred onto ice to densify the agar, and stored at 4 °C until use. Once adult, worms were incubated on a cheese-extract agar plate for 2 h and 4 h at 20 °C, with food (heat-killed *E. coli* OP50). The translocation of DAF-16::GFP was scored by assaying the presence of the GFP accumulation in the *C. elegans* cell nuclei, using a 40× magnification fluorescence microscope (Evos FL, Invitrogen, Carlsbad, CA, USA) [17].

2.11. Survival of the Worms on the Oxidative Medium

The experiment was performed to determine the potential antioxidant activity of dried milk and cheese extracts as a preventive measure. This study was performed with the wild-type *C. elegans* N2 strain for the milk and cheese extracts. The experiment was also performed with the GR1307 strain (DAF-16 loss-of-function) and the IG10 strain (TOL-1 loss-of-function), but only for the dried cheese extracts. This assay was performed according to Grompone et al. (2012) [18], with some modifications. The same agar medium as described for the longevity assay was prepared and supplemented as described above with dried milk or cheese extracts at 1% (w/v). The aliquots were supplemented with food (40 µL of HK *E. coli* OP50 suspension at 100 mg/mL) for each condition, before being poured, at 40 °C, into a 24-well plate with 0.12 mM of FUdR. The aliquots were also supplemented with amphotericin B (1.6 µg/mL) in the case of the lipid-free milk (LFM), freeze-dried cheese (FDC) and cheese-lipid extract to prevent any significant fungal development. As described above, two control conditions were used during the assay: with and without amphotericin B. After pouring, the plates were immediately transferred onto ice to densify the agar, and stored at 4 °C until use. Once adult, worms were incubated on a supplemented agar medium (or agar medium for the control condition) with ~50 worms per well, and the worms were incubated for 5 days at 20 °C. After incubation, the worms were transferred onto an agar medium with or without hydrogen peroxide (3 mM in the medium). After 3 h 30 min of contact, the worm survival rate τ was scored and expressed with the following formula:

$$\tau = \frac{\left(\dfrac{n_{\text{living worms at } t=3h30}}{n_{\text{living worms at } t=0h}} \right)_{\text{medium with } H_2O_2}}{\left(\dfrac{n_{\text{living worms at } t=3h30}}{n_{\text{living worms at } t=0h}} \right)_{\text{medium without } H_2O_2}}, \tag{1}$$

A worm was considered as dead when it did not respond to a mechanical stimulus. The effect of the LFM, FDC and the cheese-lipid extract was evaluated in comparison with their respective control condition with amphotericin B and food (CC2) whereas the effect of the other extracts was evaluated in comparison with their respective control condition with food only (CC1). This assay was performed as at least four independent experiments containing three wells per condition and conducted simultaneously with the control conditions.

2.12. Determination of the Expression of Gene of Interest
2.12.1. Incubation of the Worms

In order to determine the expression of the gene of interest (GOI), an RNA isolation and RT-quantitative PCR were performed with the wild-type *C. elegans* N2 strain. The experiment was conducted in order to evaluate the gene expression at two incubation times as determined in the longevity assay performed in our previous study [4]: at 3 days (start of the decrease in the population during the longevity assay) and at 10 days (mean lifespan of the control population during longevity assay). The worms were incubated on the medium supplemented with the cheese extracts at 1% (w/v) as described for the longevity assay, in 55 mm diameter Petri dishes (1 per replicate and per time). Once adult, worms were incubated on a supplemented agar medium (or agar medium for the control condition) with ~500 worms per well (3 days) or ~1000 worms (10 days), provided with the necessary amount of food for each time and kept at 20 °C. The freeze-dried cheese (FDC) and the cheese-lipid extract were compared to the control condition with amphotericin B and food (CC2), whereas the extracts W40, W70 and WF were compared to the control condition with food only (CC1). This assay was performed as at least three independent experiments.

2.12.2. RNA Isolation and RT-Quantitative PCR

The RNA isolation and RT-quantitative PCR were adapted from Poupet et al. [19]. After a 3- or 10 day-incubation period, the worms were collected with M9 buffer. The total RNA was extracted by adding 500 µL of TRIzol reagent. The worms were disrupted by using a Precellys (Bertin instruments, Montigny-le-Bretonneux, France) and glass beads (PowerBead Tubes Glass 0.1 mm, Mo Bio Laboratories, Carlsbad, CA, USA). The beads were removed by centrifugation at 14,000 rpm for 1 min (Eppendorf® 5415D, Hamburg, Germany), and 100 µL of chloroform was added to the supernatant. The tubes were vortexed for 30 s and incubated at room temperature for 3 min. The tubes were then centrifuged (12,000 rpm, 15 min, 4 °C) and the phenolic phase was removed. The aqueous phase was treated again with chloroform. The RNA was precipitated in the second aqueous phase by adding 250 µL of isopropanol. The tubes were incubated at room temperature for 4 min before centrifugation (12,000 rpm, 10 min, 4 °C). The supernatant was discarded, and the pellet was washed with 1000 µL of 70% ethanol. The supernatant was discarded after centrifugation (14,000 rpm, 5 min, 4 °C) and the pellet was dissolved into 20 µL of RNase-free water. Then, 2 µg of RNA was reverse-transcribed using the High-Capacity cDNA Archive kit, according to the manufacturer's instructions. For real-time qPCR assay, each reaction contained 2.5 µL of cDNA, 6.25 µL of Rotor-Gene SYBR Green Mix, 1.25 µL of 10 µM primers (Table 2) and 1.25 µL of RNase-free water. All samples were run in triplicate. Rotor-Gene Q Series Software (Qiagen GmbH, Hilden, Germany) was used for the analysis. In our study, one reference gene, Y45F10D.4, was used in all of the experimental groups. The quantification of GOI expression (E_{GOI}) was performed according to Equation (2) [20]:

$$EGOI = \frac{(GOI\ efficiency)^{\Delta Ct_{GOI}}}{(Y45F10D.4\ efficiency)^{\Delta Ct_{Y45F10D.4}}}, \tag{2}$$

Table 2. Targeted *C. elegans* genes primers for qPCR analysis.

Gene Name	Gene Type	Forward Primer (5'-3')	Reverse Primer (5'-3')	Reference
Y45F10D.4	housekeeping	CGAGAACCCGCGAAATGTCGGA	CGGTTGCCAGGGAAGATGAGGC	[19]
daf-16	GOI	TTCAATGCAAGGAGCATTTG	AGCTGGAGAAACACGAGACG	[19,21]
sek-1	GOI	GCCGATGGAAAGTGGTTTTA	TAAACGGCATCGCCAATAAT	[19,21]
pmk-1	GOI	CCGACTCCACGAGAAGGATA	AGCGAGTACATTCAGCAGCA	[19,21]

2.13. Leukocyte Viability

Blood was collected from healthy human volunteers ($n = 22$; Etablissement Français du Sang, EFS, Clermont-Ferrand, France). Donors gave their written informed consent for the use of blood samples for research purposes under EFS contract n°16-21-62 (in accordance with the following articles: L1222-1, L1222-8, L1243-4 and R1243-61 of the French Public Health Code). The whole-blood leukocytes were obtained by hemolytic shock using ammonium chloride solution (NH_4Cl, 155 µM; $NaHCO_3$ 12 µM, EDTA 0.01 µM). The leukocytes were then washed with RPMI, centrifuged ($400\times g$, 10 min) and resuspended in RPMI. The cell preparations were adjusted to 10^6 cells/mL with supplemented RPMI (FBS 10%, gentamicin 50 µg/mL and glutamine 2 mM). The cells were then placed in 96-well polystyrene plates (Cell Wells™, Corning, New-York, NY, USA), incubated with the dried extracts WF, W40, W70 or the dried cheese-lipid extract at 0, 10, 50, 100 or 200 µg/mL, PMA (1 µM) and resazurin (25 µg/mL). The extract WF was filtered at 0.22 µM to avoid any significant fungal development. Fluorescence (excitation/emission: 544/590 nm) was recorded every 30 min for 2 h using the Fluoroskan Ascent FL® apparatus (ThermoFisher Scientific, Illkirch, France).

2.14. Kinetics of ROS Production by Leukocytes

Blood was collected from healthy human volunteers (n = 22). The whole-blood leukocyte preparations were obtained and adjusted as previously described. The cells were placed in 96-well polystyrene plates, incubated with the dried extracts WF, W40, W70 or the dried cheese-lipid extract at 0, 10, 50, 100 or 200 μg/mL, and dihydrorhodamine 123 (DHR 123, 1 μM), and stimulated by 1 μM PMA for 120 min to increase the ROS production. The extract WF was filtered at 0.22 μM to avoid any significant fungal development. The fluorescence intensity of rhodamine 123, which is the reduced form of dihydrorhodamine 123 oxidation by ROS, was recorded every 5 min for 120 min (excitation/emission: 485/538 nm) using the Fluoroskan Ascent FL® apparatus.

2.15. Statistical Analysis

Results of lifespan experiments were examined by using the Kaplan–Meier method, and compared among group scoring for significance using the log-rank test with R software version 3.6.0. The differences between conditions, in the survival assay on the oxidative medium, the qPCR analysis, the cellular viability assay and the ROS production assay, were determined by using the Kruskal–Wallis test followed by an Uncorrected Dunn's test using GraphPad Prism version 8.2.1 for Windows (GraphPad Software, La Jolla, CA, USA). Differences were considered statistically significant if p-value $\leq$ 0.05.

3. Results

3.1. Implication of DAF-16 in the Capacity of the Extracts to Induce an Increase in Longevity

In order to determine the implication of the transcription factor DAF-16 (involved in the longevity phenomenon) in the mechanisms of action of the dried cheese extracts (freeze-dried cheese (FDC), cheese-lipid extract, and WF, W40 and W70), a longevity assay was carried out using the GR1307 mutant strain that did not produce the protein. If the transcription factor DAF-16 was required, the extracts would no longer be able to induce an increase in the lifespan. In this regard, no variation of the beneficial effect of the cheese extracts should be noted.

The worms incubated with the FDC did not show any variation in longevity compared with the control CC2 condition (Figure 2A). The mean and maximum lifespans were identical between the control condition and both concentrations of FDC (Table 3). However, the significant difference observed between the curves representing the FDC and the CC2 (p = 0.001 and p = 0.03 for 0.5% and 1% concentration, respectively) suggests that the FDC was responsible for a beneficial effect on the worms, allowing a larger part of the population to remain alive for a given amount of time compared with the CC2.

The cheese-lipid extract showed a variation between the two concentrations. An increase in longevity was observed for the worms incubated on the extract at 0.5%, with a significant difference between this curve and the CC2 curve (p = 0.02) (Figure 2B). The maximum lifespan was also higher, increasing from 17 (CC2) to 18 (+6%) days (Table 3). At 1%, the beneficial effects of the extract were no longer significant (p = 0.5).

The worms incubated with the dried aqueous cheese extracts WF and W40 had a significant decrease in their lifespan compared with the control CC1 (p < 0.001) (Figure 3A,B). The curves representing these extracts were significantly below the CC1 curve, with a decrease in the mean lifespan from 9 (CC1) to 8 days (WF at 1%, W40 at 0.5% and 1%). Concerning WF at 0.5%, the mean lifespan decreased from 9 to 7 days (Table 4). However, the maximum lifespan was higher for all conditions, with an increase of between 12% and 18% compared with CC1. Concerning the extract W70, no variation in lifespan was observed in comparison with CC1. The evolution of the curve was similar to that of the control condition for both concentrations (Figure 3C). An increase in the maximum lifespan was noted for a small proportion of the population of the worms (up to 5% of the population).

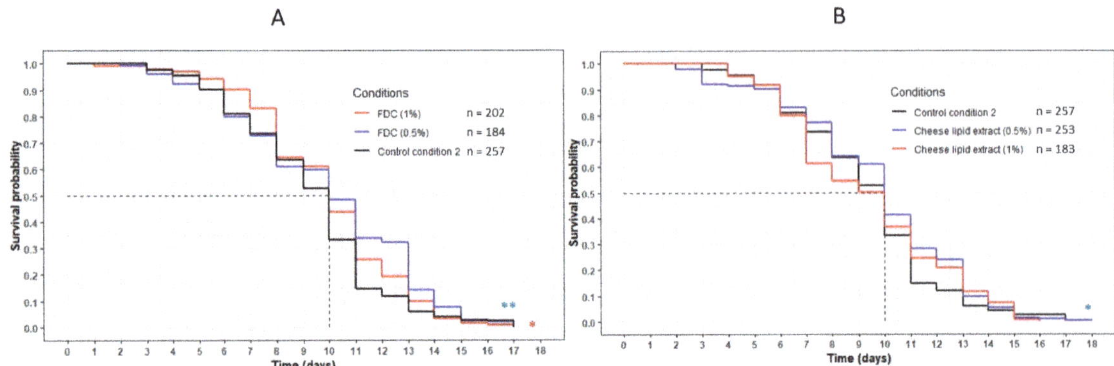

Figure 2. Influence of the FDC (freeze-dried cheese) (**A**) and the cheese-lipid extract (**B**) on the lifespan of *C. elegans* GR1307 strain. The worms were incubated on the medium supplemented with the dried extracts at day 0 and regularly fed with HK *E. coli* OP50. The conditions were considered significantly different when the *p*-value was lower than 0.05 (*) or 0.01 (**) (log-rank test). The asterisks next to the curves represent the differences with the control condition CC2. The asterisks next to the legend represent the differences between the extracts.

Table 3. Data of the longevity assay of the *C. elegans* GR1307 strain on the medium supplemented with the FDC (freezz-dried cheese) and the cheese-lipid extract, and CC2 as control. Mean lifespan, maximum lifespan and the percentage of population still alive were taken from the survival curves in Figure 2. *p*-values were calculated by comparing conditions with CC2 using the log-rank test.

Tested Conditions	Concentration (*w/v*) (%)	Mean Lifespan (Days)	Maximum Lifespan (Days)	Relative Increase in the Maximum Lifespan (%)	Percentage of Population Still Alive at 17 Days (%)	*p*-Value
CC2	-	10	17	-	0	-
Freeze-Dried	0.5	10	17	0	0	0.001
Cheese (FDC)	1	10	17	0	0	0.03
Cheese-lipid extract	0.5	10	18	+6	1	0.02
	1	10	16	−6	0	0.5

Table 4. Data of the longevity assay of the *C. elegans* GR1307 strain on the medium supplemented with the aqueous extracts WF, W40 and W70, and CC1 as a control. Mean lifespan, maximum lifespan and the percentage of population still alive were taken from the survival curves in Figure 3. *p*-values were calculated by comparing conditions with CC1 using the log-rank test.

Tested Conditions	Concentration (*w/v*) (%)	Mean Lifespan (Days)	Maximum Lifespan (Days)	Relative Increase in the Maximum Lifespan (%)	Percentage of Population Still Alive at 17 Days (%)	*p*-Value
CC1	-	9	17	-	0	-
Extract WF	0.5	7	19	+12	1	<0.0001
(40 °C)	1	8	19	+12	2	0.0008
Extract W40	0.5	8	19	+12	2	<0.0001
(40 °C)	1	8	20	+18	1	0.0002
Extract W70	0.5	10	22	+29	2	0.9
(70 °C)	1	9	22	+29	5	0.09

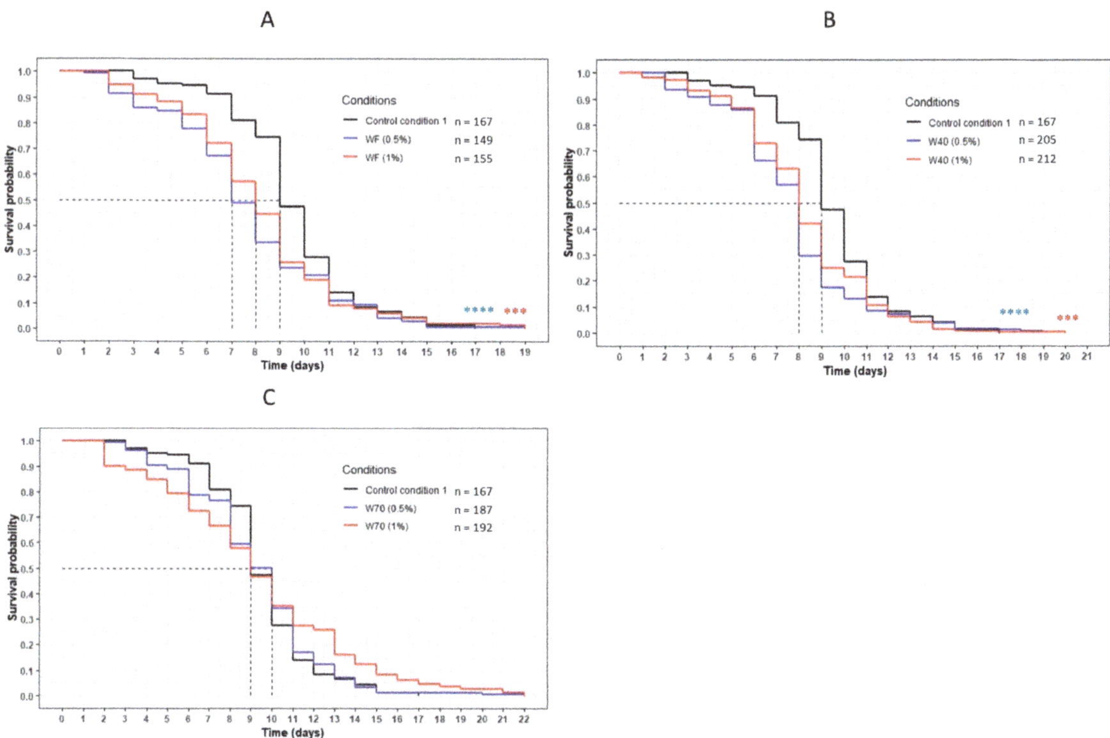

Figure 3. The influence of the aqueous extracts WF (**A**), W40 (**B**) and W70 (**C**) on the *C. elegans* GR1307 strain lifespan. The worms were incubated on the medium supplemented with the dried extracts at day 0 and regularly fed with HK *E. coli* OP50. The conditions were considered significantly different when the *p*-value was lower than 0.001 (***) or 0.0001 (****) (log-rank test). The asterisks next to the curves represent the differences with the control condition CC1. The asterisks next to the legend represent the differences between the extracts.

This result was reinforced by the comparison of the three extracts WF, W40 and W70 at the same concentration (Figure 4). For both concentrations, the curve representing W70 was significantly above the other curves corresponding to the extracts WF and W40 ($p < 0.0001$ with W40; $p = 0.0001$ with WF). These results confirmed that the transcription factor DAF-16 was required for the FDC, the cheese-lipid extract and the extracts WF, W40 and W70 in order to increase the lifespan of the nematode *C. elegans* significantly.

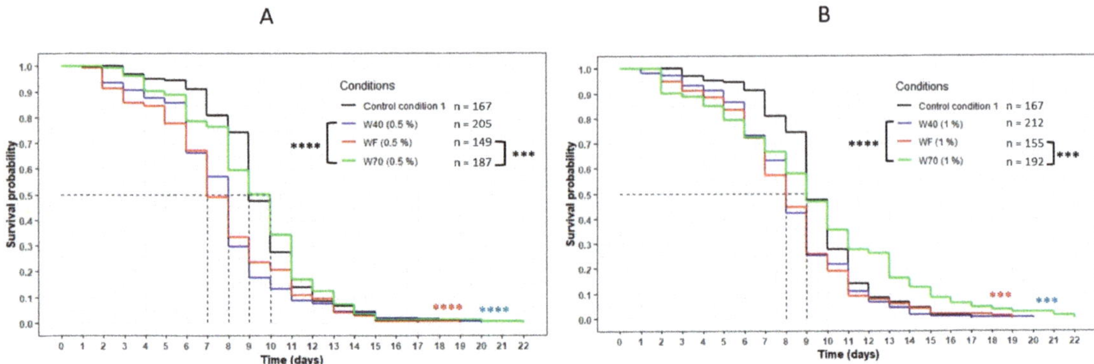

Figure 4. Comparison of the effect of the aqueous extracts WF, W40 and W70 at 0.5% (**A**) and 1% (**B**) concentration on the *C. elegans* GR1307 strain lifespan. The worms were incubated on the medium supplemented with the dried extracts obtained with water at day 0 and regularly fed with HK *E. coli* OP50. The conditions were considered significantly different when the *p*-value was lower than 0.001 (***) or 0.0001 (****) (log-rank test). The asterisks next to the curves represent the differences with the control condition CC1. The asterisks next to the legend represent the differences between the extracts.

3.2. Cellular Localisation of DAF-16::GFP

The nuclear translocation of the DAF-16/FOXO transcription factor was examined using the *C. elegans* TJ356 strain (which constitutively expresses the fusion protein DAF-16::GFP) (Figure 5). The localization of DAF-16 in the cells of the worms was established by fluorescence to determine if this signaling pathway was involved in the effects of the extracts. Figure 5 shows that the freeze-dried cheese (FDC) and the extract W70 were the only extracts that tended to induce a translocation of the transcription factor into the nuclei after 2 h of incubation. At 4 h, each extract tended to induce a translocation of DAF-16 into the nuclei. The control condition did not show any variation in DAF-16 cellular localization for the duration of the assay, suggesting that the translocation observed was induced by all of the extracts. These observations reinforced the results obtained from the longevity assays, suggesting that the transcription factor DAF-16 was involved in the mechanisms of action of the extracts.

3.3. Effect of the Dried Cheese Extracts on the Survival Rate of the Wild-Type C. elegans N2 Strain on the Oxidative Medium

The impact of the dried cheese extracts (freeze-dried cheese (FDC), cheese-lipid extract, WF, W40 and W70) on the survival rate of the wild-type *C. elegans* N2 strain on an oxidative medium was evaluated (Figure 6). The survival rate was determined by measuring the worm viability after 3 h 30 min incubation on an agar H_2O_2-medium. The results demonstrated that the cheese extracts significantly influenced the ability of the worms to survive longer on the oxidative medium. Those incubated with the FDC and the cheese-lipid extract exhibited a better resistance to the oxidizing medium compared with the CC2 worms, with a survival rate of 3 ($p = 0.0003$) and 2.4 ($p = 0.0306$), respectively (Figure 6A). The same observation was made for the worms incubated with the cheese extracts WF, W40 and W70, where the survival rate increased to 4.4 ($p = 0.0052$), 4.3 ($p = 0.0091$) and 4.1 ($p = 0.0181$), respectively, compared with the CC1 condition (Figure 6B). The cheese extracts exhibited a beneficial effect on the worms by improving their survival rate on the oxidative medium.

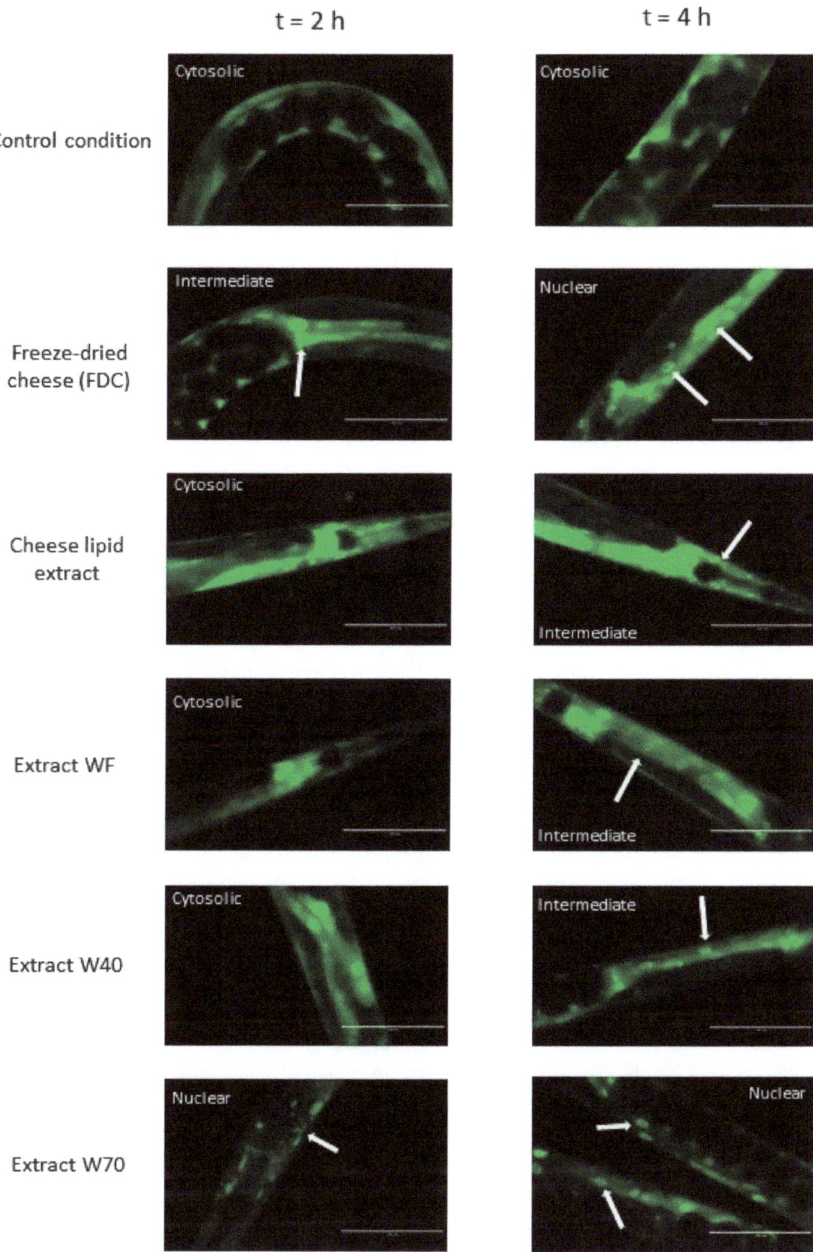

Figure 5. The effects of the FDC (freeze-dried cheese) and the extracts on DAF-16 cellular localization in *C. elegans* transgenic strain TJ356 expressing DAF-16::GFP, after 2 h and 4 h of incubation on the supplemented medium. The arrows indicate the accumulation of the transcription factor in the nuclei. Scale bar: 100 μm.

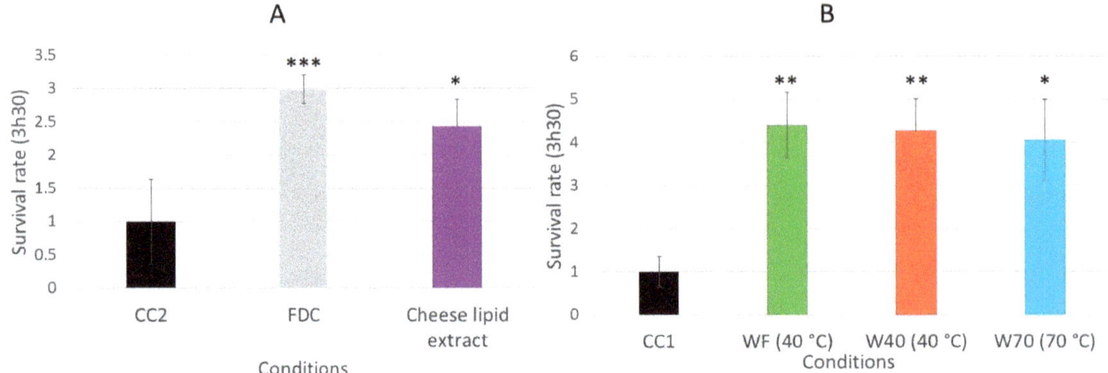

Figure 6. Relative survival rates of the wild-type *C. elegans* N2 strain on an oxidative medium after 5 days of incubation on a medium supplemented with FDC (freeze-dried cheese) and cheese-lipid extract (**A**) or aqueous extracts WF, W40 and W70 (**B**). The conditions were considered significantly different from the control when the *p*-value was lower than 0.05 (*), 0.01 (**), 0.001 (***) (Kruskal–Wallis test).

3.4. Implication of the Signaling Pathways in the Survival of the C. elegans on the Oxidative Medium

The increase of the survival rate of the worms, induced by the dried cheese extracts, may be due to an activation of the signaling pathways involved in the defense mechanisms of the worms, such as the insulin-like pathway or the p38 mitogen activated protein kinase (p38 MAPK) [8,9,22]. The same survival assay was performed with mutants to determine if these pathways are involved in the biological mechanism. The survival rate for the *C. elegans* GR1307 strain, which does not express *daf*-16, was determined by measuring the worm viability after 3 h 30 min of incubation on the oxidative medium. When incubated with the freeze-dried cheese (FDC) ($p < 0.0001$), and the cheese extracts WF ($p = 0.0057$), W40 ($p = 0.0003$) and W70 ($p = 0.0371$) (Figure 7A,B), the worms exhibited a significant resistance to the oxidative medium compared with that in their respective control conditions. The FDC increased the survival rate from 1 to 4.8. The worms incubated with WF, W40 and W70 had a survival rate between 3.5 and 4.7. The cheese-lipid extract tended to improve the worms' resistance to the oxidative medium from 1 to 2.9. However, this increase was not significant compared with that in the CC2 group. The results suggested that only the cheese-lipid extract required the presence of the protein DAF-16 to improve the survival rate of the worms.

The same experiment was conducted with the *C. elegans* IG10 strain, which does not express the gene of the receptor *tol*-1 that is linked to the p38 MAPK signaling pathways involved in the nematode immunity. The absence of the receptor may prevent the activation of the pathway by the extracts and cancel the beneficial effects observed with the N2 strain. The same observations as for the GR1307 strain were made for the IG10 mutant. The worms incubated on the FDC or the dried aqueous extracts (W40, W70, WF) demonstrated a better resistance to the oxidizing medium (Figure 7C,D). The survival rate increased from 1 to 4.4 ($p = 0.0162$) for WF and 4.6 ($p = 0.0037$) for W40 and W70. For the FDC, it increased from 1 to 2.9 ($p = 0.0005$). Once again, the cheese-lipid extract tended to increase the worms' resistance to the oxidative medium (survival rate of 2.1), but the effect observed was not significant in comparison with the CC2 condition. Based on these observations, only the cheese-lipid extract required the receptor TOL-1 in order to improve the survival rate of the worms.

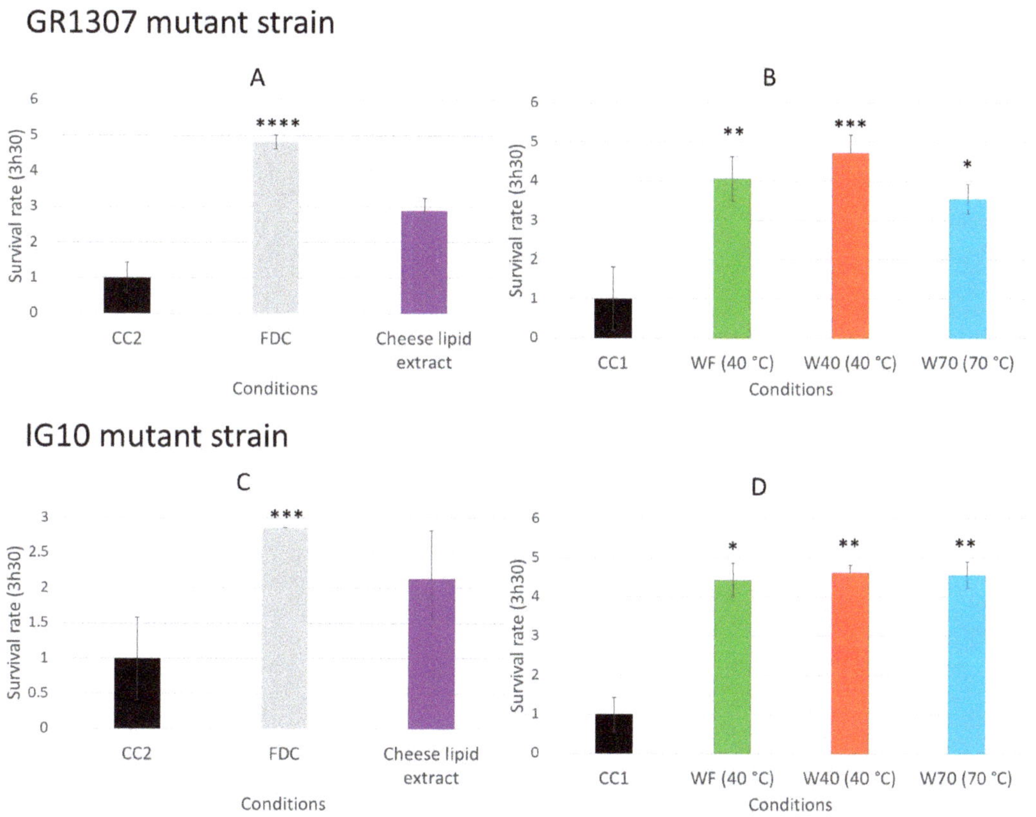

GR1307 mutant strain

IG10 mutant strain

Figure 7. The relative survival rates of the *C. elegans* GR1307 strain (DAF-16 loss-of-function) (**A**,**B**) and IG10 strain (TOL-1 loss-of-function) (**C**,**D**) on an oxidative medium after 5 days of incubation on a medium supplemented with the cheese extracts (FDC (freeze-dried cheese), cheese-lipid extract and aqueous extracts WF, W40 and W70). The conditions were considered significantly different from the control when the *p*-value was lower than 0.05 (*), 0.01 (**), 0.001 (***) or 0.0001 (****) (Kruskal–Wallis test).

3.5. Evaluation of the Expression of the Genes of Interest (GOI) daf-16, sek-1 and pmk-1

The expression of the three genes of interest (*daf*-16, *sek*-1 and *pmk*-1) was investigated as two of these are implicated in the p38 MAPK pathway (Figure 8). The experiment was conducted during two different time periods: 3 days and 10 days of incubation with the dried cheese extracts. The freeze-dried cheese (FDC) and the cheese-lipid extract did not modulate the expression of any of the genes of interest after 3 days of incubation (Table 5). At 10 days, the FDC significantly upregulated the expression of the three genes to 2.78 for *daf*-16 (*p* = 0.0039), 2.89 for *sek*-1 (*p* = 0.0297) and 2.80 for *pmk*-1 (*p* = 0.016). The cheese-lipid extract upregulated the expression of *daf*-16 and *pmk*-1 to 3.41 (*p* = 0.0019) and 2.39 (*p* = 0.0094), respectively, after 10 days of incubation (Table 5).

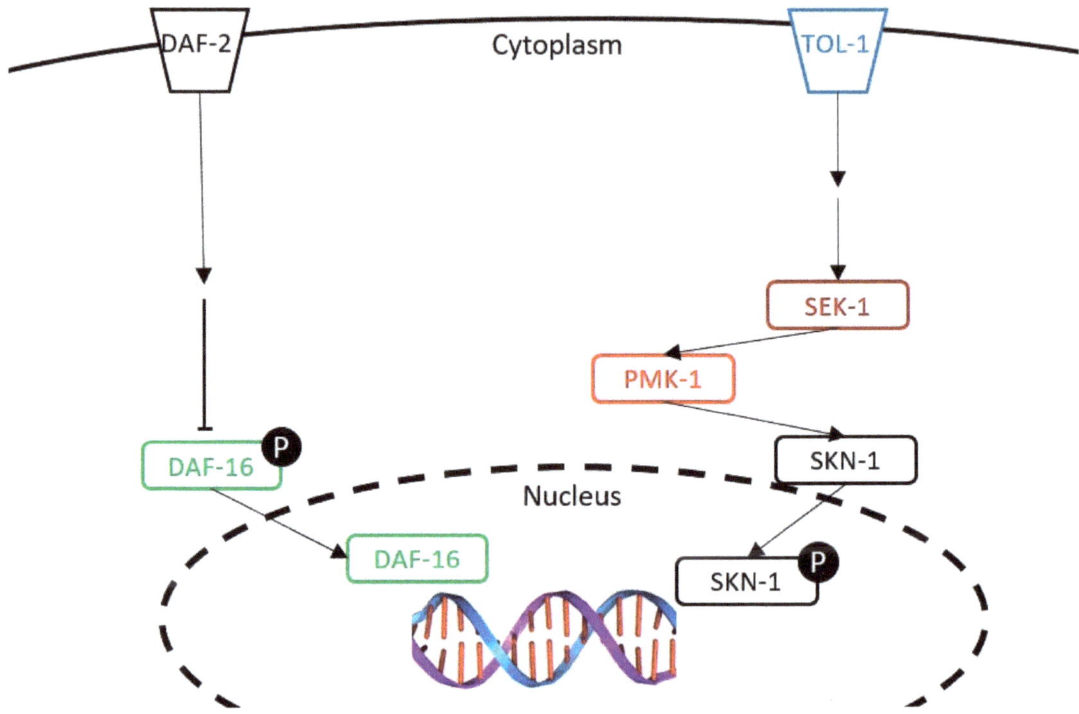

Figure 8. The representation of the insulin-like pathway and the p38 MAPK pathway studied in this article. The following genes were tested: *daf*-16 (studied with a mutant and a transcriptomic analysis), *tol*-1 (studied with a mutant), and *pmk*-1 and *sek*-1 (studied with a transcriptomic analysis).

Table 5. The relative expression of the three *C. elegans* genes of interest after 3 days or 10 days of incubation on a medium supplemented with FDC (freeze-dried cheese) or cheese-lipid extract in comparison with CC2 condition. The expressions were considered significantly different when the p-value was lower than 0.05 (*) or 0.01 (**), and simultaneous when the expression change was of at least 2 times higher or 0.5 times lower.

	Genes of Interest		
Conditions	***daf*-16**	***sek*-1**	***pmk*-1**
FDC 3 days	1.95	1.13	1.00
FDC 10 days	2.78 **	2.89 *	2.80 *
Cheese-lipid extract 3 days	1.07	1.02	1.09
Cheese-lipid extract 10 days	3.41 **	1.19	2.39 **

With regard to the different dried aqueous extracts, at 3 days, only the extract WF overexpressed *daf*-16 up to 5.52 ($p = 0.0018$) before going back to a normal expression after 10 days. The expression of *sek*-1 and *pmk*-1 was not modified by the extract whatever the duration. As for W40, the results demonstrated that the extracts did not modulate the expression of the genes after 3 days of incubation (Table 6). At 10 days, W40 significantly increased the expression of *daf*-16 to 2.93 ($p = 0.0043$) and tended to overexpress *sek*-1 to 3.39. Finally, the extract W70 did not influence the expression of the three genes of interest. The results demonstrated that, except for W70, the extracts influenced the expression of *daf*-16. Concerning the genes *sek*-1 and *pmk*-1, only the FDC and the cheese-lipid extract were able to overexpress at least one of them.

Table 6. The relative expression of the three *C. elegans* genes of interest after 3 days or 10 days of incubation on a medium supplemented with aqueous extracts WF, W40 or W70, in comparison with CC1 condition. The expressions were considered significantly different when the *p*-value was lower than 0.01 (**), and simultaneous when the expression change was of at least 2 times higher or 0.5 times lower.

	Genes of Interest		
Conditions	*daf*-16	*sek*-1	*pmk*-1
WF (40 °C) 3 days	5.52 **	1.11	1.27
WF (40 °C) 10 days	1.00	1.87	1.01
W40 (40 °C) 3 days	1.08	0.83	1.24
W40 (40 °C) 10 days	2.93 **	3.39	0.67
W70 (70 °C) 3 days	1.20	0.80	1.08
W70 (70 °C) 10 days	1.96	1.26	0.59

3.6. Production of ROS in Human Blood Leukocytes Triggered by PMA

The effect of the dried cheese extracts (cheese-lipid extract, WF, W40 and W70) on the production of the reactive oxygen species (ROS) was quantified in human blood leukocytes triggered by PMA (Figure 9A). Only two extracts exhibited the capacity to reduce the ROS production in the cells. A significant decrease was observed for W70 for the highest concentration (200 µg/mL), decreasing the ROS production by 28% ($p = 0.0029$). The other concentrations also tended to decrease the ROS production, but not significantly. Finally, the cheese-lipid extract significantly reduced the production of ROS for each concentration, with the same strength as no dose response was observed (by 23% for 10 µg/mL ($p = 0.025$) and by 24% for 50 µg/mL ($p = 0.0202$), 100 µg/mL and 200 µg/mL ($p = 0.0124$ for both concentrations)). The results obtained were not influenced by any toxic effects of the extracts, as no significant differences were observed with the leukocyte viability assay (Figure 9B).

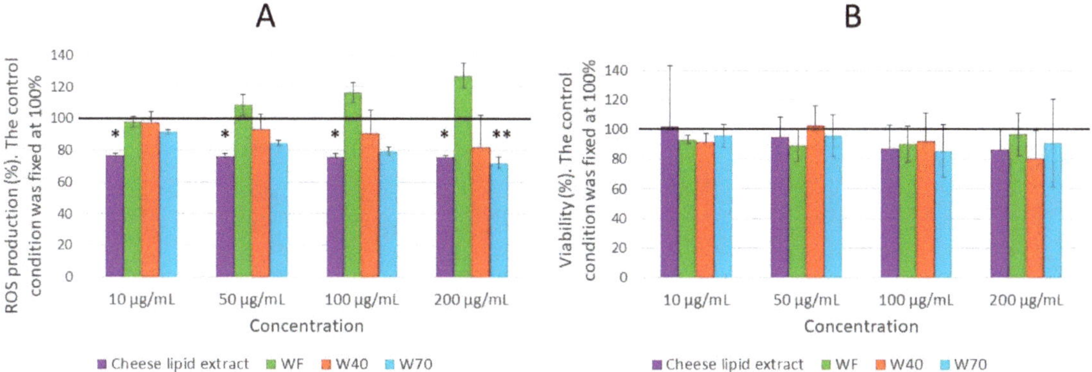

Figure 9. The effect of the cheese-lipid extract, WF, W40 and W70 on the ROS production in the leukocytes (**A**) and the viability of the leukocytes (**B**). The cells were treated with the indicated concentrations of the extract for 2 h, and measurements were made every 30 min. Data were expressed as relative production or viability in comparison with the control which was fixed at 100%. The conditions were considered significantly different from the control when the *p*-value was lower than 0.05 (*), 0.01 (**) (Kruskal–Wallis test).

3.7. Effect of the Dried Milk Extracts on the Longevity of Wild-Type C. elegans N2 Strain and Its Survival Rate on the Oxidative Medium

In order to determine if the milk could provide the same beneficial effect on the worms' longevity as the cheese, a longevity assay was conducted with the dried milk extracts, using the wild-type *C. elegans* N2 strain. Indeed, the milk used for making the cheese could contain the same metabolites and so, could exert the same beneficial effect.

Incubating the worms with the freeze-dried milk (FDM) showed a significant increase in longevity for both concentrations tested ($p < 0.0001$) (Figure 10). The mean lifespan increased by 33% and 25% for the 0.25% and 0.5% concentrations, respectively (Table 7). The maximum lifespan also increased from 23 days (maximum lifespan of CC1) to 31 days (+35%) and 26 days (+13%), and the percentage of the population still alive when the worms on the CC1 condition had died was 6% and 5%, respectively. Moreover, no differences were observed between the two concentrations of FDM when compared to each other. The same observation was made for the milk lipid extract, the milk extracts MA (obtained with dichloromethane) and MC (obtained with ethanol) and the residual solid milk (RSM). Indeed, these conditions also significantly increased the lifespan of the worms ($p < 0.0001$). The mean lifespan increased between 21% and 25% for all conditions, and the percentage of the population still alive at 23 days reached a maximum of 5%. The maximum lifespan was more variable, with an increase of up to 52% with the RSM compared to the CC1 condition. Concerning the extract MA, a significant difference was observed between the two concentration curves ($p = 0.04$), with the 0.25% curve above the 0.5% concentration curve.

Table 7. Data of the longevity assay of the wild-type *C. elegans* N2 strain on the medium supplemented with the FDM (freeze-dried milk), milk lipid extract, extracts MA (obtained with dichloromethane), MB (obtained with ethyl acetate), MC (obtained with absolute ethanol) and the RSM (residual solid milk), and CC1 as a control. Mean lifespan, maximum lifespan and the percentage of the population still alive are from the survival curves. *p*-values were calculated by comparing conditions with the CC1 using the log-rank test.

Tested Conditions	Concentration (*w/v*) (%)	Mean Lifespan (Days)	Maximum Lifespan (Days)	Relative Increase of the Maximum Lifespan (%)	Percentage of Population Still Alive at 23 Days (%)	*p*-Value
CC1	-	12	23	-	0	-
Freeze-Dried Milk (FDM)	0.25	16	31	+35	6	<0.0001
	0.5	15	26	+13	5	<0.0001
Milk Lipid Extract	0.25	14.5	26	+13	3	<0.0001
	0.5	15	30	+30	5	<0.0001
	1	15	23	0	0	<0.0001
Extract MA	0.25	15	31	+35	5	<0.0001
	0.5	14	28	+22	3	<0.0001
Extract MB	0.5	7	22	−4	0	<0.0001
Extract MC	0.25	14	26	+13	2	<0.0001
	0.5	15	28	+22	4	<0.0001
Residual Solid Milk (RSM)	0.25	15	28	+22	5	<0.0001
	0.5	15	25	+9	5	<0.0001
	1	15	35	+52	4	<0.0001

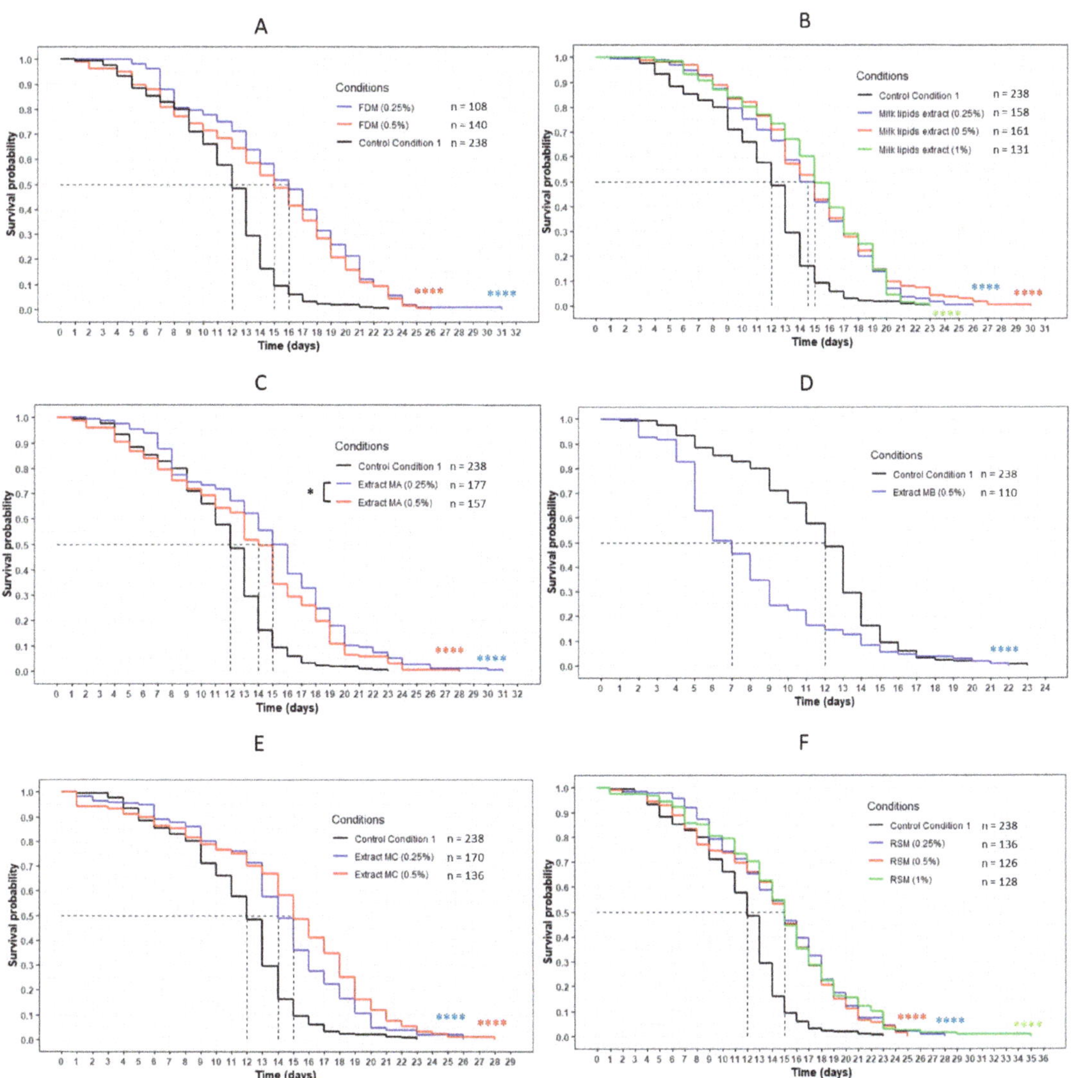

Figure 10. The influence of FDM (freeze-dried milk) (**A**), milk lipid extract (**B**), extracts MA (obtained with dichloromethane) (**C**), MB (obtained with ethyl acetate) (**D**), MC (obtained with ethanol) (**E**) and RSM (residual solid milk) (**F**) on the lifespan of wild-type *C. elegans* N2 strain. The worms were incubated on the medium supplemented with the dried extracts at day 0 and regularly fed with HK *E. coli* OP50. The conditions were considered significantly different when the *p*-value was lower than 0.05 (*) or 0.0001 (****). The asterisks next to the curves represent the differences with the control condition CC1. The asterisks next to the legend represent the differences between the extracts (log-rank test).

The extract MB, unlike the others, exhibited a significant negative effect on the lifespan of the wild-type *C. elegans* N2 strain. Indeed, the curve representing this extract was significantly below the CC1 curve ($p < 0.0001$) (Figure 10). The mean lifespan was reduced by 42% and the maximum lifespan was lower than the control, with a maximum of 22 days against 23 days (−4%) (Table 7).

The worms incubated with the lipid-free milk (LFM) presented a significant increase in their lifespan compared with the CC2 condition, with the curves of each concentration of the extract above the CC2 curve ($p < 0.0001$) (Figure 11). The mean lifespan increased by 8%, 17% and 25% for the 0.25%, 0.5% and 1% concentrations, respectively (Table 8). The maximum lifespan increased between 4% and 26%, and the percentage of the population still alive when the worms on the CC2 condition had died was between 5% and 7%.

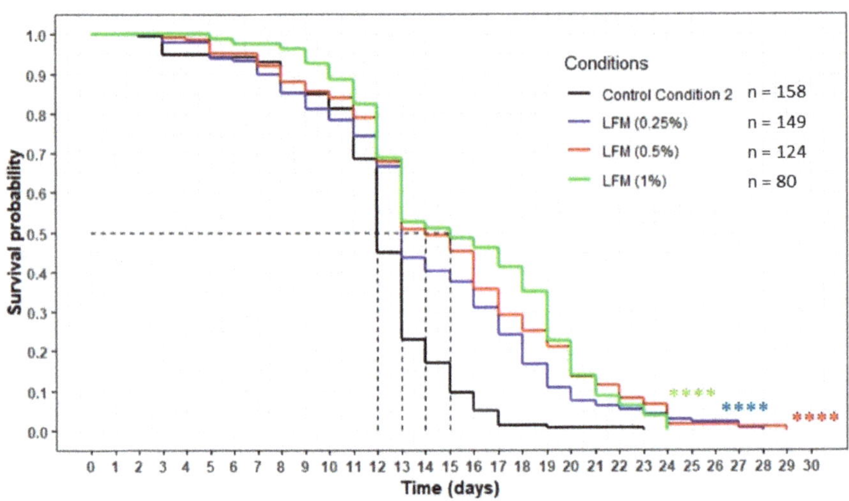

Figure 11. The influence of the LFM (lipid-free milk) on the lifespan of the wild-type *C. elegans* N2 strain. The worms were incubated on the medium supplemented with the dried extract at day 0 and regularly fed with HK *E. coli* OP50. The conditions were considered significantly different from the control conditions CC2 when the *p*-value was lower than 0.0001 (****) (log-rank test).

Table 8. Data of the longevity assay of the wild-type *C. elegans* N2 strain on the medium supplemented with the LFM (lipid-free milk) and CC2 as a control. Mean lifespan, maximum lifespan and the percentage of population still alive are from survival curves. *p*-values were calculated by comparing conditions with CC2 using the log-rank test.

Tested Conditions	Concentration (*w*/*v*) (%)	Mean Lifespan (Days)	Maximum Lifespan (Days)	Relative Increase of the Maximum Lifespan (%)	Percentage of Population Still Alive at 23 Days (%)	*p*-Value
CC2	-	12	23	-	0	-
Lipid-Free Milk (LFM)	0.25	13	28	+22	5	<0.0001
	0.5	14	29	+26	7	<0.0001
	1	15	24	+4	5	<0.0001

The effects of the milk extracts on the survival abilities of the worms on the oxidative medium were also evaluated. Only the worms incubated with the FDM and the extract MC exhibited a better resistance to the oxidizing medium compared to the CC1 condition (Figure 12A), with a survival rate increasing from 1.0 for the CC1 to 2.0 ($p = 0.0006$) and 1.8 ($p = 0.0142$), respectively. The other extracts did not show any significant effect on the survival of the worms.

With the exception of the extract MB, the results of our lifespan assay demonstrated that all of the other milk extracts (FDM, milk lipid extract, MA, MC, RSM and LFM) exerted a beneficial effect and were able to significantly increase the lifespan of the worms. However, in the survival assay on the oxidative medium, the effects of these extracts, still excepting MB, on the worms' survival rate were not as noteworthy as those of the cheese extracts (FDC, cheese-lipid extract, WF, W40, W70). Indeed, only two milk extracts (FDM and MC) exhibited an effect on the worms.

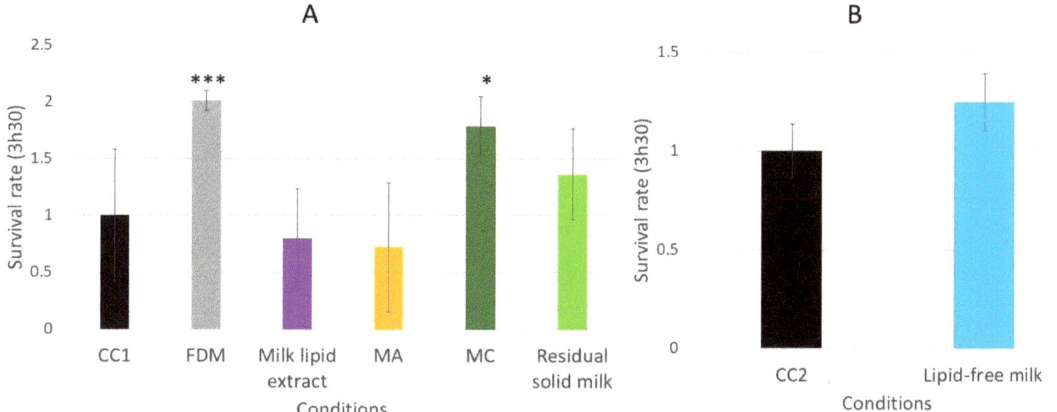

Figure 12. The relative survival rates of the wild-type *C. elegans* N2 strain on an oxidative medium after 5 days of incubation on a medium supplemented with FDM (freeze-dried milk), milk lipid extract, extract MA (obtained with dichloromethane), MC (obtained with ethanol) and RSM (residual solid milk) (**A**) or LFM (lipid-free milk) (**B**). The conditions were considered significantly different from the control when the *p*-value was lower than 0.05 (*) or 0.001 (***) (Kruskal–Wallis test).

4. Discussion

4.1. The Transcription Factor DAF-16 Is Implicated in the Mechanism by Which the Dried Cheese Extracts Increase Longevity

In our previous study, we demonstrated the beneficial effects of different dried cheese extracts (the freeze-dried cheese (FDC), the cheese-lipid extract and the three cheese extracts obtained with water) on nematode longevity [4]. To better understand the mechanisms of action, the present study was focused on the identification of the signaling pathways that may be involved in these beneficial effects. *C. elegans* has already been the subject of many investigations and is now well described [23]. The insulin-like pathway, also named DAF-2/DAF-16, has been reported to be involved in the development and in the longevity phenomenon of the worm [11,24]. Many studies have highlighted that the transcription factor DAF-16 regulates the lifespan by influencing downstream genes [10,11]. As our extracts were shown to increase longevity, an assumption was established that DAF-16 was associated with this phenomenon. Two experiments were performed to determine the involvement of DAF-16: a lifespan assay with the GR1307 mutant strain that does not express DAF-16 and the study of the cellular localization of DAF-16 with the transgenic strain TJ356 expressing DAF-16::GFP. These assays were completed with a transcriptomic analysis to determine the expression of *daf*-16 when the worms were incubated with the extracts.

It has been shown that the FDC significantly increased the lifespan of the wild-type *C. elegans* N2 strain. The longevity assay on the DAF-16 loss-of-function strain (GR1307) showed a difference in the effect of the cheese, with a disappearance of the beneficial effects. There was no variation in the mean and maximum lifespan of the worms incubated with FDC compared with the CC2 worms. The results demonstrated that FDC cannot increase longevity without DAF-16. However, for both concentrations, the FDC curve was significantly above the CC2 curve, suggesting that the FDC exercised a beneficial effect on the worms, allowing a larger part of the population to remain alive for a given amount of time compared with the CC2 condition. The hypothesis of the implication of DAF-16 was also consolidated by the observations of the cellular localization of DAF-16::GFP (Figure 5) and the analysis of gene expression. Both experiments validated the implication of DAF-16. The FDC tended to translocate the transcription factor into the nuclei after 2 h of incubation. The gene expression analysis showed that the FDC significantly upregulated the expression of *daf*-16 at 10 days. All results validated the implication of the DAF-16 transcription factor in the mechanisms of action of the FDC in increasing the lifespan of the nematode.

The cheese-lipid extract presented a dose effect during the longevity assay. A significant increase in longevity was only observed for the 0.5% (p = 0.02). The study of the DAF-16::GFP consolidated the observation made during the longevity assay, as well as the RT-qPCR analysis, for which the extract significantly overexpressed *daf*-16 at 10 days. These results validated the implication of the transcription factor DAF-16 in the mechanisms of action of the extract in increasing the longevity in the wild-type *C. elegans* N2 strain.

The monitoring of the cellular localization of DAF-16::GFP showed that the extracts W40 and WF tended to activate the signaling pathway by translocating the protein into the nuclei. The transcriptomic analysis consolidated this observation for WF and W40 which significantly up-regulated the expression of the gene at 3 days and at 10 days, respectively. These observations were reinforced by the inability of the extracts to increase the longevity of the GR1307 strain (DAF-16 loss-of-function). Even more, the curves representing the extracts were significantly below the CC1 curve (for both concentrations), suggesting that the transcription factor is necessary and required for WF and W40 to exhibit a beneficial effect. Concerning W70, no differences were observed with the CC1 condition for either concentration. Although W70 did not overexpress the gene *daf*-16, it tended to translocate the transcription factor into the nuclei. In agreement with the observation made on the wild-type *C. elegans* N2 strain the extract did not show any toxic effect on the worms. All the results obtained validated the hypothesis that the DAF-16 transcription factor is involved in the biological mechanism, allowing the longevity of the nematode to increase when incubated with the extracts. The proof of the implication of DAF-16 completes the results obtained in the previous study [4], by characterizing the mechanisms of action of the extract.

As DAF-2/DAF-16 pathway is a homologue of the insulin pathway in human, it is conceivable that the effects of the bioactive extracts on the *C. elegans* model may be similar in humans, using the same pathway. Further studies are required to investigate this hypothesis.

4.2. The Dried Cheese Extracts Influenced the Survival of the Worms on the Oxidative Medium

During aging, the defense mechanisms of the worms are weakened, making the nematode more sensitive to the oxidative stress [9]. The beneficial effects of the dried cheese extracts observed in longevity may be due to an improvement of the worms' resistance to the oxidative stress and/or to the interaction with the DAF-16 transcription factor. To test this assumption, the effect of the extracts on the survival rate of the worms on the oxidative medium was evaluated.

The assay demonstrated that the worms incubated with the dried cheese extracts exhibited a better resistance to the oxidative medium, with an increase in the survival rate of up to 4.4 (Figure 6). Following these observations, the investigation of the signaling pathways implicated in the mechanisms of action was performed with the GR1307 (which does not express DAF-16) and IG10 (which does not express the receptor TOL-1) mutant strains. The insulin-like pathway is also described to be involved in immunity and stress resistance [22,25]. The results of the longevity assays suggested that DAF-16 is implicated in the beneficial effect of the extracts. The same survival assay on an oxidative medium was performed with the GR1307 strain. In absence of the protein DAF-16, no variation in the effect of the cheese extracts was observed in the worms, as evidenced by the survival rates, except for the cheese-lipid extract for which the increased survival rate was not significant compared with the control. This extract requires the expression of the protein DAF-16 to exhibit an improvement in the worms' resistance to the oxidative stress.

Another signaling pathway was investigated to describe the mechanisms of action of the cheese extracts. The p38 MAPK is a pathway involved in the human immune system [26]. This pathway is highly conserved in *C. elegans*, and is also implicated in the worms' resistance to oxidative stress by the synthesis of glutathione [8,22,27]. The receptor TOL-1, which is connected to the p38 MAPK, is a homolog of the toll-like receptor family in humans, which is implicated in immunity [9] and may play a role in the effect of the extracts. The survival assay on the oxidative medium performed with the TOL-1-deficient mutant strain showed the same results as the GR1307 strain. Indeed, the lipid extract alone could not significantly increase the survival rate of the worms without the presence of TOL-1, suggesting that the lipid extract also needs the p38 MAPK as well to exert its effect. The transcriptomic analysis revealed that the expression of two downstream genes of the p38 MAPK pathway, *sek*-1 and *pmk*-1, was modulated by some cheese extracts. The FDC significantly upregulated *sek*-1 and *pmk*-1, and the cheese-lipid extract, only influenced the expression of *pmk*-1, suggesting that the two extracts exert an effect via the p38 MAPK signaling pathway. These results suggest that the cheese-lipid extract requires the activation of the two signaling pathways to improve the resistance of the worms. With regard to WF, W40 and W70, they did not influence the expression of these genes. The results from the assays suggest that they may use another pathway for increasing the survival rate of the worms. Further mechanistic studies are needed to better understand their mechanisms of action.

4.3. Influence of the Dried Cheese Extracts on ROS Production in Human Leukocytes

The FDC (freeze-dried cheese), the cheese-lipid extract and the cheese extracts WF, W40 and W70 demonstrated the ability to increase the longevity of the wild-type *C. elegans* N2 strain, as well as its survival rate on an oxidative medium, suggesting an action on the worms' resistance to the oxidative stress. In keeping with the investigation of the effects of the extract on this stress, an in vitro assay was conducted to determine the capacity of four extracts (cheese-lipid extract, WF, W40 and W70) to reduce the ROS production in human leukocytes. Only two extracts reduced the amount of ROS produced by the cells when triggered with PMA. The cheese-lipid extract seemed to exhibit an effect without any dose response, suggesting that the maximum efficient concentration was reached. Further studies with lower concentrations should give more information on the activity of the cheese-lipid extract on the ROS production. Conversely, the extract W70 only demonstrated a beneficial effect for the highest concentration (200 µg/mL). For the lower concentrations, the extract tended to decrease the ROS production but without significant differences, suggesting that the effect of W70 increased with the concentration. This assay allowed us to make the correlation between the findings observed in *C. elegans* and in humans. The two effective extracts were able to improve the resistance of the worms to the oxidative medium and to reduce the ROS production in the human lymphocyte cells. Further studies are required in order to determine if the mechanisms of action involved in human cells are similar to those in *C. elegans*.

4.4. The Milk and Its Extracts Exert a Lower Beneficial Effect on the Wild-Type C. elegans N2 Strain Compared with the Cheese and Its Extracts

The goal of this assay was to determine if the milk could exert the same effect as the cheese on the longevity and survival rate of the *C. elegans*. The results suggested that, except for the extract MB (obtained with ethyl acetate as a solvent), all of the dried milk extracts significantly increased the longevity of the wild-type *C. elegans* N2 strain. However, this beneficial effect seems lower in comparison with the effect of the dried cheese extracts as determined in our previous study [4]. Indeed, the comparison between the milk extracts and the equivalent cheese extracts, at the same concentration, revealed that milk was less efficient in improving the longevity of the nematode *C. elegans* than the cheese. The increase of the mean lifespan was higher with the freeze-dried milk (FDM) at 0.5% than the freeze-dried cheese (FDC) at 0.5% (+25% for the FDM and +18% for the FDC), but the maximum lifespan was lower. The FDM increased the maximum lifespan by 13%, whereas the FDC increased it by 63% at the same concentration. Moreover, the part of population still alive on the conditions, when the worms of the control had died, was higher with the FDC, with a maximum of 13% of the population still alive, whereas only a maximum of 5% was observed with the worms on the FDM. The same observation was made when the cheese-lipid extract was compared to the milk-lipid extract: a mean lifespan higher with the milk-lipid extract (+30% against +18%, for the same 0.5% concentration) but the maximum lifespan was lower than the cheese-lipid extract (+30% against +37%, for the same 0.5% concentration). The differences observed in their biological response may be due to a variation in the metabolite composition and/or concentration between these two dairy foods, induced by the microorganism's activities (fermentation, etc.) of the milk during the cheese-making process. This observation was confirmed by the fact that the cheese extracts A (extracted with dichloromethane) and B (extracted with ethyl acetate) killed the worms instantly, whereas survival curves from the longevity study with the equivalent extracts of milk (MA and MB) were recovered, confirming the variation in the metabolite composition, or their concentration, between the two dairy foods.

The same observations were noted in the survival assay on the oxidative medium. The results suggested that the milk extracts were less beneficial than the cheese extracts. Indeed, only two milk extracts (freeze-dried milk (FDM) and extract MC) improved the survival of the worms, whereas the five cheese extracts (FDC, cheese-lipid extract, WF, W40 and W70) exerted a beneficial effect on the wild-type *C. elegans* N2 strain. Moreover, the survival rate was higher for the worms incubated with the cheese extracts. The survival assay on the oxidative medium, as well as the longevity assay, validated that the milk extracts had a less favorable impact on the *C. elegans* nematode.

5. Conclusions

This study allowed us to deepen the understanding of the biological effects of dried cheese extracts on longevity and resistance to oxidative stress in *C. elegans*. To our knowledge, this investigation had never been performed. The study of the signaling pathways involved in the mechanisms of action of the cheese extracts revealed, for the first time, that the insulin-like pathway is implicated, via DAF-16, to increase longevity. The extracts also revealed the capacity to increase the survival rate of the worms that were incubated on an oxidative medium. However, no evidence of the implication of the DAF-2/DAF-16 and/or the p38 MAPK in this mechanism has been highlighted. Only the cheese-lipid extract seemed to require these two pathways to improve the worms' resistance. The beneficial effects of the cheese extracts on the wild-type *C. elegans* N2 strain were correlated with their effect in human leukocytes. Indeed, two extracts (cheese-lipid extract and W70) decreased the ROS production, confirming the link between the results obtained with the nematode and human cells. These results allow us to hypothesis that the benefits of the raw-milk cheese could be similar in the aging process in humans. However, further studies on human cells are needed to pursue the investigation of the action of the extracts and understand the mechanisms implied in this phenomenon. Finally, the comparison of the effects of the milk and cheese extracts on longevity and the survival rate of the worms demonstrated a less favorable effect of the milk compared with the cheese, suggesting that the bioactive metabolites were only present, or were at a higher concentration, in the fermented food. The investigation of the bioactive metabolites in the goat cheese should continue, by subfractionating the interesting extracts and evaluating the biological effects of the resulting subfractions on the *C. elegans* and human cells in order to deepen our knowledge of the biological activity of the cheese. Alongside, a comparison of the beneficial effects of the goat cheese from this study and other raw-milk cheeses should give more information regarding the bioactive metabolites in cheese and its potential applications concerning human health.

Author Contributions: Conceptualization, G.C., P.C., S.B. and L.R.; methodology, G.C., I.R., C.P., M.B., P.V., P.C., S.B. and L.R.; validation, I.R., P.C., S.B. and L.R.; formal analysis, G.C.; investigation, G.C., C.P., M.B. and P.V.; data curation, G.C.; writing—original draft preparation, G.C.; writing—review and editing, I.R., C.P., M.B., P.V., P.C., J.P., S.B. and L.R.; supervision, P.C., A.C., E.S., J.P., S.B. and L.R. All authors have read and agreed to the published version of the manuscript.

Funding: This work was supported by VetAgro Sup by way of grants to G.C. Dômes Pharma provided support in the form of salaries for authors A.C., E.S. and J.P. Dômes Pharma also contributed by providing research materials for this study.

Data Availability Statement: All relevant data are within this manuscript.

Conflicts of Interest: The funders (VetAgro Sup and Dômes Pharma) had no role in the design of the study; in the collection, analyses, or interpretation of data; in the writing of the manuscript; or in the decision to publish the results.

References

1. Phelan, M.; Kerins, D. The potential role of milk-derived peptides in cardiovascular disease. *Food Funct.* **2011**, *2*, 153–167. [CrossRef]
2. Muro Urista, C.; Álvarez Fernández, R.; Riera Rodriguez, F.; Arana Cuenca, A.; Téllez Jurado, A. Review: Production and functionality of active peptides from milk. *Food Sci. Technol. Int.* **2011**, *17*, 293–317. [CrossRef] [PubMed]
3. Yasuda, S.; Kuwata, H.; Kawamoto, K.; Shirakawa, J.; Atobe, S.; Hoshi, Y.; Yamasaki, M.; Nishiyama, K.; Tachibana, H.; Yamada, K.; et al. Effect of highly lipolyzed goat cheese on HL-60 human leukemia cells: Antiproliferative activity and induction of apoptotic DNA damage. *J. Dairy Sci.* **2012**, *95*, 2248–2260. [CrossRef] [PubMed]
4. Cardin, G.; Ripoche, I.; Poupet, C.; Bonnet, M.; Veisseire, P.; Chalard, P.; Chauder, A.; Saunier, E.; Priam, J.; Bornes, S.; et al. Development of an innovative methodology combining chemical fractionation and in vivo analysis to investigate the biological properties of cheese. *PLoS ONE* **2020**, *15*, e0242370. [CrossRef]
5. Wilson, M.A.; Shukitt-Hale, B.; Kalt, W.; Ingram, D.K.; Joseph, J.A.; Wolkow, C.A. Blueberry polyphenols increase lifespan and thermotolerance in *Caenorhabditis elegans*. *Aging Cell* **2006**, *5*, 59–68. [CrossRef]
6. Clark, L.C.; Hodgkin, J. Commensals, probiotics and pathogens in the *Caenorhabditis elegans* model. *Cell. Microbiol.* **2014**, *16*, 27–38. [CrossRef] [PubMed]
7. Park, M.R.; Yun, H.S.; Son, S.J.; Oh, S.; Kim, Y. Short communication: Development of a direct in vivo screening model to identify potential probiotic bacteria using *Caenorhabditis elegans*. *J. Dairy Sci.* **2014**, *97*, 6828–6834. [CrossRef]
8. Poupet, C.; Chassard, C.; Nivoliez, A.; Bornes, S. *Caenorhabditis elegans*, a host to reveal and investigate the probiotic properties of beneficial microorganisms. *Front. Nutr.* **2020**, *7*, 135. [CrossRef]
9. Roselli, M.; Schifano, E.; Guantario, B.; Zinno, P.; Uccelletti, D.; Devirgiliis, C. *Caenorhabditis elegans* and Probiotics Interactions from a Prolongevity Perspective. *Int. J. Mol. Sci.* **2019**, *20*, 5020. [CrossRef]
10. Murphy, C.T.; McCarroll, S.A.; Bargmann, C.I.; Fraser, A.; Kamath, R.S.; Ahringer, J.; Li, H.; Kenyon, C. Genes that act downstream of DAF-16 to influence the lifespan of *Caenorhabditis elegans*. *Nature* **2003**, *424*, 277–284. [CrossRef]
11. Lee, S.S.; Kennedy, S.; Tolonen, A.C.; Ruvkun, G. DAF-16 target genes that control *C. elegans* Life-span and metabolism. *Science* **2003**, *300*, 644–647. [CrossRef]
12. Kampkötter, A.; Pielarski, T.; Rohrig, R.; Timpel, C.; Chovolou, Y.; Wätjen, W.; Kahl, R. The *Ginkgo biloba* extract EGb761 reduces stress sensitivity, ROS accumulation and expression of catalase and glutathione S-transferase 4 in *Caenorhabditis elegans*. *Pharmacol. Res.* **2007**, *55*, 139–147. [CrossRef]
13. Brenner, S. The genetics of *Caenorhabditis elegans*. *Genetics* **1974**, *77*, 71–94. [CrossRef] [PubMed]
14. Poupet, C.; Saraoui, T.; Veisseire, P.; Bonnet, M.; Dausset, C.; Gachinat, M.; Camarès, O.; Chassard, C.; Nivoliez, A.; Bornes, S. *Lactobacillus rhamnosus* Lcr35 as an effective treatment for preventing *Candida albicans* infection in the invertebrate model *Caenorhabditis elegans*: First mechanistic insights. *PLoS ONE* **2019**, *14*, e0216184. [CrossRef] [PubMed]
15. Porta-de-la-Riva, M.; Fontrodona, L.; Villanueva, A.; Cerón, J. Basic Caenorhabditis elegans methods: Synchronization and observation. *J. Vis. Exp.* **2012**, *64*, e4019. [CrossRef]
16. Breger, J.; Fuchs, B.B.; Aperis, G.; Moy, T.I.; Ausubel, F.M.; Mylonakis, E. Antifungal Chemical Compounds Identified Using a *C. elegans* Pathogenicity Assay. *PLoS Pathog.* **2007**, *3*, 168–178. [CrossRef] [PubMed]
17. Fatima, S.; Haque, R.; Jadiya, P.; Kumar, L.; Nazir, A. Ida-1, the *Caenorhabditis elegans* Orthologue of Mammalian Diabetes Autoantigen IA-2, Potentially Acts as a Common Modulator between Parkinson's Disease and Diabetes: Role of Daf-2/Daf-16 Insulin Like Signalling Pathway. *PLoS ONE* **2014**, *9*, e113986. [CrossRef]
18. Grompone, G.; Martorell, P.; Llopis, S.; González, N.; Genovés, S.; Mulet, A.P.; Fernández-Calero, T.; Tiscornia, I.; Bollati-Fogolín, M.; Chambaud, I.; et al. Anti-Inflammatory *Lactobacillus rhamnosus* CNCM I-3690 Strain Protects against Oxidative Stress and Increases Lifespan in *Caenorhabditis elegans*. *PLoS ONE* **2012**, *7*, e52493. [CrossRef] [PubMed]
19. Poupet, C.; Veisseire, P.; Bonnet, M.; Camarès, O.; Gachinat, M.; Dausset, C.; Chassard, C.; Nivoliez, A.; Bornes, S. Curative treatment of candidiasis by the live biotherapeutic microorganism *Lactobacillus rhamnosus* lcr35® in the invertebrate model *Caenorhabditis elegans*: First mechanistic insights. *Microorganisms* **2020**, *8*, 34. [CrossRef] [PubMed]
20. Hellemans, J.; Mortier, G.; De Paepe, A.; Speleman, F.; Vandesompele, J. qBase relative quantification framework and software for management and automated analysis of real-time quantitative PCR data. *Genome Biol.* **2007**, *8*, R19. [CrossRef]
21. Nakagawa, H.; Shiozaki, T.; Kobatake, E.; Hosoya, T.; Moriya, T.; Sakai, F.; Taru, H.; Miyazaki, T. Effects and mechanisms of prolongevity induced by *Lactobacillus gasseri* SBT2055 in *Caenorhabditis elegans*. *Aging Cell* **2016**, *15*, 227–236. [CrossRef] [PubMed]
22. Millet, A.C.M.; Ewbank, J.J. Immunity in *Caenorhabditis elegans*. *Curr. Opin. Immunol.* **2004**, *16*, 4–9. [CrossRef]
23. The *C. elegans* Sequencing Consortium Genome Sequence of the Nematode *C. elegans*: A Platform for Investigating Biology. *Science* **1998**, *282*, 2012–2018.
24. Finch, C.E.; Ruvkun, G. The genetics of aging. *Annu. Rev. Genomics Hum. Genet.* **2001**, *2*, 435–462. [CrossRef]

25. Kondo, M.; Yanase, S.; Ishii, T.; Hartman, P.S.; Matsumoto, K.; Ishii, N. The p38 signal transduction pathway participates in the oxidative stress-mediated translocation of DAF-16 to *Caenorhabditis elegans* nuclei. *Mech. Ageing Dev.* **2005**, *126*, 642–647. [CrossRef]
26. Nakahara, T.; Moroi, Y.; Uchi, H.; Furue, M. Differential role of MAPK signaling in human dendritic cell maturation and Th1/Th2 engagement. *J. Dermatol. Sci.* **2006**, *42*, 1–11. [CrossRef] [PubMed]
27. Pastuhov, S.I.; Hisamoto, N.; Matsumoto, K. MAP kinase cascades regulating axon regeneration in *C. elegans*. *Proc. Jpn. Acad. Ser. B Phys. Biol. Sci.* **2015**, *91*, 63–75. [CrossRef]

Review

Differences in the Effects of Anthocyanin Supplementation on Glucose and Lipid Metabolism According to the Structure of the Main Anthocyanin: A Meta-Analysis of Randomized Controlled Trials

Risa Araki [1,2], Akira Yada [1,3], Hirotsugu Ueda [1], Kenichi Tominaga [1,3] and Hiroko Isoda [1,2,4,5,*]

[1] Open Innovation Laboratory for Food and Medicinal Resource Engineering (FoodMed-OIL), National Institute of Advanced Industrial Science and Technology (AIST), 1-1-1 Tennodai, Tsukuba 305-8577, Japan; raar51835@gmail.com (R.A.); a-yada@aist.go.jp (A.Y.); ueda.hirotsugu@aist.go.jp (H.U.); k-tominaga@aist.go.jp (K.T.)

[2] R&D Center for Tailor-Made QOL, University of Tsukuba, 1-2 Kasuga, Tsukuba 305-8550, Japan

[3] Interdisciplinary Research Center for Catalytic Chemistry, National Institute of Advanced Industrial Science and Technology (AIST), Tsukuba Central 5, 1-1-1 Higashi, Tsukuba 305-8565, Japan

[4] Alliance for Research on the Mediterranean and North Africa (ARENA), University of Tsukuba, 1-1-1 Tennodai, Tsukuba 305-8572, Japan

[5] Faculty of Life and Environmental Sciences, University of Tsukuba, 1-1-1 Tennodai, Tsukuba 305-8572, Japan

* Correspondence: isoda.hiroko.ga@u.tsukuba.ac.jp; Tel.: +81-298-53-5775

Citation: Araki, R.; Yada, A.; Ueda, H.; Tominaga, K.; Isoda, H. Differences in the Effects of Anthocyanin Supplementation on Glucose and Lipid Metabolism According to the Structure of the Main Anthocyanin: A Meta-Analysis of Randomized Controlled Trials. *Nutrients* **2021**, *13*, 2003. https://doi.org/10.3390/nu13062003

Academic Editor: Yoshinori Katakura

Received: 15 May 2021
Accepted: 6 June 2021
Published: 10 June 2021

Publisher's Note: MDPI stays neutral with regard to jurisdictional claims in published maps and institutional affiliations.

Copyright: © 2021 by the authors. Licensee MDPI, Basel, Switzerland. This article is an open access article distributed under the terms and conditions of the Creative Commons Attribution (CC BY) license (https://creativecommons.org/licenses/by/4.0/).

Abstract: The effectiveness of anthocyanins may differ according to their chemical structures; however, randomized clinical controlled trials (RCTs) or meta-analyses that examine the consequences of these structural differences have not been reported yet. In this meta-analysis, anthocyanins in test foods of 18 selected RCTs were categorized into three types: cyanidin-, delphinidin-, and malvidin-based. Delphinidin-based anthocyanins demonstrated significant effects on triglycerides (mean difference (MD): -0.24, $p < 0.01$), low-density lipoprotein cholesterol (LDL-C) (MD: -0.28, $p < 0.001$), and high-density lipoprotein cholesterol (HDL-C) (MD: 0.11, $p < 0.01$), whereas no significant effects were observed for cyanidin- and malvidin-based anthocyanins. Although non-significant, favorable effects on total cholesterol (TC) and HDL-C were observed for cyanidin- and malvidin-based anthocyanins, respectively (both $p < 0.1$). The ascending order of effectiveness on TC and LDL-C was delphinidin-, cyanidin-, and malvidin-based anthocyanins, and the differences among the three groups were significant (both $p < 0.05$). We could not confirm the significant effects of each main anthocyanin on glucose metabolism; however, insulin resistance index changed positively and negatively with cyanidin- and delphinidin-based anthocyanins, respectively. Therefore, foods containing mainly unmethylated anthocyanins, especially with large numbers of OH groups, may improve glucose and lipid metabolism more effectively than those containing methylated anthocyanins.

Keywords: anthocyanins; structure; glucose and lipid metabolism; human health; meta-analysis

1. Introduction

Anthocyanidins are water-soluble pigments consisting of three ring structures: a double benzoyl ring A and B, and a heterocyclic C ring. They are classified, based on the number of hydroxyl and methoxyl groups attached to the B ring, into six types based: cyanidin, delphinidin, pelargonidin, peonidin, malvidin, and petunidin (Figure 1), the percentage distributions of which in red to purplish-blue-colored foods are 50%, 12%, 12%, 12%, 7%, and 7%, respectively [1,2]. Anthocyanidins are normally present in foods in their glycoside forms, called anthocyanins. Anthocyanins may have organic acids, such as succinic acid and malonic acid, bound to them, apart from sugars, and can be acylated [3]. After absorption, glycosides of cyanidin, delphinidin, pelargonidin, peonidin, malvidin, and petunidin are metabolized to protocatechuic acid (PCA), gallic acid, 4-hydroxybenzoic acid, syringic acid, and vanillic acid, respectively [4].

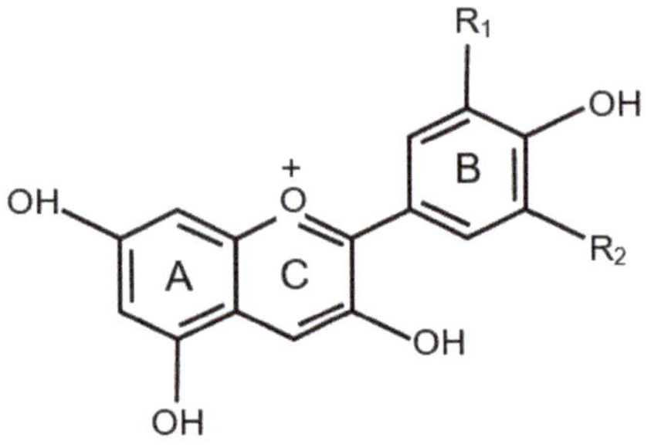

Anthocyanidin	R_1	R_2
Pelargonidin	-H	-H
Cyanidin	-OH	-H
Delphinidin	-OH	-OH
Peonidin	$-OCH_3$	-H
Petunidin	$-OCH_3$	-OH
Malvidin	$-OCH_3$	$-OCH_3$

Figure 1. Molecular structures of anthocyanidins.

Of the six compounds, delphinidin has the highest polarity [5], making it more soluble in water than malvidin [6], which has the lowest polarity. Studies have indicated that addition of methoxyl groups to the B ring may reduce solubility [1]. Pelargonidin-based anthocyanins are more easily absorbed than cyanidin-based anthocyanins, and this is suggested to be attributed to the structure of pelargonidin, which cannot undergo methylation due to only one hydroxyl group on B ring and may be more available for glucuronidation [7]. The higher the number of hydroxyl groups, the stronger the blue color of the anthocyanins, and replacing the hydroxyl groups on the B ring with methoxyl groups results in a reddish color. In particular, pelargonidin with 4'-hydroxyl groups, cyanidin with 3',4'-dihydroxyl groups, and delphinidin with 3',4',5'-trihydroxyl groups have orange, reddish-purple, and blue color, respectively [8]. Methoxyl groups in the B ring improve the stability of the anthocyanins in the digestive process, whereas hydroxyl groups reduce it [9]. Moreover, delphinidin-based anthocyanins exhibit higher biophysical interaction than cyanidin- and pelargonidin-based anthocyanins [10].

Anthocyanins are effective antioxidants [11], and their structure, with a positively charged oxygen atom in the C ring [1], lends them their antioxidant activity [12]. Oxidative damage is involved in the onset and progression of various diseases [13], and many health-promoting effects of anthocyanins have been reported in in vitro and clinical studies [1–3].

Seeram et al. [14] reported that the antioxidant activities of anthocyanidins increase with the increasing number of hydroxyl groups in the B ring, as evidenced by the following order of decreasing activity: delphinidin with 3',4',5'-trihydroxyl groups (70%) > cyanidin with 3',4'-dihydroxyl groups (60%) > pelargonidin with 4'-hydroxyl groups (40%). They also confirmed a decrease in antioxidant activity on substitution of the hydroxyl groups with methoxyl groups, with the activity in the following order: peonidin with 3'-methoxyl groups (45%) > malvidin with 3',5'-dimethoxyl groups (43%). Furthermore, the number of glycosyl groups in the A and C rings [15] and the acylation of glycosides [16] have been suggested as factors that reduce the antioxidant activity of anthocyanins.

Excessive intracellular oxidation is considered a risk factor for type 2 diabetes as it leads to inflammation, which in turn induces pancreatic β-cell damage and insulin secretion disorder [17]. Therefore, the differences in the antioxidant activity of anthocyanins owing to their structural differences may variably affect the ability of the pancreas to secrete insulin. Studies have reported a positive association between the ability to secrete insulin and the number of hydroxyl groups in the B rings of anthocyanins [18]. The inhibitory activity of aldose reductase in anthocyanins is attenuated upon methylation of the substituent in the B ring; hence, cyanidin-based anthocyanins show higher activity than peonidin-based anthocyanins [19]. The inhibitory activity of the intestinal and pancreatic glycolytic enzymes increases when glucose is added at the 3-O position but decreases when it is added at the 5-O position [20]. Cell culture studies have reported decreased triglyceride accumulation and mRNA expression of the fatty acid synthase (FAS) and sterol regulatory element-binding protein-1c (SREBP-1c) in 3T3-L1 adipocytes upon treatment with cyanidin compared to that with malvidin [21]. In another experimental study, vascular endothelial growth factor (VEGF) release in vascular smooth muscle cells by platelet-derived growth factor AB ($PDGF_{AB}$) stimulation, one of the risk factors for arteriosclerosis, was inhibited by cyanidin and delphinidin but not by malvidin and peonidin [22]. Additionally, Skemiene et al. has suggested that ischemia-induced activation of the apoptosis-promoting factor caspase depends on the reduction ability of cytosolic cytochrome c, and this ability was the highest in delphinidin-3-glucoside followed by cyanidin-3-glucoside while it was lower in pelargonidin-, malvidin-, and peonidin-3-glucoside [23].

Similarly, the extent to which anthocyanins improve glucose and lipid metabolism among humans may vary depending on their structure and have not yet been investigated using clinical trials or meta-analyses. Therefore, we conducted a meta-analysis of randomized controlled trials (RCTs) to estimate the differences in the effects of anthocyanins on glucose and lipid metabolism based on the structure of the main compound of their sources.

2. Materials and Methods

We conducted this meta-analysis based on the Preferred Reporting Items for Systematic Reviews and Meta-Analyses (PRISMA) statement [24].

2.1. Search Strategy and Trial Selection

A literature search was conducted up to 5 March 2021, using the databases of PubMed, Cochrane Library, and Web of Science. First, articles were screened according to their titles and abstracts for trials meeting the following inclusion criteria: (1) participants aged ≥18 years; (2) randomized, parallel-group, placebo-controlled clinical trials that compared purified anthocyanins or anthocyanin-rich extracts or anthocyanin-rich food as test foods to placebo or appropriate controls (not therapeutic agents); (3) intervention ≥4 weeks; (4) availability of data for blood triglyceride (TG), total cholesterol (TC), low-density lipoprotein cholesterol (LDL-C), high-density lipoprotein cholesterol (HDL-C), glucose or insulin, or HbA1c levels after fasting or homeostatic model assessment of insulin resistance (HOMA-IR) index; and (5) independent studies written in English.

Trials were excluded if (1) the participants were aged <18 years; (2) a control group was not included or a therapeutic agent was used as a control; (3) anthocyanin supplementation was combined with other functional ingredients; (4) intervention duration was <4 weeks; (5) the intervention was combined with exercise loading; (6) details of the sources of anthocyanins and/or daily dosage of anthocyanins were not described; (7) appropriate data extraction details not provided; (8) the language of publication was not English; or (9) the publication was not an independent research article (e.g., review, conference reports, editorial, and clinical trials registration).

The search terms for PubMed were ("anthocyanin" OR "anthocyanins" OR "cyanidin" OR "delphinidin" OR "malvidin" OR "peonidin" OR "petunidin" OR "pelargonidin") AND ("randomized clinical controlled trial" OR "randomized controlled trial" OR "RCT") AND ("metabolic" OR "lipids" OR "TG" OR "triglyceride" OR "TC" OR "LDL" OR "HDL" OR "Glucose" OR "Insulin" OR "HbA1c" OR "HOMA") NOT ("in vitro" OR "review" OR "bioavailability" OR "kinetics" OR "excretion" OR "children" OR "postprandial" OR "acute"). The search terms for Cochrane Library were ("randomized controlled study" OR "randomized controlled" OR "RCT") NOT ("review" OR "meta-analysis" OR "postprandial" OR "acute" OR "children" OR "bioavailability" OR "excretion" OR "kinetics" OR "animal" OR in "vitro") AND ("anthocyanin" OR "anthocyanins" OR "cyanidin" OR "delphinidin" OR "malvidin" OR "peonidin" OR "petunidin" OR "pelargonidin") AND ("metabolic" OR "cardiovascular" OR "lipids" OR "triglyceride" OR "cholesterol" OR "TC" OR "LDL-C" OR "HDL-C" OR "Glucose" OR "Insulin" OR "HOMA" OR "HbA1c"). The search terms for Web of Science were: #1 (TS = (randomized controlled clinical trial* OR

RCT) AND TS = (anthocyanin* OR cyanidin* OR delphinidin* OR malvidin* OR peonidin* OR petunidin* OR pelargonidin) NOT TS = (in vitro* OR review* OR meta-analysis* OR children* OR bioavailability* OR acute* OR postprandial)) AND #2 (TS = (metabolic* OR cardiovascular* OR lipids* OR triglyceride* OR cholesterol* OR LDL* or HDL* OR Glucose* OR insulin* OR HOMA* OR HbA1c* OR TG) AND language: (English) AND document type: (Article)).

As shown in Figure 2, a total of 183 articles were obtained in the initial search, of which 130 papers were assessed after removing 53 duplicate articles. Forty-three articles were reviewed in full after reviewing the title and abstract. Of these, we excluded three trials that used the same population as the selected studies. Finally, 18 eligible studies (22 trials) were included in our meta-analysis.

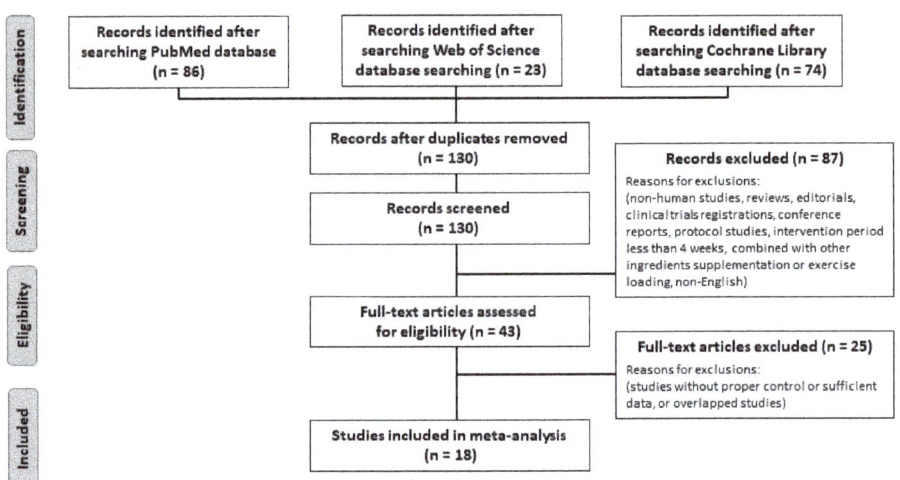

Figure 2. Flow diagram showing the trial selection process.

2.2. Outcomes

The primary outcomes of this study were changes in parameters of lipid metabolism (TG, TC, LDL-C, HDL-C) and glucose metabolism (fasting glucose, insulin, HbA1c, HOMA-IR).

2.3. Data Extraction and Risk of Bias Assessment

The following information were extracted from each trial: first author, years of data collection, year of publication, clinical characteristics of participants, sample size, duration of intervention, sources of anthocyanins, main anthocyanin in the test foods, daily dosage of anthocyanins, and the means and standard deviations (SD) of glucose and lipid profile at baseline and at the end of the intervention in test and control groups or change by group. If the trial presented the data distribution as standard errors (SE), it was converted to SD by multiplying SE by the square root of the sample size. If the trial presented the data as medians and ranges, it was converted to means and SD using the formula proposed by Hozo et al. [25]. The unit was standardized to mmol/L for TG, TC, LDL, HDL-C, and glucose, μIU/mL for insulin, and %National Glycohemoglobin Standardization Program (NGSP) for HbA1c.

The anthocyanins in the test foods used in all trials were mixtures of different classes of anthocyanins rather than single compounds. Therefore, we identified the main anthocyanin in the test food as the one mentioned in the articles or its references as the "most abundant" or "predominant", constituting >50% of the total anthocyanin content, or having the highest content among the six compounds. If such information could not be obtained, we defined the main anthocyanin based on the component composition shown in a related review paper or product information.

We judged the risk of bias of the selected studies as low, high, or unclear according to the Cochrane Handbook for Systematic Reviews of Interventions for the following seven domains: (1) random sequence generation; (2) allocation concealment; (3) blinding of participants and personnel; (4) blinding of outcome assessment; (5) incomplete outcome data; (6) selective reporting; and (7) other potential threats to validity [26]. We assessed the evidence using GRADE criteria (risk of bias; inconsistency of results; indirectness of evidence; imprecision; and publication bias) [27] and created a summary of findings table.

These were conducted by two authors independently, and disagreements were resolved through discussion with all authors.

2.4. Data Analysis

This meta-analysis was conducted using RevMan 5.4 (the Cochrane Collaboration, Odense, Denmark) and StatsDirect statistics software (StatsDirect Ltd., Birkenhead, UK).

The weighted mean differences (MD) for net change and 95% confidence intervals (CI) were used to estimate the effect of anthocyanins on parameters of glucose and lipid metabolism. For trials with ≥ 2 treatment groups, we compared each treatment group with the control group.

Statistical heterogeneity across studies were assessed using Cochran's Q test and calculating the I^2 value, and $p < 0.1$ and/or I^2 value ≥ 50 was considered significant [28]. If heterogeneity in either whole food group or each compound group was significant, a random effect model was applied and if none of the heterogeneities were significant, a fixed effects statistical model was applied. To evaluate the robustness of the results and sources of heterogeneity, we performed sensitivity analyses using leave-one-out method. If the significances of the effects on each parameter could not be observed or sensitivity analyses could not be improved for high heterogeneity, subgroup analyses of further subdivisions were performed. Potential publication bias was assessed visually and through Egger's test [29,30]. Except for heterogeneity, $p < 0.05$ was judged to be statistically significant.

3. Results

3.1. Study Characteristics

A summary of the characteristics of the 22 trials included in this study are presented in Table 1.

The target population were patients with dyslipidemia in six trials, those with prediabetes and/or type 2 diabetes in three trials, healthy individuals in six trials, overweight or obese individuals in three trials, patients with metabolic syndrome in two trials, and others in two trials. Out of the 22 trials, 20 targeted both genders, and the remaining two trials targeted only men. The source of anthocyanins were various berries, black soybean, and whole purple wheat. The main anthocyanin in the test foods was delphinidin in 13 trials, cyanidin in five trials, and malvidin in four trials. The dosage of anthocyanins ranged from 1.65 to 320 mg/d (mean 160 mg/d), the ratio of main anthocyanin to total anthocyanins ranged from 34% to 98.4%, and the treatment duration ranged from 4 to 24 weeks.

Table 1. Characteristics of selected trials.

Study ID	Participants	Sample Size (n)		Gender	Duration (Week)	Intervention Material	Type of Anthocyanin Source	Dosage (mg/d)	Main Anthocyanins	
		Test Group	Control Group						Compounds [Most Abundant Type]	% of Total [$]
Bakuradze, T., 2019 [31]	Healthy	30	27	Men	8	Anthocyanin-rich fruit juice	Extract	205.5	malvidin [mv-3-glc]	35.5
Davinelli, S., 2015 [32]	Overweight	26	16	Both	4	Maqui berry extract	Extract	162	delphinidin [unknown]	80
Gamel, T., 2020 [33]	Overweight or obese	15	13	Both	8	Whole purple wheat bars	Extract	1.65	cyanidin [cy-3-glc]	83.7 [34]
Hansen, A.S., 2015-a [35]	Healthy	15	18	Both	4	Red grape extract	Extract	24–36	malvidin [mv-3-glc]	41.2–41.3
Hansen, A.S., 2015-b [35]	Healthy	17	18	Both	4	Red grape extract	Extract	48–71	malvidin [mv-3-glc]	38.1–38.7
Johnson, S.A., 2020 [36]	Metabolic syndrome	9	10	Both	12	Tart cherry juice	Extract	88	cyanidin [cy-3-glc]	42
Khan, F., 2014-a [37]	Healthy	22	21	Both	6	Blackcurrant juice	Extract	40	delphinidin [dp-3-rut]	54.6 [38]
Khan, F., 2014-b [37]	Healthy	21	21	Both	6	Blackcurrant juice	Extract	143	delphinidin [dp-3-rut]	54.6 [38]
Kianbakht, S., 2014 [39]	Dyslipidemia	40	40	Both	8	*Vaccinium arctostaphylos* fruit hydroalcoholic extract	Extract	7.35	delphinidin [unknown]	41.0 [40]
Kim, H., 2018 [41]	Metabolic syndrome	19	18	Both	12	Açaí beverage	Extract	216	cyanidin [cy-3-rut]	98.4 [42]
Krikorian, R., 2012 [43]	Mild cognitive impairment	10	11	Both	16	Concord grape juice	Extract	142–208	delphinidin [dp-3-glc]	40.1 [44]
Lee, M., 2016 [45]	Overweight or obese	32	31	Both	8	Black soybean testa extract	Extract	31.45	cyanidin [cy-3-glc]	68.3
Li, D., 2015 [46]	Type 2 diabetes	29	29	Both	24	Purified anthocyanins from bilberry and blackcurrant	Purified anthocyanins	320	delphinidin [dp-3-glc]	58.0 [47]

Table 1. Cont.

Study ID	Participants	Sample Size (n)		Gender	Duration (Week)	Intervention Material	Type of Anthocyanin Source	Dosage (mg/d)	Main Anthocyanins	
		Test Group	Control Group						Compounds [Most Abundant Type]	% of Total $
Lynn, A., 2014 [48]	Healthy	24	19	Both	6	Tart cherry juice	Extract	273.5	cyanidin [cy-3-rut]	80.0 [49]
Soltani, R., 2014 [50]	Dyslipidemia	25	25	Both	4	Vaccinium arctostaphylos L. fruit extract	Extract	90	delphinidin [unknown]	41.0 [51]
Stote, K.S., 2020 [52]	Type 2 diabetes	26	26	Men	8	Freeze-dried blueberries	Others	261.8	malvidin [unknown]	34 [53]
Xu, Z., 2014-a [54]	Dyslipidemia	45	46	Both	12	Purified anthocyanins from bilberry and blackcurrant	Purified anthocyanins	40	delphinidin [dp-3-glc]	58.0 [47]
Xu, Z., 2014-b [54]	Dyslipidemia	42	46	Both	12	Purified anthocyanins from bilberry and blackcurrant	Purified anthocyanins	80	delphinidin [dp-3-glc]	58.0 [47]
Xu, Z., 2014-c [54]	Dyslipidemia	43	46	Both	12	Purified anthocyanins from bilberry and blackcurrant	Purified anthocyanins	320	delphinidin [dp-3-glc]	58.0 [47]
Yang, L., 2017 [55]	Prediabetes and early untreated diabetes	80	80	Both	12	Purified anthocyanins from bilberry and blackcurrant	Purified anthocyanins	320	delphinidin [dp-3-glc]	58.0 [47]
Zhang, P.W., 2015 [56]	NAFLD	37	37	Both	12	Purified anthocyanins from bilberry and blackcurrant	Purified anthocyanins	320	delphinidin [dp-3-glc]	58.0 [47]
Zhang, X., 2016 [57]	Dyslipidemia	73	73	Both	24	Purified anthocyanins from bilberry and blackcurrant	Purified anthocyanins	320	delphinidin [dp-3-glc]	58.0 [47]

The characteristics of 22 trials (18 studies) are shown. The study by Hansen et al. [35], Khan et al. [37], and Xu et al. [54] had two or more intervention groups. $: If the values were not described in the original papers, we applied the values of the references indicated in this column. Cy-3-glc, cyanidin-3-glucoside; dp-3-glc, delphinidin-3-glucoside; mv-3-glc, malvidin-3-glucoside; cy-3-rut, cyanidin-3-rutinoside; dp-3-rut, delphinidin-3-rutinoside.

3.2. Risk of Bias Assessment

The results of the risk of bias assessment are shown in Figure 3. Of the 22 selected trials, random sequence generation was conducted in 12 trials. It was confirmed that the allocation concealment was properly performed in eight trials, while the concealment was unclear in 13 trials and was not executed in one trial. Both participants and personnel were blinded, except in three trials (single-blinding in two trials and unblinded in one trial). With respect to incomplete outcome data and selective reporting, all trials were judged to be low-risk. In the trial by Johnson et al. [36], although there was no difference in any outcome, the age at the baseline was significantly different between the test and control groups. Therefore, other risks of bias were evaluated as unclear.

3.3. Publication Bias Assessment

Based on Egger's tests, a significant small-study effect was observed on HDL-C ($p = 0.004$). However, there was no significant evidence of small-study effect on TG ($p = 0.236$), TC ($p = 0.749$), LDL-C ($p = 0.437$), glucose ($p = 0.962$), insulin ($p = 0.386$), HOMA-IR ($p = 0.825$), and HbA1c ($p = 0.695$).

The characteristics of 22 trials (18 studies) are shown. The study by Hansen et al. [35], Khan et al. [37], and Xu et al. [54] had two or more intervention groups.

3.4. Meta-Analysis

3.4.1. Effects of Anthocyanins and Main Anthocyanin in the Test Foods on TG

Favorable effects of anthocyanins were observed (MD: -0.20, 95% CI: -0.33 to -0.07, $p < 0.01$) with moderate heterogeneity ($I^2 = 34\%$, $p = 0.10$), when all the trials that reported TG levels (n = 15) were pooled. When the trials were classified according to the main anthocyanin in the test foods, significant effects were observed only in the trials using delphinidin-based anthocyanins (MD: -0.24, 95% CI: -0.41 to -0.07, $p < 0.01$), with moderate but significant heterogeneity ($I^2 = 45\%$, $p = 0.07$) (Figure 4).

In the sensitivity analysis, it was observed that the trial by Soltani et al. [50] greatly affected the overall heterogeneity and that of trials using delphinidin-based anthocyanins; after this trial was removed from the dataset, the heterogeneities for all trials and trials using delphinidin-based anthocyanins became non-significant (all trials: $I^2 = 4\%$, $p = 0.40$, trials using delphinidin-based anthocyanins: $I^2 = 12\%$, $p = 0.34$). These overall effects remained significant as shown: all trials (MD: -0.20, 95% CI: -0.30 to -0.11, $p < 0.0001$), and trials using delphinidin-based anthocyanins (MD: -0.23, 95% CI: -0.36 to -0.10, $p < 0.001$).

The overall effect of anthocyanins was significant in the subgroups "Anthocyanin dosage above 160 mg/d" (MD: -0.18, 95% CI: -0.31 to -0.04, $p < 0.05$), "Anthocyanin extract" (MD: -0.22, 95% CI: -0.43 to -0.02, $p < 0.05$), and "Baseline TG above 1.7 mmol/L" (MD: -0.27, 95% CI: -0.42 to -0.11, $p < 0.001$). Although with significant heterogeneity, the overall effect of trials using delphinidin-based anthocyanins were also significant in the subgroups "Anthocyanin extract" (MD: -0.41, 95% CI: -0.78 to -0.04, $p < 0.05$) and "Baseline TG above 1.7 mmol/L" (MD: -0.25, 95% CI: -0.44 to -0.07, $p < 0.01$). In addition, delphinidin-based anthocyanins were favored in the subgroups "Main anthocyanin to total anthocyanin above 50%" (MD: -0.16, 95% CI: -0.32 to -0.01, $p = 0.06$) with no heterogeneity, and "Dyslipidemia" (MD: -0.27, 95% CI: -0.55 to 0.00, $p = 0.05$) with large heterogeneity ($I^2 = 58\%$, $p = 0.05$) (Table S1).

3.4.2. Effects of Anthocyanins and Main Anthocyanin in the Test Foods on TC

The effects of anthocyanins on TC levels were not significant (MD: -0.19, 95% CI: -0.42 to 0.04, $p = 0.10$) with very high heterogeneity ($I^2 = 83\%$, $p < 0.00001$) when all the data were pooled (n = 20). Based on the classification of trials according to the main anthocyanin in the test foods, favorable effects were observed in those of cyanidin-based anthocyanins (MD: -0.24, 95% CI: -0.53 to 0.04, $p = 0.09$) with insignificant heterogeneity ($I^2 = 0\%$, $p = 0.44$) and delphinidin-anthocyanins (MD: -0.30, 95% CI: -0.64 to 0.04, $p = 0.08$) with very high heterogeneity ($I^2 = 87\%$, $p < 0.00001$) (Figure 5).

Study	Random sequence generation (selection bias)	Allocation concealment (selection bias)	Blinding of participants and personnel (performance bias)	Blinding of outcome assessment (detection bias)	Incomplete outcome data (attrition bias)	Selective reporting (reporting bias)	Other bias
Bakuradze, T., 2019					+	+	+
Davinelli, S., 2015		+	+	+	+	+	+
Gamel, T., 2020	+	+		+	+	+	+
Hansen, AS., 2005-a			+		+	+	+
Hansen, AS., 2005-b			+		+	+	+
Johnson, SA., 2020	+				+	+	
Khan, F., 2014-a			+		+	+	+
Khan, F., 2014-b			+		+	+	+
Kianbakht, S., 2014	+	+	+	+	+	+	+
Kim, H., 2018			+		+	+	+
Krikorian, R., 2012			+		+	+	+
Lee, M., 2016	+		+	+	+	+	+
Li, D., 2015			+	+	+	+	+
Lynn, A., 2014	+	−	−		+	+	+
Soltani, R., 2014	+		+	+	+	+	+
Stote, KS., 2020	+		+		+	+	+
Xu, Z., 2020-a	+	+	+	+	+	+	+
Xu, Z., 2020-b	+	+	+	+	+	+	+
Xu, Z., 2020-c	+	+	+	+	+	+	+
Yang, L., 2017	+	+	+	+	+	+	+
Zhang, PW., 2015	+	+	+	+	+	+	+
Zhang, X., 2016			+		+	+	+

Figure 3. Risk of bias summary. Low risk is indicated as green, high risk as red, and unknown risk as blank.

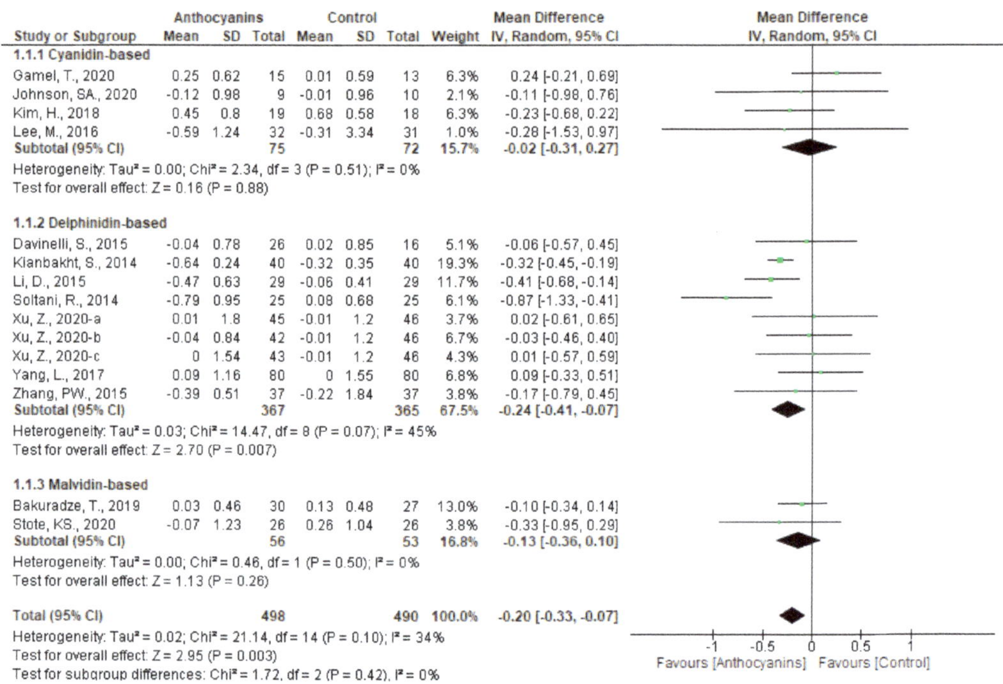

Study or Subgroup	Anthocyanins			Control			Weight	Mean Difference IV, Random, 95% CI
	Mean	SD	Total	Mean	SD	Total		
1.1.1 Cyanidin-based								
Gamel, T., 2020	0.25	0.62	15	0.01	0.59	13	6.3%	0.24 [-0.21, 0.69]
Johnson, SA., 2020	-0.12	0.98	9	-0.01	0.96	10	2.1%	-0.11 [-0.98, 0.76]
Kim, H., 2018	0.45	0.8	19	0.68	0.58	18	6.3%	-0.23 [-0.68, 0.22]
Lee, M., 2016	-0.59	1.24	32	-0.31	3.34	31	1.0%	-0.28 [-1.53, 0.97]
Subtotal (95% CI)			75			72	15.7%	-0.02 [-0.31, 0.27]
Heterogeneity: Tau² = 0.00; Chi² = 2.34, df = 3 (P = 0.51); I² = 0%								
Test for overall effect: Z = 0.16 (P = 0.88)								
1.1.2 Delphinidin-based								
Davinelli, S., 2015	-0.04	0.78	26	0.02	0.85	16	5.1%	-0.06 [-0.57, 0.45]
Kianbakht, S., 2014	-0.64	0.24	40	-0.32	0.35	40	19.3%	-0.32 [-0.45, -0.19]
Li, D., 2015	-0.47	0.63	29	-0.06	0.41	29	11.7%	-0.41 [-0.68, -0.14]
Soltani, R., 2014	-0.79	0.95	25	0.08	0.68	25	6.1%	-0.87 [-1.33, -0.41]
Xu, Z., 2020-a	0.01	1.8	45	-0.01	1.2	46	3.7%	0.02 [-0.61, 0.65]
Xu, Z., 2020-b	-0.04	0.84	42	-0.01	1.2	46	6.7%	-0.03 [-0.46, 0.40]
Xu, Z., 2020-c	0	1.54	43	-0.01	1.2	46	4.3%	0.01 [-0.57, 0.59]
Yang, L., 2017	0.09	1.16	80	0	1.55	80	6.8%	0.09 [-0.33, 0.51]
Zhang, PW., 2015	-0.39	0.51	37	-0.22	1.84	37	3.8%	-0.17 [-0.79, 0.45]
Subtotal (95% CI)			367			365	67.5%	-0.24 [-0.41, -0.07]
Heterogeneity: Tau² = 0.03; Chi² = 14.47, df = 8 (P = 0.07); I² = 45%								
Test for overall effect: Z = 2.70 (P = 0.007)								
1.1.3 Malvidin-based								
Bakuradze, T., 2019	0.03	0.46	30	0.13	0.48	27	13.0%	-0.10 [-0.34, 0.14]
Stote, KS., 2020	-0.07	1.23	26	0.26	1.04	26	3.8%	-0.33 [-0.95, 0.29]
Subtotal (95% CI)			56			53	16.8%	-0.13 [-0.36, 0.10]
Heterogeneity: Tau² = 0.00; Chi² = 0.46, df = 1 (P = 0.50); I² = 0%								
Test for overall effect: Z = 1.13 (P = 0.26)								
Total (95% CI)			498			490	100.0%	-0.20 [-0.33, -0.07]
Heterogeneity: Tau² = 0.02; Chi² = 21.14, df = 14 (P = 0.10); I² = 34%								
Test for overall effect: Z = 2.95 (P = 0.003)								
Test for subgroup differences: Chi² = 1.72, df = 2 (P = 0.42), I² = 0%								

Figure 4. Effects of anthocyanins and main anthocyanin in the test foods on TG.

In the sensitivity analyses, it was observed that the trial by Kianbakht et al. [39] highly affected the heterogeneity of all trials and trials using delphinidin-based anthocyanins. After that trial was removed, these heterogeneities became insignificant (all trials: I^2 = 30%, p = 0.11, trials of delphinidin-based anthocyanins: I^2 = 16%, p = 0.29). The significance of the overall effect of all trials (MD: −0.08, 95% CI: −0.20 to 0.04, p = 0.19), and trials using delphinidin-based anthocyanins (MD: −0.14, 95% CI: −0.27 to 0.00, p = 0.06) were not changed.

The results of the subgroup analyses are shown in Table S2. In the subgroup "Main anthocyanin to total anthocyanin above 50%", the heterogeneity of all trials (I^2 = 0%, p = 0.95) and trials using delphinidin-based anthocyanins (I^2 = 0%, p = 0.99) reduced to zero. The overall effect of all the trials included became favorable (MD: −0.11, 95% CI: −0.23 to 0.01, p = 0.07), whereas the favorable tendency observed for trials using delphinidin-based anthocyanins disappeared (MD: −0.08, 95% CI: −0.21 to 0.05, p = 0.21). Similar results were also observed in the subgroup "Anthocyanin dosage above 160 mg/d". In contrast, in the subgroup "Baseline TC above 5.17 mmol/L", favorable effects were observed only in the trials using delphinidin-based anthocyanins (MD: −0.40, 95% CI: −0.83 to 0.03, p = 0.07) with very high heterogeneity (I^2 = 90%, p < 0.00001). Furthermore, in the subgroup "Anthocyanin extract", overall effects were favorable only according to the pooled data of cyanidin-based anthocyanin trials (MD: −0.24, 95% CI: −0.53 to 0.04, p = 0.09). The MD of cyanidin- and delphinidin-based anthocyanin trials showed negative values, while that of malvidin showed positive values. Therefore, a significant difference was observed among the three main anthocyanin trials (p < 0.05).

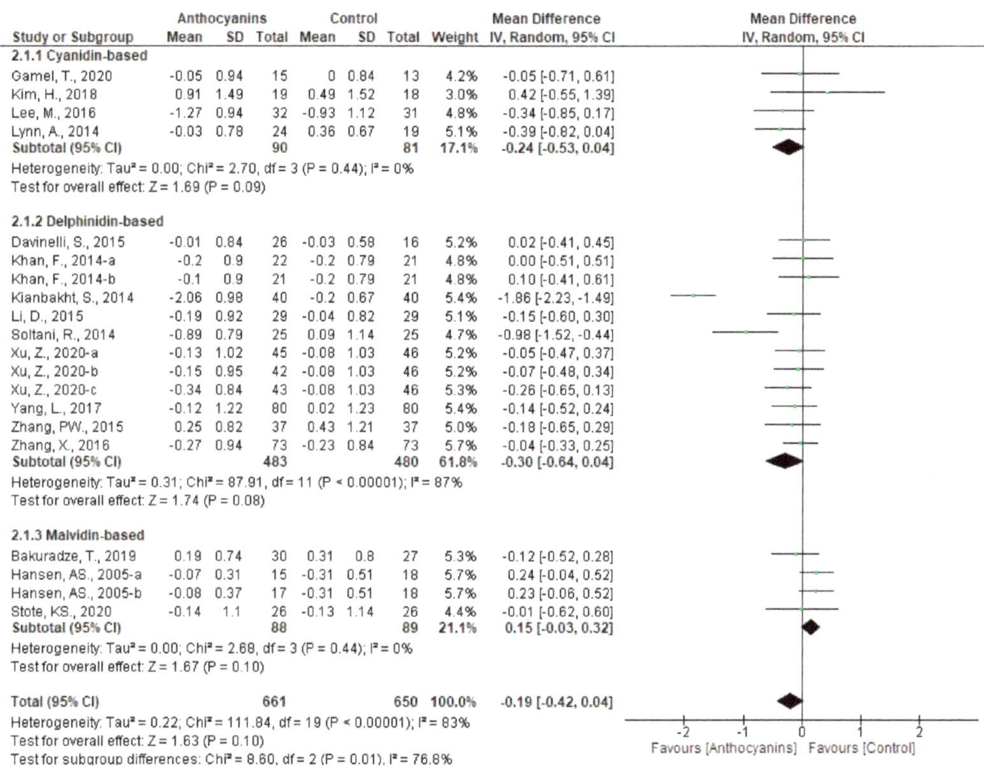

Study or Subgroup	Anthocyanins Mean	SD	Total	Control Mean	SD	Total	Weight	Mean Difference IV, Random, 95% CI
2.1.1 Cyanidin-based								
Gamel, T., 2020	-0.05	0.94	15	0	0.84	13	4.2%	-0.05 [-0.71, 0.61]
Kim, H., 2018	0.91	1.49	19	0.49	1.52	18	3.0%	0.42 [-0.55, 1.39]
Lee, M., 2016	-1.27	0.94	32	-0.93	1.12	31	4.8%	-0.34 [-0.85, 0.17]
Lynn, A., 2014	-0.03	0.78	24	0.36	0.67	19	5.1%	-0.39 [-0.82, 0.04]
Subtotal (95% CI)			90			81	17.1%	-0.24 [-0.53, 0.04]
Heterogeneity: Tau² = 0.00; Chi² = 2.70, df = 3 (P = 0.44); I² = 0%								
Test for overall effect: Z = 1.69 (P = 0.09)								
2.1.2 Delphinidin-based								
Davinelli, S., 2015	-0.01	0.84	26	-0.03	0.58	16	5.2%	0.02 [-0.41, 0.45]
Khan, F., 2014-a	-0.2	0.9	22	-0.2	0.79	21	4.8%	0.00 [-0.51, 0.51]
Khan, F., 2014-b	-0.1	0.9	21	-0.2	0.79	21	4.8%	0.10 [-0.41, 0.61]
Kianbakht, S., 2014	-2.06	0.98	40	-0.2	0.67	40	5.4%	-1.86 [-2.23, -1.49]
Li, D., 2015	-0.19	0.92	29	-0.04	0.82	29	5.1%	-0.15 [-0.60, 0.30]
Soltani, R., 2014	-0.89	0.79	25	0.09	1.14	25	4.7%	-0.98 [-1.52, -0.44]
Xu, Z., 2020-a	-0.13	1.02	45	-0.08	1.03	46	5.2%	-0.05 [-0.47, 0.37]
Xu, Z., 2020-b	-0.15	0.95	42	-0.08	1.03	46	5.2%	-0.07 [-0.48, 0.34]
Xu, Z., 2020-c	-0.34	0.84	43	-0.08	1.03	46	5.3%	-0.26 [-0.65, 0.13]
Yang, L., 2017	-0.12	1.22	80	0.02	1.23	80	5.4%	-0.14 [-0.52, 0.24]
Zhang, PW., 2015	0.25	0.82	37	0.43	1.21	37	5.0%	-0.18 [-0.65, 0.29]
Zhang, X., 2016	-0.27	0.94	73	-0.23	0.84	73	5.7%	-0.04 [-0.33, 0.25]
Subtotal (95% CI)			483			480	61.8%	-0.30 [-0.64, 0.04]
Heterogeneity: Tau² = 0.31; Chi² = 87.91, df = 11 (P < 0.00001); I² = 87%								
Test for overall effect: Z = 1.74 (P = 0.08)								
2.1.3 Malvidin-based								
Bakuradze, T., 2019	0.19	0.74	30	0.31	0.8	27	5.3%	-0.12 [-0.52, 0.28]
Hansen, AS., 2005-a	-0.07	0.31	15	-0.31	0.51	18	5.7%	0.24 [-0.04, 0.52]
Hansen, AS., 2005-b	-0.08	0.37	17	-0.31	0.51	18	5.7%	0.23 [-0.06, 0.52]
Stote, KS., 2020	-0.14	1.1	26	-0.13	1.14	26	4.4%	-0.01 [-0.62, 0.60]
Subtotal (95% CI)			88			89	21.1%	0.15 [-0.03, 0.32]
Heterogeneity: Tau² = 0.00; Chi² = 2.68, df = 3 (P = 0.44); I² = 0%								
Test for overall effect: Z = 1.67 (P = 0.10)								
Total (95% CI)			661			650	100.0%	-0.19 [-0.42, 0.04]
Heterogeneity: Tau² = 0.22; Chi² = 111.84, df = 19 (P < 0.00001); I² = 83%								
Test for overall effect: Z = 1.63 (P = 0.10)								
Test for subgroup differences: Chi² = 8.60, df = 2 (P = 0.01), I² = 76.8%								

Figure 5. Effects of anthocyanins and main anthocyanin in the test foods on TC.

3.4.3. Effects of Anthocyanins and Main Anthocyanin in the Test Foods on LDL-C

The favorable effects of anthocyanins on LDL-C levels were significant (MD: −0.19, 95% CI: −0.31 to −0.06, $p < 0.01$) with high heterogeneity ($I^2 = 53\%$, $p < 0.01$) upon pooling data from all trials. When classified according to the main anthocyanin in the test foods, a significant effect was observed only in delphinidin-based anthocyanins (MD: −0.28, 95% CI: −0.42 to −0.13, $p < 0.001$) with large heterogeneity ($I^2 = 52\%$, $p < 0.05$). The differences among the three main anthocyanin groups were also significant ($p < 0.05$) (Figure 6).

In the sensitivity analyses, the trial by Kianbakht et al. [39] affected the heterogeneity of all trials and trials using delphinidin-based anthocyanins. After that trial was removed, heterogeneities of all trials and trials using delphinidin-based anthocyanins became non-significant (all trials: $I^2 = 15\%$, $p = 0.28$; trials using delphinidin-based anthocyanins: $I^2 = 0\%$, $p = 0.70$). These overall effects remained significant in all trials (MD: −0.16, 95% CI: −0.25 to −0.07, $p < 0.001$) and trials using delphinidin-based anthocyanins (MD: −0.25, 95% CI: −0.34 to −0.15, $p < 0.00001$). The significance of the differences among the three main anthocyanin groups was also retained ($p < 0.05$).

The overall effects of anthocyanins were significant in the subgroups "Main anthocyanin to total anthocyanin above 50%" (MD: −0.24, 95% CI: −0.34 to −0.14, $p < 0.00001$) with no heterogeneity ($I^2 = 0\%$, $p = 0.71$), "Anthocyanin dosage above 160 mg/d" (MD: −0.25, 95% CI: −0.35 to −0.15, $p < 0.00001$) with no heterogeneity ($I^2 = 0\%$, $p = 0.62$), and "Baseline LDL-C levels below 3.6 mmol/L" (MD: −0.16, 95% CI: −0.27 to −0.05, $p < 0.01$) with moderate heterogeneity ($I^2 = 30\%$, $p = 0.14$). The overall effects of delphinidin-based anthocyanins were also significant in the subgroups "Main anthocyanin to total anthocyanin above 50%" (MD: −0.24, 95% CI: −0.34 to −0.14, $p < 0.00001$), "Anthocyanin dosage above 160 mg/d" (MD: −0.26, 95% CI: −0.37 to −0.16, p < 0.00001), and "Baseline LDL-C levels below 3.6 mmol/L" (MD: −0.27, 95% CI: −0.38 to 0.17, p < 0.00001), none of which showed heterogeneity (all I^2 = 0%, p > 0.1). Furthermore, delphinidin-based anthocyanins had significant effects in the subgroups "Purified anthocyanins" (MD: −0.25, 95% CI: −0.35 to −0.15, p < 0.00001), "Prediabetes and/or type 2 diabetes" (MD: −0.23, 95% CI: −0.41 to −0.06, p < 0.05), "Dyslipidemia" (MD: −0.36, 95% CI: −0.60 to −0.11, p < 0.01), and "Baseline BMI above 25.0 kg/m^2" (MD: −0.29, 95% CI: −0.43 to −0.14, p < 0.0001). Except for in the "Dyslipidemia" subgroup (I^2 = 66%, p < 0.05), heterogeneity was not significant (all I^2 < 50%, p > 0.1) in the other subgroups (Table S3).

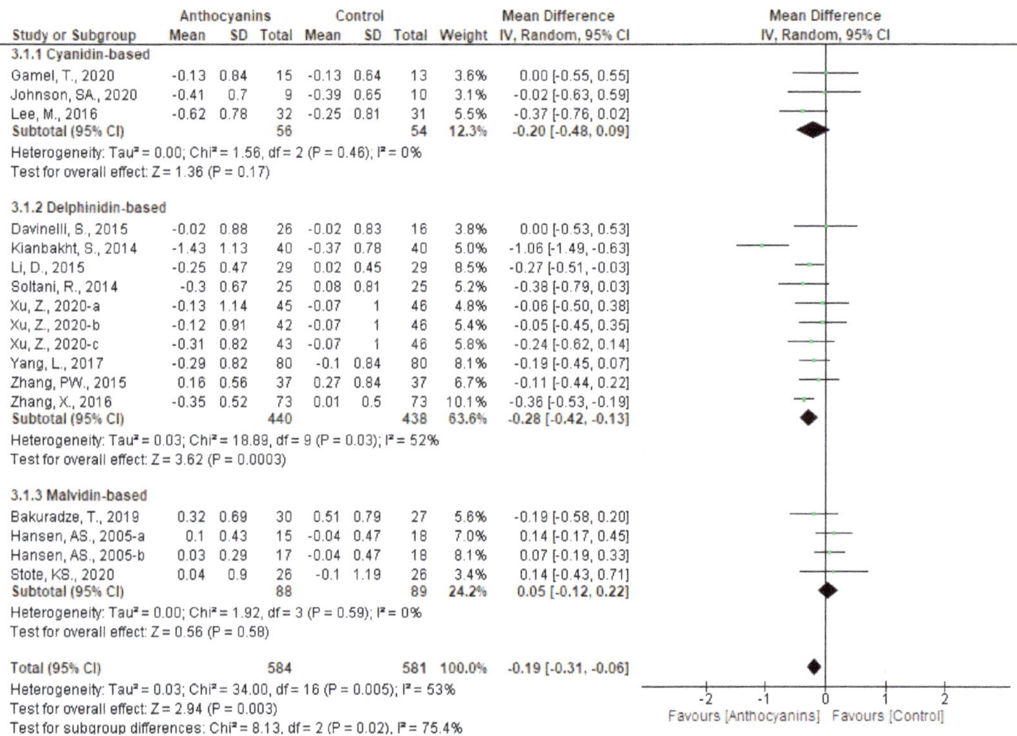

Figure 6. Effects of anthocyanins and main anthocyanin in the test foods on LDL-C.

3.4.4. Effects of Anthocyanins and Main Anthocyanin in the Test Foods on HDL-C

The favorable effects of anthocyanins on HDL-C levels were significant (MD: 0.09, 95% CI: 0.02 to 0.15, p < 0.01) with very high heterogeneity (I^2 = 76%, p < 0.00001) when data from all 14 trials were pooled. When classified according to the main anthocyanin in the test foods, significant effects were observed in the trials of delphinidin-based anthocyanins (MD: 0.11, 95% CI: 0.04 to 0.19, p < 0.01) with very high heterogeneity (I^2 = 81%, p < 0.00001), and favorable effects were observed in the trials of malvidin-based anthocyanins (MD: 0.08, 95% CI: −0.01 to 0.17, p = 0.08) with no heterogeneity (I^2 = 0%, p = 0.41) (Figure 7).

In the sensitivity analyses, it was observed that the exclusion of each trial from the dataset did not influence the heterogeneity and overall effect of all trials, as well as that of trials involving cyanidin- and delphinidin-based anthocyanins. However, the overall effect of malvidin-based anthocyanins trials became significant only after removing the trial conducted by Bakuradze et al. [31] (MD: 0.12, 95% CI: 0.01 to 0.23, p = 0.03).

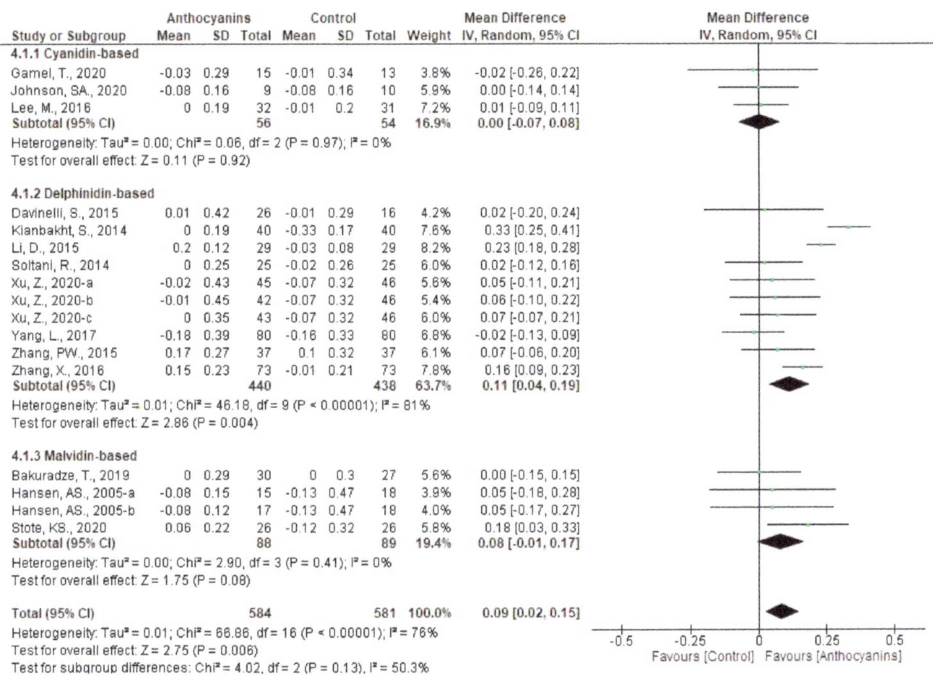

| | Anthocyanins | | | Control | | | | Mean Difference |
Study or Subgroup	Mean	SD	Total	Mean	SD	Total	Weight	IV, Random, 95% CI
4.1.1 Cyanidin-based								
Gamel, T., 2020	-0.03	0.29	15	-0.01	0.34	13	3.8%	-0.02 [-0.26, 0.22]
Johnson, SA., 2020	-0.08	0.16	9	-0.08	0.16	10	5.9%	0.00 [-0.14, 0.14]
Lee, M., 2016	0	0.19	32	-0.01	0.2	31	7.2%	0.01 [-0.09, 0.11]
Subtotal (95% CI)			56			54	16.9%	0.00 [-0.07, 0.08]
Heterogeneity: Tau² = 0.00; Chi² = 0.06, df = 2 (P = 0.97); I² = 0%								
Test for overall effect: Z = 0.11 (P = 0.92)								
4.1.2 Delphinidin-based								
Davinelli, S., 2015	0.01	0.42	26	-0.01	0.29	16	4.2%	0.02 [-0.20, 0.24]
Kianbakht, S., 2014	0	0.19	40	-0.33	0.17	40	7.6%	0.33 [0.25, 0.41]
Li, D., 2015	0.2	0.12	29	-0.03	0.08	29	8.2%	0.23 [0.18, 0.28]
Soltani, R., 2014	0	0.25	25	-0.02	0.26	25	6.0%	0.02 [-0.12, 0.16]
Xu, Z., 2020-a	-0.02	0.43	45	-0.07	0.32	46	5.6%	0.05 [-0.11, 0.21]
Xu, Z., 2020-b	-0.01	0.45	42	-0.07	0.32	46	5.4%	0.06 [-0.10, 0.22]
Xu, Z., 2020-c	0	0.35	43	-0.07	0.32	46	6.0%	0.07 [-0.07, 0.21]
Yang, L., 2017	-0.18	0.39	80	-0.16	0.33	80	6.8%	-0.02 [-0.13, 0.09]
Zhang, PW., 2015	0.17	0.27	37	0.1	0.32	37	6.1%	0.07 [-0.06, 0.20]
Zhang, X., 2016	0.15	0.23	73	-0.01	0.21	73	7.8%	0.16 [0.09, 0.23]
Subtotal (95% CI)			440			438	63.7%	0.11 [0.04, 0.19]
Heterogeneity: Tau² = 0.01; Chi² = 46.18, df = 9 (P < 0.00001); I² = 81%								
Test for overall effect: Z = 2.86 (P = 0.004)								
4.1.3 Malvidin-based								
Bakuradze, T., 2019	0	0.29	30	0	0.3	27	5.6%	0.00 [-0.15, 0.15]
Hansen, AS., 2005-a	-0.08	0.15	15	-0.13	0.47	18	3.9%	0.05 [-0.18, 0.28]
Hansen, AS., 2005-b	-0.08	0.12	17	-0.13	0.47	18	4.0%	0.05 [-0.17, 0.27]
Stote, KS., 2020	0.06	0.22	26	-0.12	0.32	26	5.8%	0.18 [0.03, 0.33]
Subtotal (95% CI)			88			89	19.4%	0.08 [-0.01, 0.17]
Heterogeneity: Tau² = 0.00; Chi² = 2.90, df = 3 (P = 0.41); I² = 0%								
Test for overall effect: Z = 1.75 (P = 0.08)								
Total (95% CI)			584			581	100.0%	0.09 [0.02, 0.15]
Heterogeneity: Tau² = 0.01; Chi² = 66.86, df = 16 (P < 0.00001); I² = 76%								
Test for overall effect: Z = 2.75 (P = 0.006)								
Test for subgroup differences: Chi² = 4.02, df = 2 (P = 0.13), I² = 50.3%								

Figure 7. Effects of anthocyanins and main anthocyanin in the test foods on HDL-C.

In the subgroup "Main anthocyanin to total anthocyanins above 50%", the heterogeneity of all the trials included and that of trials using delphinidin-based anthocyanins showed a slight decline but remained high, nevertheless, in all trials ($I^2 = 73\%$, $p < 0.001$) and trials using delphinidin-based anthocyanins ($I^2 = 72\%$, $p < 0.001$). Similar results were observed in the following subgroups: "Anthocyanin dosage above 160 mg/d", "Purified anthocyanin", and "Baseline HDL-C below 1.4 mmol/L", while in the subgroup "Baseline BMI above 25.0 kg/m^2", all trials ($I^2 = 22\%$, $p = 0.24$) and trials involving delphinidin-based anthocyanins ($I^2 = 32\%$, $p = 0.22$) was insignificant. Additionally, the overall effect in all trials (MD: 0.07, 95% CI: 0.02 to 0.12, $p < 0.01$) and in trials involving delphinidin-based anthocyanins (MD: 0.10, 95% CI: 0.02 to 0.17, $p < 0.05$) remained significant. In this subgroup, the favorable effects of trials involving malvidin-based anthocyanins (MD: 0.12, 95% CI: 0.01 to 0.23, $p < 0.05$) were also significant (Table S4).

3.4.5. Effects of Anthocyanins and Main Anthocyanin in the Test Foods on Glucose

The favorable effects of anthocyanins on glucose levels (MD: −0.17, 95% CI: −0.31 to −0.03, $p < 0.05$) without heterogeneity ($I^2 = 0\%$, $p = 0.96$) were observed when all eight trials were pooled (Figure 8). When classified according to the main anthocyanin in the test foods, favorable effects were observed in the trials involving cyanidin-based anthocyanins (MD: −0.25, 95% CI: −0.53 to 0.03, $p = 0.08$) and delphinidin-based anthocyanins (MD: −0.15, 95% CI: −0.31 to 0.02, $p = 0.09$), with no heterogeneity (all $I^2 = 0\%$, $p > 0.1$).

Because only one trial was applicable, the analysis limited to malvidin-based anthocyanins was omitted.

In sensitivity analyses, the exclusion of each trial did not influence the heterogeneity. However, the overall effect of all trials (MD: −0.13, 95% CI: −0.29 to −0.03, $p = 0.10$) and the trials using cyanidin-based anthocyanins (MD: 0.14, 95% CI: −0.68 to 0.96, $p = 0.74$) became insignificant only after excluding the trial conducted by Gamel et al. [33].

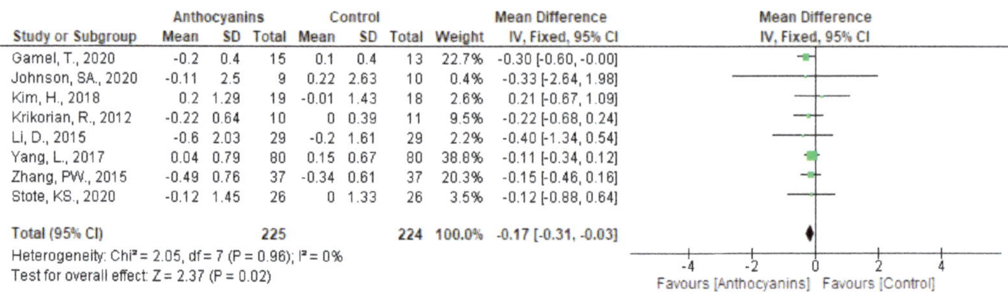

Figure 8. Effects of anthocyanins and main anthocyanin in the test foods on glucose.

The overall effects of anthocyanins were significant in the subgroups "Main anthocyanin to total anthocyanins above 50%" (MD: −0.17, 95% CI: −0.32 to −0.02, $p < 0.05$), and "Baseline glucose below 7.0 mmol/L" (MD: −0.17, 95% CI: −0.31 to 0.64, $p < 0.05$), none of which showed heterogeneity (all $I^2 = 0\%$, $p > 0.1$).

3.4.6. Effects of Anthocyanins and Main Anthocyanin in the Test Foods on Insulin

The effects of anthocyanins on insulin levels were not significant (MD: −0.28, 95% CI: −0.87 to 0.30, $p = 0.34$), with no heterogeneity ($I^2 = 0\%$, $p = 0.59$) when data from all eight trials were pooled (Figure 9). Based on the classification of trials according to the main anthocyanin in the test foods, both cyanidin-based anthocyanins (MD: 0.80, 95% CI: −2.75 to 4.35, $p = 0.66$) and delphinidin-based anthocyanins (MD: −0.34, 95% CI: −0.93 to 0.26, $p = 0.27$) did not show any significant effects on insulin levels. Because only one trial was applicable, the analysis limited to malvidin-based anthocyanins was omitted.

Study or Subgroup	Anthocyanins Mean	SD	Total	Control Mean	SD	Total	Weight	Mean Difference IV, Fixed, 95% CI
Gamel, T., 2020	0.66	6.73	15	2.14	6.83	13	1.3%	-1.48 [-6.52, 3.56]
Johnson, SA., 2020	1.44	6.34	9	0.3	6.32	10	1.1%	1.14 [-4.56, 6.84]
Kim, H., 2018	7.68	12.42	19	-1.74	19.07	18	0.3%	9.42 [-1.01, 19.85]
Krikorian, R., 2012	1	4	10	3	8.19	11	1.2%	-2.00 [-7.44, 3.44]
Li, D., 2015	-0.8	4.15	29	0.1	3.96	29	7.8%	-0.90 [-2.99, 1.19]
Yang, L., 2017	-0.41	6.35	80	-0.1	6.46	80	8.7%	-0.31 [-2.29, 1.67]
Zhang, PW., 2015	0.02	1.38	37	0.28	1.5	37	79.2%	-0.26 [-0.92, 0.40]
Stote, KS., 2020	4	10.24	26	-0.2	19.44	26	0.5%	4.20 [-4.25, 12.65]
Total (95% CI)			225			224	100.0%	-0.28 [-0.87, 0.30]

Heterogeneity: Chi² = 5.59, df = 7 (P = 0.59); I² = 0%
Test for overall effect: Z = 0.95 (P = 0.34)

Figure 9. Effects of anthocyanins and main anthocyanin in the test foods on insulin.

In the sensitivity analyses, the exclusion of each trial did not influence the heterogeneity and overall effect. No significant effects of anthocyanins or each main anthocyanin on insulin levels were demonstrated through the subgroup analyses (Table S6).

3.4.7. Effects of Anthocyanins and Main Anthocyanin in the Test Foods on HOMA-IR

The effects of anthocyanins on HOMA-IR levels were not significant (MD: −0.04, 95% CI: −0.11 to 0.02, $p = 0.17$) with moderate heterogeneity ($I^2 = 49\%$, $p = 0.10$) when data from all eight trials were pooled (Figure 10).

Although no significant effect was observed in the analyses based on main anthocyanins of the test foods, HOMA-IR changed positively by cyanidin-based anthocyanins (MD: 0.43, 95% CI: −0.37 to 1.22, $p = 0.29$), whereas negatively by delphinidin-based anthocyanins (MD: −0.18, 95% CI: −0.50 to 0.13, $p = 0.25$). Trials using malvidin-based anthocyanins were not included in the HOMA-IR dataset.

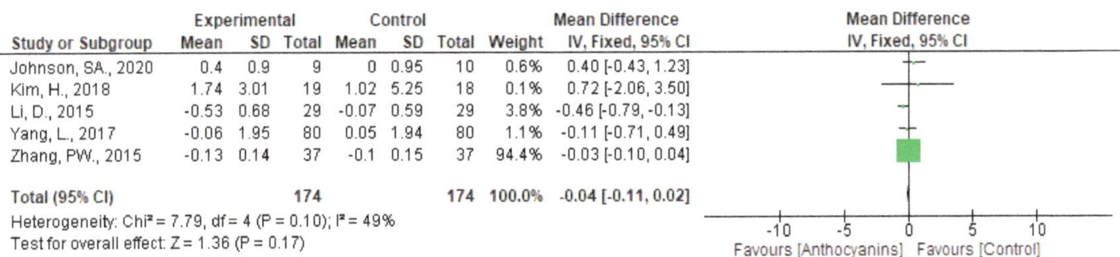

Figure 10. Effects of anthocyanins and main anthocyanin in the test foods on HOMA-IR.

In the sensitivity analyses, the exclusion of each trial from the dataset did not influence the heterogeneity and overall effect.

The overall effect was significant (MD: -0.38, 95% CI: -0.67 to -0.99, $p < 0.05$) with no heterogeneity ($I^2 = 0\%$, $p = 0.32$) in the subgroup "Prediabetes and/or type 2 diabetes" (two delphinidin-based trials). Moreover, similar results were confirmed in the subgroups "Baseline HOMA-IR levels above 2.5" and "Baseline BMI levels below 25.0 kg/m^2" (Table S7).

3.4.8. Effects of Anthocyanins and Main Anthocyanin in the Test Foods on HbA1c

The effects of anthocyanins on HbA1c levels were not significant (MD: -0.14, 95% CI: -0.30 to 0.03, $p = 0.11$), with no heterogeneity ($I^2 = 0\%$, $p = 0.94$), when data from all five trials were pooled (Figure 11). Moreover, the effect of delphinidin-based anthocyanins was not significant (MD: -0.13, 95% CI: -0.31 to 0.04, $p = 0.14$). Because there was only one trial each using cyanidin- and malvidin-based anthocyanins, analyses of all main anthocyanins except delphinidin-based anthocyanins were omitted.

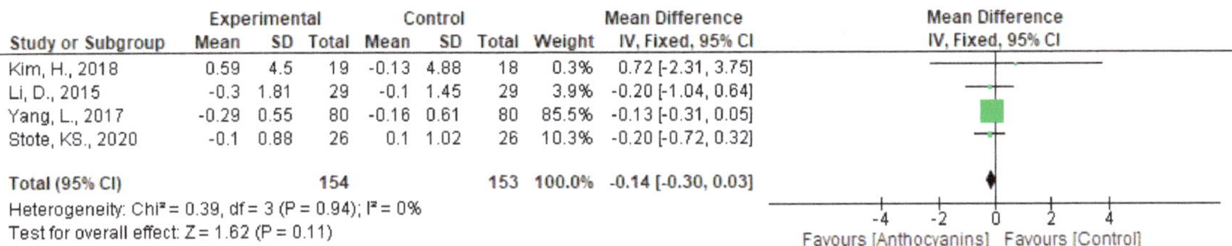

Figure 11. Effects of anthocyanins and main anthocyanin in the test foods on HbA1c.

In the sensitivity analyses, the exclusion of each trial from the dataset did not influence the heterogeneity and overall effect. As shown in Table S8, no subgroup showed a significant trend in terms of overall effect in either all trials or trials using each main anthocyanin.

3.4.9. Summary of Findings

We have summarized our main results in the summary of findings table (Table 2). We judged the quality of the evidence on the outcome of TG, LDL-C, and glucose to be moderate, insulin, HOMA-IR, and HbA1c to be low, and TC and HDL-C to be very low.

Table 2. Summary of findings for main comparison.

Outcomes	No of Participants (Trials)	Effect Estimates (95% CI)	Quality of the Evidence (GRADE)
TG (mmol/L)	988 (15 trials)	0.20 lower (0.33 lower to 0.07 lower)	⊕⊕⊕◯ Moderate [a]
TC (mmol/L)	1311 (20 trials)	0.19 lower (0.42 lower to 0.04 higher)	⊕◯◯◯ Very low [b,c]
LDL-C (mmol/L)	1165 (17 trials)	0.19 lower (0.31 lower to 0.06 lower)	⊕⊕⊕◯ Moderate [a]
HDL-C (mmol/L)	1165 (17 trials)	0.09 higher (0.02 higher to 0.15 higher)	⊕◯◯◯ Very low [c,d]
Glucose (mmol/L)	449 (8 trials)	0.17 lower (0.31 lower to 0.03 lower)	⊕⊕⊕◯ Moderate [b]
Insulin (μU/mL)	449 (8 trials)	0.28 lower (0.87 lower to 0.03 higher)	⊕⊕◯◯ Low [b,e]
HOMA-IR	348 (5 trials)	0.04 lower (0.11 lower to 0.02 higher)	⊕⊕◯◯ Low [b,e]
HbA1c (%)	307 (4 trials)	0.14 lower (0.30 lower to 0.03 higher)	⊕⊕◯◯ Low [b,e]

CI, confidence interval; MD, mean difference; TG, triglyceride; TC, total cholesterol; LDL-C, low-density lipoprotein-cholesterol; HDL-C, high-density lipoprotein-cholesterol; HOMA-IR, homeostatic model assessment of insulin resistance; HbA1c; hemoglobin A1c.

GRADE Working Group grades of evidence

High quality: Further research is very unlikely to change our confidence in the estimate of effect.

Moderate quality: Further research is likely to have an important impact on our confidence in the estimate of effect and may change the estimate.

Low quality: Further research is very likely to have an important impact on our confidence in the estimate of effect and is likely to change the estimate.

Very low quality: We are very uncertain about the estimate.

[a]: Important inconsistency (substantial heterogeneity ($I^2 > 50\%$)), [b]: Important imprecision (95% CI close to or crossing the line of no effect), [c]: Very serious inconsistency (considerable heterogeneity ($I^2 > 75\%$)), [d]: The possibility of publication bias (Egger's test was significant ($p < 0.05$)), [e]: Sparse data (total number of trials < 10).

4. Discussion

In this meta-analysis of 18 RCTs, significant improvements of TG, LDL-C, HDL-C, and glucose by anthocyanins were confirmed. The anthocyanins used in the test foods in these RCTs could be classified into three types: cyanidin-, delphinidin-, and malvidin-based, and overall, the TG, LDL-C, and HDL-C levels were significantly improved only in the trials involving delphinidin-based anthocyanins. The tendency of delphinidin-based anthocyanins to improve TC and glucose levels were also observed. In contrast, the trials involving cyanidin-based anthocyanins only showed improvement in TC and glucose, and malvidin-based anthocyanins only showed improvement in HDL-C. The MD of TC was −0.30, −0.24, and 0.15 for delphinidin-, cyanidin-, and malvidin-based anthocyanins, respectively, and a significant difference was observed among the three groups. The result for LDL-C was similar. Sensitivity analyses indicated that, except for the effects of cyanidin-based anthocyanins on glucose and of malvidin-based anthocyanins on HDL-C, the robustness of the effects of anthocyanins or main anthocyanin in the test foods on glucose and lipid metabolism remains consistent. From these results, the effectiveness of anthocyanins in improving lipid metabolism seemed to be in the following order: delphinidin- > cyanidin- > and malvidin-based. In the present study, it was not possible to examine the extent of the effects of malvidin-based anthocyanins on four parameters related to glucose metabolism and of cyanidin-based anthocyanins on HbA1c because of insufficient data. However, the MD for the four glucose metabolism-related parameters was negative in all trials using delphinidin-based anthocyanins, whereas it was positive in all or some trials using cyanidin-based anthocyanins. Therefore, it was presumed that the effects of each anthocyanin compound on glucose metabolism might be similar to those observed in our meta-analysis of lipid metabolism.

In previous animal studies which confirmed that anthocyanin supplementation decreased fat accumulation and serum lipid levels, the activation of 5′ adenosine monophosphate-activated protein kinase (AMPK) by *Acanthopanax senticosus* [58], suppression of FAS and 3-hydroxy-3-methylglutaryl coenzyme A reductase expression by mulberry water extracts [59], and decreased expression of lipid metabolism-related genes such as peroxisome proliferator-activated receptor γ (PPARγ) and SREBP-1c by *Aronia melanocarpa* extract [60] were observed. Similarly, in cell culture studies, the elevation of phosphorylated AMPK levels [61] and reduction of lipid accumulation and PPARγ protein levels [62] in 3T3-L1 cells by anthocyanins have been reported. Regarding the HDL-C increasing effects of anthocyanins, enhancing cholesterol efflux by activating Liver X receptor α (LXRα) pathway [63] and paraoxonase 1 (PON1) [64] were suggested as related factors. Activation of AMPK by anthocyanins are associated not only with the downregulation of adipogenesis but also the upregulation of glucose transporter type 4 gene expression, glucose uptake elevation [65], and insulin sensitivity improvement [66]. Anthocyanins also act as inhibitors of glycolytic enzymes such as α-glucosidase [67] and sucrase [20].

Furthermore, it has been reported that lipid accumulation and mRNA expression of FAS and SREBP1-c in 3T3-L1 adipocytes were decreased when treated with cyanidin rather than malvidin [21], and the inhibitory effects of anthocyanidins on α-glucosidase were in the order of delphinidin > cyanidin > malvidin [68]. In a study by Aboufarrag et al., gallic acid, a delphinidin-3-glucoside metabolite, increased the lactonase activity of PON1 R192R phenotypes significantly, which did not change with PCA, a cyanidin-3-glucoside metabolite, or syringic acid, a malvidin-3-glucoside metabolite [69].

These reports suggest that the activity of unmethylated anthocyanins, especially those with large number of OH groups in the B ring, is higher than that of methylated anthocyanins, which supports our results. However, the effects of anthocyanins in this meta-analysis were compared based not on a single compound alone, but on the most abundant anthocyanins in each test food. Therefore, it cannot be denied that the effects observed in this study may have been affected by other compounds coexisting with the main anthocyanins as proved in the animal study by Grace et al. [70]. Moreover, the data of RCTs using pelargonidin-, peonidin-, and petunidin-based anthocyanins were not included in this meta-analysis. The percentages of the main anthocyanin with respect to the total anthocyanins, anthocyanin type, and target population were not completely consistent among trials using cyanidin-, delphinidin-, and malvidin-based anthocyanins. Although we performed subgroup analyses to confirm the effects of confounding factors, they could not be analyzed conclusively due to the small sample size of the selected trials especially for glucose metabolism-related parameters. This may be a reason for the unremarkable effects of anthocyanins on glucose metabolism in our study. In addition, some meta-analyses indicated significant effects of anthocyanins on glucose, and HbA1c [71,72] while no effects of anthocyanins were observed (except for HOMA-IR) in other meta-analyses [73]. Thus, the effects of anthocyanins on glucose metabolism should be judged carefully. Besides, the overall effect of anthocyanins on HDL-C may have a publication bias and low robustness.

Despite the several limitations mentioned above, to our knowledge, we revealed, for the first time, the possibility that the effects of anthocyanins on glucose and lipid metabolism in humans may be affected by the extent of hydroxylation and methylation of the B ring. In addition, the differences in the effects of the three main anthocyanins were influenced by the type of anthocyanin supplementation and the clinical characteristics of the participants, and it was considered that the conditions under which the differences became remarkable may differ for each parameter. Further studies including RCTs comparing the effects of anthocyanins with different structures will be expected to further the development of personalized medicines.

5. Conclusions

In this meta-analysis, it was suggested that foods which contain delphinidin-, cyanidin-, and malvidin-based anthocyanins may be effective for the improvement of lipid metabolism in humans in that order, due to their chemical structure. The differences in the effects of the three main anthocyanins were particularly noticeable for TC and HDL-C. While the effects on the glucose metabolism parameters were not remarkable, they seemed to be more favorable for glucose, insulin, and HOMA-IR in trials using delphinidin-based anthocyanins compared with trials using cyanidin-based anthocyanins. Additionally, it was also suggested that the optimal conditions for achieving the effects of each compound easily, such as the types of anthocyanin source and target population, may differ. The possibility that these findings will contribute to the elucidation of anthocyanin intake to prevent lifestyle-related diseases and atherosclerosis more effectively was indicated.

Supplementary Materials: The following are available online at https://www.mdpi.com/article/10.3390/nu13062003/s1, Table S1: Subgroup analyses for effects of anthocyanins and main anthocyanin in the test foods on TG, Table S2: Subgroup analyses for effects of anthocyanins and main anthocyanin in the test foods on TC, Table S3: Subgroup analyses for effects of anthocyanins and main anthocyanin in the test foods on LDL-C, Table S4: Subgroup analyses for effects of anthocyanins and main anthocyanin in the test foods on HDL-C, Table S5: Subgroup analyses for effects of anthocyanins and main anthocyanin in the test foods on Glucose, Table S6: Subgroup analyses for effects of anthocyanins and main anthocyanin in the test foods on insulin, Table S7: Subgroup analyses for effects of anthocyanins and main anthocyanin in the test foods on HOMA-IR, Table S8: Subgroup analyses for effects of anthocyanins and main anthocyanin in the test foods on HbA1c.

Author Contributions: Performed the systematic review and risk of bias assessment, R.A., A.Y., H.U., K.T. and H.I.; formal analysis and writing—original draft preparation, R.A.; writing—review and editing, R.A., A.Y., H.U., K.T. and H.I. All authors have read and agreed to the published version of the manuscript.

Funding: This research received no external funding.

Institutional Review Board Statement: Not applicable.

Informed Consent Statement: Not applicable.

Data Availability Statement: Data are available upon the corresponding author after reasonable request.

Conflicts of Interest: The authors declare no conflict of interest.

References

1. Khoo, H.E.; Azlan, A.; Tang, S.T.; Lim, S.M. Anthocyanidins and anthocyanins: Colored pigments as food, pharmaceutical ingredients, and the potential health benefits. *Food Nutr. Res.* **2017**, *61*, 1361779. [CrossRef]
2. Castañeda-Ovando, A.; de Lourdes Pacheco-Hernández, M.; Páez-Hernández, M.E.; Rodríguez, J.A.; Galán-Vidal, C.A. Chemical studies of anthocyanins: A review. *Food Chem.* **2009**, *113*, 859–871. [CrossRef]
3. de Pascual-Teresa, S.; Sanchez-Ballesta, M.T. Anthocyanins: From plant to health. *Phytochem. Rev.* **2008**, *7*, 281–299. [CrossRef]
4. Hidalgo, M.; Oruna-Concha, M.J.; Kolida, S.; Walton, G.E.; Kallithraka, S.; Spencer, J.P.E.; de Pascual-Teresa, S. Metabolism of anthocyanins by human gut microflora and their influence on gut bacterial growth. *J. Agric. Food Chem.* **2012**, *60*, 3882–3890. [CrossRef]
5. Amić, D.; Davidović-Amić, D.; Trinajstić, N. Application of topological indices to chromatographic data: Calculation of the retention indices of anthocyanins. *J. Chromatogr. A* **1993**, *653*, 115–121. [CrossRef]
6. Cheynier, V.; Arellano, I.H.; Souquet, J.M.; Moutounet, M. Estimation of the oxidative changes in phenolic compounds of Carignane during winemaking. *Am. J. Enol. Viticult.* **1997**, *48*, 225–228.
7. Felgines, C.; Texier, O.; Besson, C.; Lyan, B.; Lamaison, J.L.; Scalbert, A. Strawberry pelargonidin glycosides are excreted in urine as intact glycosides and glucuronidated pelargonidin derivatives in rats. *Br. J. Nutr.* **2007**, *98*, 1126–1131. [CrossRef]
8. Habtemariam, S. Chapter 7—Bilberries and blueberries as potential modulators of type 2 diabetes and associated diseases. In *Medicinal Foods as Potential Therapies for Type-2 Diabetes and Associated Diseases*; Habtemariam, S., Ed.; Academic Press: Cambridge, MA, USA, 2019; pp. 135–175.
9. Liu, Y.; Zhang, D.; Wu, Y.; Wang, D.; Wei, Y.; Wu, J.; Ji, B. Stability and absorption of anthocyanins from blueberries subjected to a simulated digestion process. *Int. J. Food Sci. Nutr.* **2014**, *65*, 440–448. [CrossRef]

10. Damián-Medina, K.; Salinas-Moreno, Y.; Milenkovic, D.; Figueroa-Yáñez, L.; Marino-Marmolejo, E.; Higuera-Ciapara, I.; Vallejo-Cardona, A.; Lugo-Cervantes, E. In silico analysis of antidiabetic potential of phenolic compounds from blue corn (*Zea mays* L.) and black bean (*Phaseolus vulgaris* L.). *Heliyon* **2020**, *6*, e03632. [CrossRef]

11. Pojer, E.; Mattivi, F.; Johnson, D.; Stockley, C.S. The case for anthocyanin consumption to promote human health: A review. *Compr. Rev. Food Sci. Food Saf.* **2013**, *12*, 483–508. [CrossRef]

12. Kong, J.M.; Chia, L.S.; Goh, N.K.; Chia, T.F.; Brouillard, R. Analysis and biological activities of anthocyanins. *Phytochemistry* **2003**, *64*, 923–933. [CrossRef]

13. Hajhashemi, V.; Vaseghi, G.; Pourfarzam, M.; Abdollahi, A. Are antioxidants helpful for disease prevention? *Res. Pharm. Sci.* **2010**, *5*, 1–8. [PubMed]

14. Seeram, N.P.; Nair, M.G. Inhibition of lipid peroxidation and structure-activity-related studies of the dietary constituents anthocyanins, anthocyanidins, and catechins. *J. Agric. Food Chem.* **2002**, *50*, 5308–5312. [CrossRef]

15. Sadilova, E.; Stintzing, F.C.; Carle, R. Anthocyanins, colour and antioxidant properties of eggplant (*Solanum melongena* L.) and violet pepper (*Capsicum annuum* L.) peel extracts. *Z. Naturforsch. C J. Biosci.* **2006**, *61*, 527–535. [CrossRef]

16. Lachman, J.; Hamouz, K.; Šulc, M.; Orsák, M.; Pivec, V.; Hejtmánková, A.; Dvořák, P.; Čepl, J. Cultivar differences of total anthocyanins and anthocyanidins in red and purple-fleshed potatoes and their relation to antioxidant activity. *Food Chem.* **2009**, *114*, 836–843. [CrossRef]

17. Tangvarasittichai, S. Oxidative stress, insulin resistance, dyslipidemia and type 2 diabetes mellitus. *World J. Diabetes* **2015**, *6*, 456–480. [CrossRef]

18. Belwal, T.; Nabavi, S.F.; Nabavi, S.M.; Habtemariam, S. Dietary anthocyanins and insulin resistance: When food becomes a medicine. *Nutrients* **2017**, *9*, 1111. [CrossRef]

19. Yawadio, R.; Tanimori, S.; Morita, N. Identification of phenolic compounds isolated from pigmented rices and their aldose reductase inhibitory activities. *Food Chem.* **2007**, *101*, 1616–1625. [CrossRef]

20. Akkarachiyasit, S.; Charoenlertkul, P.; Yibchok-Anun, S.; Adisakwattana, S. Inhibitory activities of cyanidin and its glycosides and synergistic effect with acarbose against intestinal α-glucosidase and pancreatic α-amylase. *Int. J. Mol. Sci.* **2010**, *11*, 3387–3396. [CrossRef]

21. Park, S.; Kang, S.; Jeong, D.Y.; Jeong, S.Y.; Park, J.J.; Yun, H.S. Cyanidin and malvidin in aqueous extracts of black carrots fermented with *Aspergillus oryzae* prevent the impairment of energy, lipid and glucose metabolism in estrogen-deficient rats by AMPK activation. *Genes Nutr.* **2015**, *10*, 455. [CrossRef]

22. Oak, M.-H.; Bedoui, J.E.; Madeira, S.V.F.; Chalupsky, K.; Schini-Kerth, V.B. Delphinidin and cyanidin inhibit $PDGF_{AB}$-induced VEGF release in vascular smooth muscle cells by preventing activation of p38 MAPK and JNK. *Br. J. Pharmacol.* **2006**, *149*, 283–290. [CrossRef]

23. Skemiene, K.; Rakauskaite, G.; Trumbeckaite, S.; Liobikas, J.; Brown, G.C.; Borutaite, V. Anthocyanins block ischemia-induced apoptosis in the perfused heart and support mitochondrial respiration potentially by reducing cytosolic cytochrome c. *Int. J. Biochem. Cell Biol.* **2013**, *45*, 23–29. [CrossRef]

24. Moher, D.; Liberati, A.; Tetzlaff, J.; Altman, D.G.; PRISMA Group. Preferred reporting items for systematic reviews and meta-analyses: The PRISMA statement. *PLoS Med.* **2009**, *6*, e1000097. [CrossRef]

25. Hozo, S.P.; Djulbegovic, B.; Hozo, I. Estimating the mean and variance from the median, range, and the size of a sample. *BMC Med. Res. Methodol.* **2005**, *5*, 13. [CrossRef] [PubMed]

26. Mathie, R.T.; Lloyd, S.M.; Legg, L.A.; Clausen, J.; Moss, S.; Davidson, J.R.T.; Ford, I. Randomised placebo-controlled trials of individualised homeopathic treatment: Systematic review and meta-analysis. *Syst. Rev.* **2014**, *3*, 142. [CrossRef] [PubMed]

27. Guyatt, G.H.; Oxman, A.D.; Vist, G.E.; Kunz, R.; Falck-Ytter, Y.; Alonso-Coello, P.; Schünemann, H.J.; GRADE Working Group. GRADE: An emerging consensus on rating quality of evidence and strength of recommendations. *BMJ* **2008**, *336*, 924–926. [CrossRef]

28. Higgins, J.P.T.; Thompson, S.G.; Deeks, J.J.; Altman, D.G. Measuring inconsistency in meta-analyses. *BMJ* **2003**, *327*, 557–560. [CrossRef]

29. Egger, M.; Smith, G.D.; Schneider, M.; Minder, C. Bias in meta-analysis detected by a simple, graphical test. *BMJ* **1997**, *315*, 629–634. [CrossRef]

30. Sterne, J.A.C.; Sutton, A.J.; Ioannidis, J.P.A.; Terrin, N.; Jones, D.R.; Lau, J.; Carpenter, J.; Rücker, G.; Harbord, R.M.; Schmid, C.H.; et al. Recommendations for examining and interpreting funnel plot asymmetry in meta-analyses of randomised controlled trials. *BMJ* **2011**, *343*, d4002. [CrossRef]

31. Bakuradze, T.; Tausend, A.; Galan, J.; Groh, I.A.M.; Berry, D.; Tur, J.A.; Marko, D.; Richling, E. Antioxidative activity and health benefits of anthocyanin-rich fruit juice in healthy volunteers. *Free Radic. Res.* **2019**, *53*, 1045–1055. [CrossRef]

32. Davinelli, S.; Bertoglio, J.C.; Zarrelli, A.; Pina, R.; Scapagnini, G. A randomized clinical trial evaluating the efficacy of an anthocyanin-maqui berry extract (Delphinol®) on oxidative stress biomarkers. *J. Am. Coll. Nutr.* **2015**, *34*, 28–33. [CrossRef]

33. Gamel, T.H.; Abdel-Aal, E.S.M.; Tucker, A.J.; Pare, S.M.; Faughnan, K.; O'Brien, C.D.; Dykun, A.; Rabalski, I.; Pickard, M.; Wright, A.J. Consumption of whole purple and regular wheat modestly improves metabolic markers in adults with elevated high-sensitivity C-reactive protein: A randomised, single-blind parallel-arm study. *Br. J. Nutr.* **2020**, *124*, 1179–1189. [CrossRef]

34. Abdel-Aal, E.-S.M.; Hucl, P.; Rabalski, I. Compositional and antioxidant properties of anthocyanin-rich products prepared from purple wheat. *Food Chem.* **2018**, *254*, 13–19. [CrossRef] [PubMed]

35. Hansen, A.S.; Marckmann, P.; Dragsted, L.O.; Nielsen, I.L.F.; Nielsen, S.E.; Grønbaek, M. Effect of red wine and red grape extract on blood lipids, haemostatic factors, and other risk factors for cardiovascular disease. *Eur. J. Clin. Nutr.* **2005**, *59*, 449–455. [CrossRef]
36. Johnson, S.A.; Navaei, N.; Pourafshar, S.; Jaime, S.J.; Akhavan, N.S.; Alvarez-Alvarado, S.; Proaño, G.V.; Litwin, N.S.; Clark, E.A.; Foley, E.M.; et al. Effects of Montmorency tart cherry juice consumption on cardiometabolic biomarkers in adults with metabolic syndrome: A randomized controlled pilot trial. *J. Med. Food* **2020**, *23*, 1238–1247. [CrossRef] [PubMed]
37. Khan, F.; Ray, S.; Craigie, A.M.; Kennedy, G.; Hill, A.; Barton, K.L.; Broughton, J.; Belch, J.J.F. Lowering of oxidative stress improves endothelial function in healthy subjects with habitually low intake of fruit and vegetables: A randomized controlled trial of antioxidant- and polyphenol-rich blackcurrant juice. *Free Radic. Biol. Med.* **2014**, *72*, 232–237. [CrossRef] [PubMed]
38. Bordonaba, J.G.; Terry, L.A. Biochemical profiling and chemometric analysis of seventeen UK-grown black currant cultivars. *J. Agric. Food Chem.* **2008**, *56*, 7422–7430. [CrossRef]
39. Kianbakht, S.; Abasi, B.; Dabaghian, F.H. Improved lipid profile in hyperlipidemic patients taking *Vaccinium arctostaphylos* fruit hydroalcoholic extract: A randomized double-blind placebo-controlled clinical trial. *Phytother. Res.* **2014**, *28*, 432–436. [CrossRef]
40. Nickavar, B.; Amin, G. Anthocyanins from *Vaccinium arctostaphylos* berries. *Pharm. Biol.* **2004**, *42*, 289–291. [CrossRef]
41. Kim, H.; Simbo, S.Y.; Fang, C.; McAlister, L.; Roque, A.; Banerjee, N.; Talcott, S.T.; Zhao, H.; Kreider, R.B.; Mertens-Talcott, S.U. Açaí (*Euterpe oleracea* Mart.) beverage consumption improves biomarkers for inflammation but not glucose- or lipid-metabolism in individuals with metabolic syndrome in a randomized, double-blinded, placebo-controlled clinical trial. *Food Funct.* **2018**, *9*, 3097–3103. [CrossRef]
42. Schauss, A.G.; Wu, X.; Prior, R.L.; Ou, B.; Patel, D.; Huang, D.; Kababick, J.P. Phytochemical and nutrient composition of the freeze-dried Amazonian palm berry, *Euterpe oleraceae* Mart. (acai). *J. Agric. Food Chem.* **2006**, *54*, 8598–8603. [CrossRef] [PubMed]
43. Krikorian, R.; Boespflug, E.L.; Fleck, D.E.; Stein, A.L.; Wightman, J.D.; Shidler, M.D.; Sadat-Hossieny, S. Concord grape juice supplementation and neurocognitive function in human aging. *J. Agric. Food Chem.* **2012**, *60*, 5736–5742. [CrossRef]
44. Stalmach, A.; Edwards, C.A.; Wightman, J.D.; Crozier, A. Identification of (poly)phenolic compounds in concord grape juice and their metabolites in human plasma and urine after juice consumption. *J. Agric. Food Chem.* **2011**, *59*, 9512–9522. [CrossRef] [PubMed]
45. Lee, M.; Sorn, S.R.; Park, Y.; Park, H.-K. Anthocyanin rich-black soybean testa improved visceral fat and plasma lipid profiles in overweight/obese Korean adults: A randomized controlled trial. *J. Med. Food* **2016**, *19*, 995–1003. [CrossRef] [PubMed]
46. Li, D.; Zhang, Y.; Liu, Y.; Sun, R.; Xia, M. Purified anthocyanin supplementation reduces dyslipidemia, enhances antioxidant capacity, and prevents insulin resistance in diabetic patients. *J. Nutr.* **2015**, *145*, 742–748. [CrossRef] [PubMed]
47. Qin, Y.; Xia, M.; Ma, J.; Hao, Y.T.; Liu, J.; Mou, H.Y.; Cao, L.; Ling, W.H. Anthocyanin supplementation improves serum LDL- and HDL-cholesterol concentrations associated with the inhibition of cholesteryl ester transfer protein in dyslipidemic subjects. *Am. J. Clin. Nutr.* **2009**, *90*, 485–492. [CrossRef]
48. Lynn, A.; Mathew, S.; Moore, C.T.; Russell, J.; Robinson, E.; Soumpasi, V.; Barker, M.E. Effect of a tart cherry juice supplement on arterial stiffness and inflammation in healthy adults: A randomised controlled trial. *Plant Foods Hum. Nutr.* **2014**, *69*, 122–127. [CrossRef]
49. Bonerz, D.; Würth, K.; Dietrich, H.; Will, F. Analytical characterization and the impact of ageing on anthocyanin composition and degradation in juices from five sour cherry cultivars. *Eur. Food Res. Technol.* **2007**, *224*, 355–364. [CrossRef]
50. Soltani, R.; Hakimi, M.; Asgary, S.; Ghanadian, S.M.; Keshvari, M.; Sarrafzadegan, N. Evaluation of the effects of *Vaccinium arctostaphylos* L. fruit extract on serum lipids and hs-CRP Levels and oxidative stress in adult patients with hyperlipidemia: A randomized, double-blind, placebo-controlled clinical trial. *Evid. Based Complement. Alternat. Med.* **2014**, *2014*, 217451. [CrossRef] [PubMed]
51. Lätti, A.K.; Kainulainen, P.S.; Hayirlioglu-Ayaz, S.; Ayaz, F.A.; Riihinen, K.R. Characterization of anthocyanins in caucasian blueberries (*Vaccinium arctostaphylos* L.) native to Turkey. *J. Agric. Food Chem.* **2009**, *57*, 5244–5249. [CrossRef]
52. Stote, K.S.; Wilson, M.M.; Hallenbeck, D.; Thomas, K.; Rourke, J.M.; Sweeney, M.I.; Gottschall-Pass, K.T.; Gosmanov, A.R. Effect of blueberry consumption on cardiometabolic health parameters in men with Type 2 diabetes: An 8-week, double-blind, randomized, placebo-controlled trial. *Curr. Dev. Nutr.* **2020**, *4*, nzaa030. [CrossRef]
53. Overall, J.; Bonney, S.A.; Wilson, M.; Beermann, A.; Grace, M.H.; Esposito, D.; Lila, M.A.; Komarnytsky, S. Metabolic effects of berries with structurally diverse anthocyanins. *Int. J. Mol. Sci.* **2017**, *18*, 422. [CrossRef]
54. Xu, Z.; Xie, J.; Zhang, H.; Pang, J.; Li, Q.; Wang, X.; Xu, H.; Sun, X.; Zhao, H.; Yang, Y.; et al. Anthocyanin supplementation at different doses improves cholesterol efflux capacity in subjects with dyslipidemia-a randomized controlled trial. *Eur. J. Clin. Nutr.* **2021**, *75*, 345–354. [CrossRef]
55. Yang, L.; Ling, W.; Yang, Y.; Chen, Y.; Tian, Z.; Du, Z.; Chen, J.; Xie, Y.; Liu, Z.; Yang, L. Role of purified anthocyanins in improving cardiometabolic risk factors in Chinese men and women with prediabetes or early untreated diabetes-a randomized controlled trial. *Nutrients* **2017**, *9*, 1104. [CrossRef] [PubMed]
56. Zhang, P.-W.; Chen, F.X.; Li, D.; Ling, W.H.; Guo, H.H. A CONSORT-compliant, randomized, double-blnd, placebo-controlled pilot trial of purified anthocyanin in patients with nonalcoholic fatty liver disease. *Medicine* **2015**, *94*, e758. [CrossRef] [PubMed]
57. Zhang, X.; Zhu, Y.; Song, F.; Yao, Y.; Ya, F.; Li, D.; Ling, W.; Yang, Y. Effects of purified anthocyanin supplementation on platelet chemokines in hypocholesterolemic individuals: A randomized controlled trial. *Nutr. Metab.* **2016**, *13*, 86. [CrossRef] [PubMed]

58. Saito, T.; Nishida, M.; Saito, M.; Tanabe, A.; Eitsuka, T.; Yuan, S.-H.; Ikekawa, N.; Nishida, H. The fruit of *Acanthopanax senticosus* (Rupr. et Maxim.) Harms improves insulin resistance and hepatic lipid accumulation by modulation of liver adenosine monophosphate-activated protein kinase activity and lipogenic gene expression in high-fat diet-fed obese mice. *Nutr. Res.* **2016**, *36*, 1090–1097.
59. Chan, K.C.; Ho, H.-H.; Lin, M.C.; Wu, C.H.; Huang, C.N.; Chang, W.C.; Wang, C.J. Mulberry water extracts inhibit rabbit atherosclerosis through stimulation of vascular smooth muscle cell apoptosis via activating p53 and regulating both intrinsic and extrinsic pathways. *J. Agric. Food Chem.* **2014**, *62*, 5092–5101. [CrossRef]
60. Lim, S.M.; Lee, H.S.; Jung, J.I.; Kim, S.M.; Kim, N.Y.; Seo, T.S.; Bae, J.S.; Kim, E.J. Cyanidin-3-*O*-galactoside-enriched *Aronia melanocarpa* extract attenuates weight gain and adipogenic pathways in high-fat diet-induced obese C57BL/6 mice. *Nutrients* **2019**, *11*, 1190. [CrossRef] [PubMed]
61. Han, M.H.; Kim, H.J.; Jeong, J.W.; Park, C.; Kim, B.W.; Choi, Y.H. Inhibition of adipocyte differentiation by anthocyanins isolated from the fruit of *Vitis coignetiae* Pulliat. is associated with the activation of AMPK signaling pathway. *Toxicol. Res.* **2018**, *34*, 13–21. [CrossRef]
62. Muscarà, C.; Molonia, M.S.; Speciale, A.; Bashllari, R.; Cimino, F.; Occhiuto, C.; Saija, A.; Cristani, M. Anthocyanins ameliorate palmitate-induced inflammation and insulin resistance in 3T3-L1 adipocytes. *Phytother. Res.* **2019**, *33*, 1888–1897. [CrossRef]
63. Du, C.; Shi, Y.; Ren, Y.; Wu, H.; Yao, F.; Wei, J.; Wu, M.; Hou, Y.; Duan, H. Anthocyanins inhibit high-glucose-induced cholesterol accumulation and inflammation by activating LXRα pathway in HK-2 cells. *Drug Des. Devel. Ther.* **2015**, *9*, 5099–5113. [PubMed]
64. Farrell, N.; Norris, G.; Lee, S.G.; Chun, O.K.; Blesso, C.N. Anthocyanin-rich black elderberry extract improves markers of HDL function and reduces aortic cholesterol in hyperlipidemic mice. *Food Funct.* **2015**, *6*, 1278–1287. [CrossRef] [PubMed]
65. Nizamutdinova, I.T.; Jin, Y.C.; Chung, J.I.; Shin, S.C.; Lee, S.J.; Seo, H.G.; Lee, J.H.; Chang, K.C.; Kim, H.J. The anti-diabetic effect of anthocyanins in streptozotocin-induced diabetic rats through glucose transporter 4 regulation and prevention of insulin resistance and pancreatic apoptosis. *Mol. Nutr. Food Res.* **2009**, *53*, 1419–1429. [CrossRef] [PubMed]
66. Sasaki, R.; Nishimura, N.; Hoshino, H.; Isa, Y.; Kadowaki, M.; Ichi, T.; Tanaka, A.; Nishiumi, S.; Fukuda, I.; Ashida, H.; et al. Cyanidin 3-glucoside ameliorates hyperglycemia and insulin sensitivity due to downregulation of retinol binding protein 4 expression in diabetic mice. *Biochem. Pharmacol.* **2007**, *74*, 1619–1627. [CrossRef] [PubMed]
67. Zhang, L.; Li, J.; Hogan, S.; Chung, H.; Welbaum, G.E.; Zhou, K. Inhibitory effect of raspberries on starch digestive enzyme and their antioxidant properties and phenolic composition. *Food Chem.* **2010**, *119*, 592–599. [CrossRef]
68. Adisakwattana, S.; Yibchok-Anun, S.; Charoenlertkul, P.; Wongsasiripat, N. Cyanidin-3-rutinoside alleviates postprandial hyperglycemia and its synergism with acarbose by inhibition of intestinal α-glucosidase. *J. Clin. Biochem. Nutr.* **2011**, *49*, 36–41. [CrossRef]
69. Aboufarrag, H.T.; Needs, P.W.; Rimbach, G.; Kroon, P.A. The effects of anthocyanins and their microbial metabolites on the expression and enzyme activities of paraoxonase 1, an important marker of HDL function. *Nutrients* **2019**, *11*, 2872. [CrossRef]
70. Grace, M.H.; Ribnicky, D.M.; Kuhn, P.; Poulev, A.; Logendra, S.; Yousef, G.G.; Raskin, I.; Lila, M.A. Hypoglycemic activity of a novel anthocyanin-rich formulation from lowbush blueberry, *Vaccinium angustifolium* Aiton. *Phytomedicine* **2009**, *16*, 406–415. [CrossRef] [PubMed]
71. Fallah, A.A.; Sarmast, E.; Jafari, T. Effect of dietary anthocyanins on biomarkers of glycemic control and glucose metabolism: A systematic review and meta-analysis of randomized clinical trials. *Food Res. Int.* **2020**, *137*, 109379. [CrossRef]
72. Yang, L.; Ling, W.; Du, Z.; Chen, Y.; Li, D.; Deng, S.; Liu, Z.; Yang, L. Effects of Anthocyanins on Cardiometabolic Health: A Systematic Review and Meta-Analysis of Randomized Controlled Trials. *Adv. Nutr.* **2017**, *8*, 684–693. [CrossRef] [PubMed]
73. Daneshzad, E.; Shab-Bidar, S.; Mohammadpour, Z.; Djafarian, K. Effect of anthocyanin supplementation on cardio-metabolic biomarkers: A systematic review and meta-analysis of randomized controlled trials. *Clin. Nutr.* **2019**, *38*, 1153–1165. [CrossRef] [PubMed]

Article

Exosomes Derived from Fisetin-Treated Keratinocytes Mediate Hair Growth Promotion

Mizuki Ogawa [1], Miyako Udono [2], Kiichiro Teruya [1,2], Norihisa Uehara [3] and Yoshinori Katakura [1,2,*]

1   Graduate School of Bioresources and Bioenvironmental Sciences, Kyushu University, Fukuoka 819-0395, Japan; ogawa.mizuki.908@s.kyushu-u.ac.jp (M.O.); kteruya@grt.kyushu-u.ac.jp (K.T.)
2   Faculty of Agriculture, Kyushu University, Fukuoka 819-0395, Japan; mudono@grt.kyushu-u.ac.jp
3   Department of Molecular Cell Biology and Oral Anatomy, Faculty of Dental Science, Kyushu University, Fukuoka 812-8582, Japan; ueharan@dent.kyushu-u.ac.jp
*   Correspondence: katakura@grt.kyushu-u.ac.jp; Tel.: +81-92-802-4727

Abstract: Enhanced telomerase reverse transcriptase (TERT) levels in dermal keratinocytes can serve as a novel target for hair growth promotion. Previously, we identified fisetin using a system for screening food components that can activate the TERT promoter in HaCaT cells (keratinocytes). In the present study, we aimed to clarify the molecular basis of fisetin-induced hair growth promotion in mice. To this end, the dorsal skin of mice was treated with fisetin, and hair growth was evaluated 12 days after treatment. Histochemical analyses of fisetin-treated skin samples and HaCaT cells were performed to observe the effects of fisetin. The results showed that fisetin activated HaCaT cells by regulating the expression of various genes related to epidermogenesis, cell proliferation, hair follicle regulation, and hair cycle regulation. In addition, fisetin induced the secretion of exosomes from HaCaT cells, which activated β-catenin and mitochondria in hair follicle stem cells (HFSCs) and induced their proliferation. Moreover, these results revealed the existence of exosomes as the molecular basis of keratinocyte-HFSC interaction and showed that fisetin, along with its effects on keratinocytes, caused exosome secretion, thereby activating HFSCs. This is the first study to show that keratinocyte-derived exosomes can activate HFSCs and consequently induce hair growth.

Keywords: telomerase reverse transcriptase; keratinocyte–hair follicle stem cell interaction; exosomes; telogen–anagen transition; hair cycle regulation

Citation: Ogawa, M.; Udono, M.; Teruya, K.; Uehara, N.; Katakura, Y. Exosomes Derived from Fisetin-Treated Keratinocytes Mediate Hair Growth Promotion. *Nutrients* **2021**, *13*, 2087. https://doi.org/10.3390/nu13062087

Academic Editor: Jean Christopher Chamcheu

Received: 18 May 2021
Accepted: 16 June 2021
Published: 18 June 2021

Publisher's Note: MDPI stays neutral with regard to jurisdictional claims in published maps and institutional affiliations.

Copyright: © 2021 by the authors. Licensee MDPI, Basel, Switzerland. This article is an open access article distributed under the terms and conditions of the Creative Commons Attribution (CC BY) license (https://creativecommons.org/licenses/by/4.0/).

1. Introduction

Telomerase reverse transcriptase (TERT) maintains telomere length and is known to be active in stem cells as well as cancer cells. Sarin et al. (2005) reported that when TERT is transgenically introduced into mouse skin, the hair follicle cycle shifts from the telogen phase, the quiescent phase, to the anagen phase, the active phase, resulting in hair growth [1]. The overexpression of TERT induces the proliferation of hair follicle stem cells (HFSCs) in the bulge region, leading to hair growth. In addition, Choi et al. (2008) showed that the presence of TERT in skin keratinocytes activates dormant HFSCs, which in turn activates hair follicles and promotes hair growth [2]. These studies suggest that TERT overexpression in dermal keratinocytes could be a target to promote hair growth. In a previous study, we constructed a system for screening food components that can activate the TERT promoter in keratinocytes and identified fisetin, a type of polyphenol. Fisetin activated keratinocytes, promoted hair growth, increased skin thickness, activated β-catenin, and increased the proliferation of HFSCs [3].

Recent studies have reported that exosomes derived from hair follicle tissues are involved in hair growth promotion and function as the molecular basis of cell–cell interactions [4–9]. In the present study, we aimed to determine whether fisetin activates keratinocytes, resulting in the activation of keratinocyte–HFSC interaction, which leads to hair growth. In addition, we sought to confirm the presence of exosomes as the molecular basis of this interaction.

2. Materials and Methods

2.1. Cell Line and Reagent

HaCaT cell line (human keratinocyte), obtained from Riken Bioresource Center (Tsukuba, Japan), was cultured in Dulbecco's modified Eagle's medium (Nissui, Tokyo, Japan) supplemented with 10% fetal bovine serum (Life Technologies, Gaithersburg, MD, USA) at 37 °C in a 5% CO_2 atmosphere. Human HFSCs were obtained from Celprogen (Torrance, CA, USA) and cultured in human hair follicle stem cell complete growth medium with serum (Celprogen) at 37 °C in a 5% CO_2 atmosphere. Fisetin (>96.0% purity; Tokyo Chemical Industry Co., Ltd., Tokyo, Japan) was purchased from

2.2. Quantitative Reverse Transcriptase-Polymerase Chain Reaction

RNA was extracted from cells and skin samples using the High Pure RNA Isolation Kit (Roche Diagnostics GmbH, Mannheim, Germany) and RNeasy Fibrous Tissue Mini Kit (Qiagen, Hilden, Germany), respectively, according to the manufacturers' instructions. cDNA synthesis and quantitative reverse transcription-polymerase chain reaction (qRT-PCR) were performed, as described previously [3,10]. Samples were analyzed in triplicate, and gene expression levels were normalized to the corresponding β-actin gene level. The PCR primer sequences were described previously [3].

2.3. Investigation of Hair Growth in Experimental Animals

Six-week-old female C57BL/6 mice were obtained from Clea Japan (Tokyo, Japan) and acclimated with food and water ad libitum for one week. Nine mice were used for control and fisetin treatment, respectively. At 7 weeks of age, the dorsal hair of the mice was trimmed using electric hair clippers (Thrive Model 2100, Daito Electric Machine Industry, Osaka, Japan), shaved using an electric shaver (Panasonic, Osaka, Japan), and removed using a hair removal cream (CBS, Reckitt Benckiser, Tokyo, Japan), and all the hair follicles were synchronized in the telogen stage, as described previously [11,12]. This was followed by the daily topical application of 0.05 mL of 50% ethanol containing fisetin (1% *w/v*) using a spatula for 12 days, and hair growth was evaluated. Animal experiments were conducted in accordance with the "Guide for the Care and Use of Laboratory Animals" and approved by the Ethics Committee on Animal Experimentation of Kyushu University (approval number: A28-187-0).

2.4. Immunohistochemistry

Skin tissue sample preparations and treatments were described previously [3]. Paraffin-embedded hair follicles were cut out to a thickness of 5 micrometers. Immunohistochemical analysis was performed as previously described [3]. The tissues were first stained with primary antibodies (anti-Ki67, #12202, Cell Signaling Technology, Danvers, MA, USA; anti-β-catenin, #8480, Cell Signaling Technology; and anti-CD34, sc-74499, Santa Cruz Biotechnology, Santa Cruz, CA, USA) and then with secondary antibodies (Alexa Fluor 555 anti-rabbit IgG or Alexa Fluor 488 anti-mouse IgG, Thermo Fisher Scientific KK, Tokyo, Japan). After staining with Vectashield mounting medium (Vector Laboratories, Burlingame, CA, USA), the tissue samples were observed with the EVOS M5000 Imaging System (Thermo Fisher Scientific).

2.5. Evaluation of Mitochondrial Characteristics

Cells were stained with 250 nM of MitoTracker Red CMXRos (Thermo Fisher Scientific) and incubated at 37 °C for 30 min. Next, these cells were stained with 200 nM MitoTracker Green FM (Thermo Fisher Scientific) and incubated at 37 °C for 30 min. Finally, the cells were stained with Hoechst 33342 (Dojindo, Kumamoto, Japan), followed by incubation at 37 °C for 30 min. Then, the stained cells were analyzed using the IN Cell Analyzer 2200

(GE Healthcare, Amersham Place, UK), and the number, area, and activity of mitochondria were quantitatively determined using the IN Cell Investigator high-content image analysis software (GE Healthcare).

2.6. mRNA Microarray Assay

Total RNA was extracted from HaCaT cells treated with fisetin using Isogen II (Nippon Gene, Tokyo, Japan). Microarray analysis was performed, as described previously [10].

We identified genes with altered expression, as previously described [13]. Then, we established the following criteria for significantly upregulated or downregulated genes: upregulated genes, Z-score $\geq$ 2.0 and ratio $\geq$ 1.5-fold; downregulated genes, Z-score $\leq$ −2.0 and ratio $\leq$ 0.66-fold. To determine significantly over-represented gene ontology (GO) categories and significantly enriched pathways, we used the tools and data provided by the Database for Annotation, Visualization, and Integrated Discovery (http://david.abcc.ncifcrf.gov, 2 February 2021) [14,15].

2.7. Exosome Isolation

The MagCapure Exosome Isolation Kit PS (Fujifilm Wako Pure Chemical Corp., Osaka, Japan) was used to isolate exosomes from the medium of HaCaT cells treated with 10 μM fisetin, according to the manufacturer's instructions [10,16].

2.8. Immunocytochemistry

The cells were fixed in 4% paraformaldehyde at 25 °C for 15 min. After washing with phosphate-buffered saline (PBS), the cells were blocked with blocking buffer (5% goat serum and 0.3% Triton X-100 in PBS) at room temperature for 1 h. After removing the buffer, the cells were incubated with primary antibodies (anti-Ki67, #12202; anti-active-β-catenin, #05-665, Merck Millipore, Billerica, MA, USA; and anti-TOMM20, ab186735, Abcam, Cambridge, UK) at 4 °C overnight. After washing with PBS, the cells were incubated with secondary antibodies (Alexa Fluor 555 anti-rabbit IgG, Alexa Fluor 488 anti-rabbit IgG, or Alexa Fluor 555 anti-mouse IgG) at 25 °C for 2 h. Then, the cells were washed and stained with Hoechst 33342 at room temperature for 20 min and observed under a fluorescence microscope (BZ-X800, Keyence, Osaka, Japan) [3].

2.9. Statistical Analysis

All experiments were performed at least three times, and the representative data are presented. The results are shown as the mean $\pm$ standard deviation. Statistical significance was determined using a two-sided Student's *t*-test. Statistical significance was defined as $p < 0.05$ (* $p < 0.05$; ** $p < 0.01$; *** $p < 0.001$).

3. Results

3.1. Fisetin Augments the Expression of TERT in Keratinocytes

In our previous study, we observed that fisetin augmented the expression of TERT in a human keratinocyte cell line and the dorsal skin cells of mice [3]. In the present study, we evaluated the expression of TERT in dorsal skin cells of mice treated with fisetin. The results clearly showed that fisetin augmented TERT expression in the dorsal skin cells of mice (Figure 1A).

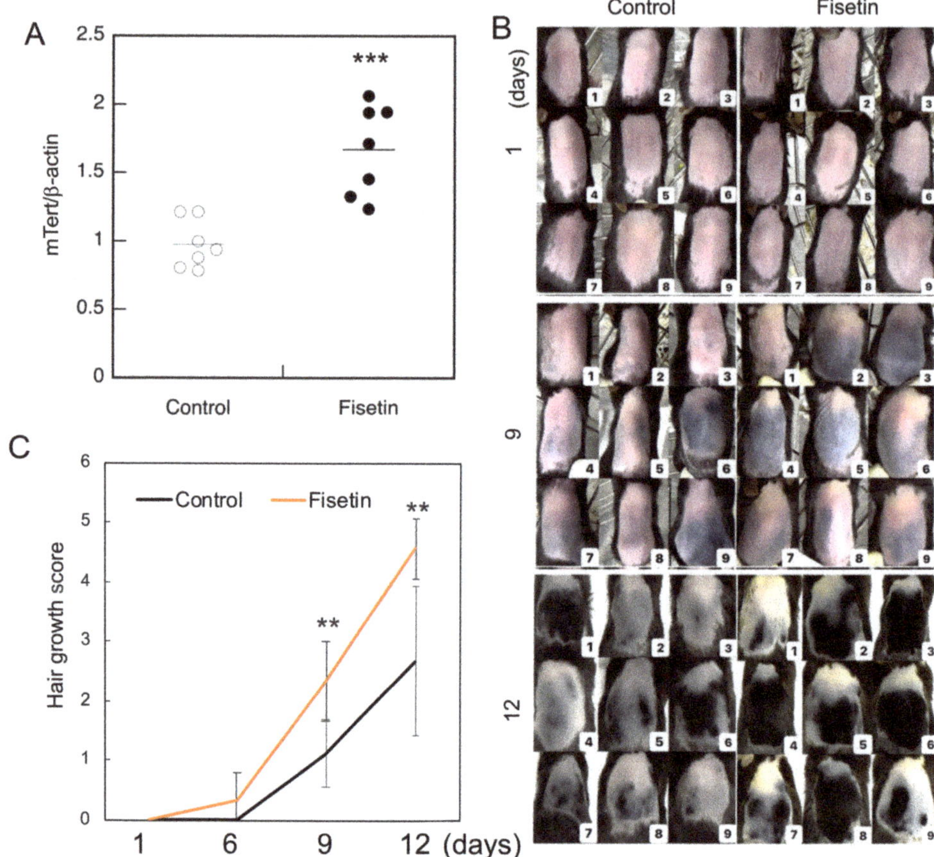

Figure 1. Hair growth-promoting effect of fisetin. (**A**) The expression of *mTert* in the dorsal skin cells of fisetin-treated mice and investigated using qRT-PCR. (**B**) Evaluation of the effect of fisetin treatment on the dorsal skin of C57BL/6 mice and hair growth after 12 days. (**C**) Hair growth score of fisetin-treated and control groups (the entire shaved back is pink = 0; part of the shaved back is blue = 1; part of the shaved back is gray = 2; the entire shaved back is gray = 3; part of the shaved back is black = 4; the entire shaved back is black = 5). Statistical significance was determined using a two-sided Student's t-test. Statistical significance was defined as $p < 0.05$ (** $p < 0.01$; *** $p < 0.001$).

3.2. Fisetin Promotes Hair Growth in Mice

Since the functions and properties of the skin change significantly with the hair cycle, it is generally considered important to control the hair cycle in animal experiments on hair growth. To obtain a uniform hair cycle, it is widely known that hair removal creams and electric shavers can be used to simultaneously remove hair within a certain range of the telogen phase and stimulate them to induce a uniform anagen phase [12]. Hair growth was strongly promoted in mice in the fisetin group compared to those in the control group (Figure 1B). The degree of hair growth was scored using skin color and hair growth as indicators, and the results showed that hair growth was significantly enhanced in the fisetin group compared to the control group (Figure 1C).

3.3. Fisetin Activates HFSCs In Vivo

After the hair growth test was completed, the condition of the HFSCs was observed using paraffin-embedded sections of the skin using antibodies against β-catenin, CD34 (a stem cell marker), and Ki67 (a cell proliferation marker). The results showed that β-catenin and Ki67 were strongly expressed in CD34$^+$ cells in the vicinity of hair follicles on the dorsal side of the skin of fisetin-treated mice (Figure 2). This suggested that the proliferation of HFSCs in the vicinity of the dorsal skin of fisetin-treated mice was activated and that fisetin activated the interaction between keratinocytes and HFSCs, resulting in the promotion of hair growth.

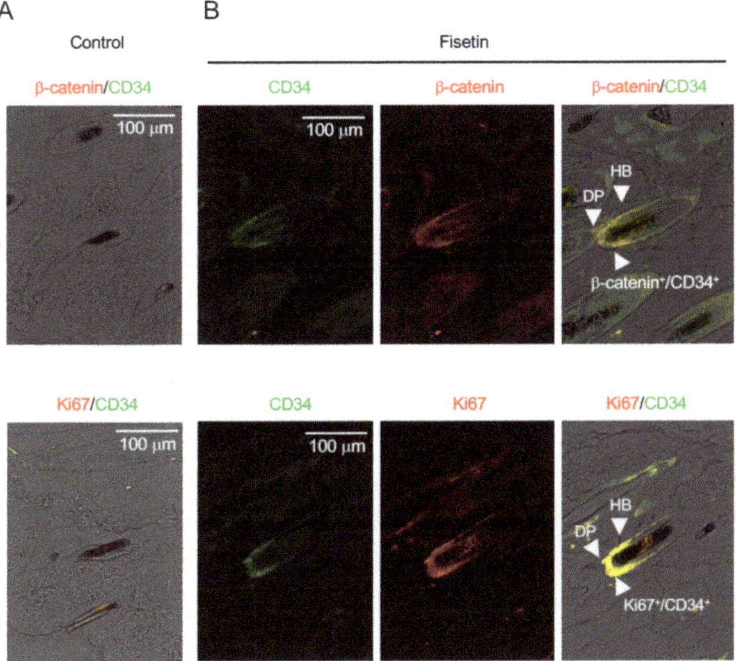

Figure 2. Effects of fisetin treatment on the expression of marker proteins in dorsal skin sections. Immunohistochemistry analysis of skin sections was performed using anti-β-catenin, anti-Ki67, and anti-CD34 antibodies ((**A**), Control; (**B**), Fisetin) (DP, dermal papilla; HB, hair bulge).

3.4. Fisetin Activates HaCaT Cells

First, we examined the effects of fisetin on keratinocytes. The results showed that fisetin augmented the expression of SIRT1 and further increased and activated the mitochondria in HaCaT cells (Figure 3A,B). In addition, fisetin regulated the expression of various genes, such as those encoding secretory factors and those involved in epidermogenesis, cell proliferation, hair follicle regulation, and hair cycle regulation (Tables 1 and 2). These results suggest that fisetin activates HaCaT cells.

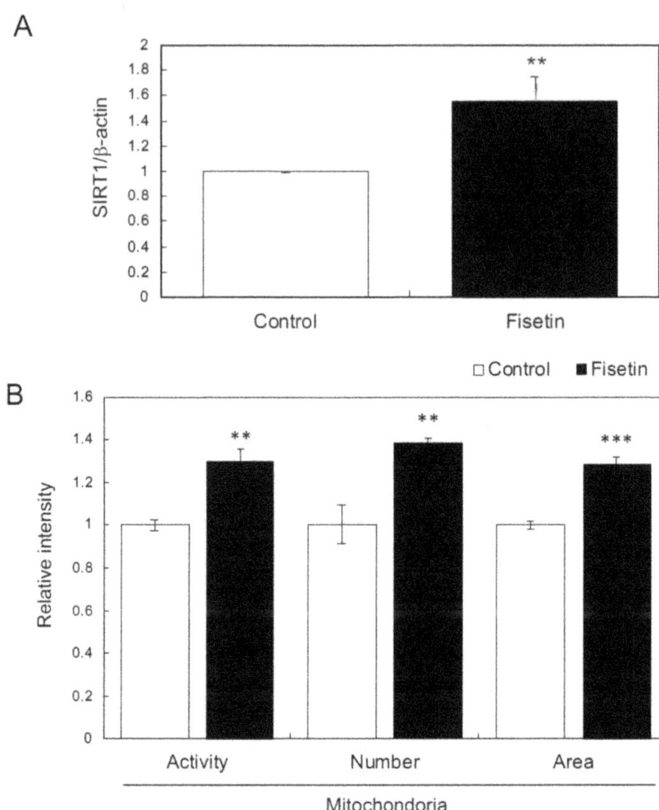

Figure 3. Effects of fisetin on keratinocytes. (**A**) The expression of *SIRT1* in HaCaT cells was evaluated using qRT-PCR. (**B**) Effects of fisetin on mitochondria evaluated using mitochondrion-specific probes (MitoTracker Red CMXRos and MitoTracker Green FM) and IN Cell Analyzer 2200. Statistical significance was determined using a two-sided Student's t-test. Statistical significance was defined as $p < 0.05$ (** $p < 0.01$; *** $p < 0.001$).

Table 1. Functional annotation.

Annotation Cluster	Term	Count	p-Value
1	Oxidoreductase	50	4.90×10^{-6}
	Oxidation-reduction process	49	9.90×10^{-5}
2	Secreted	122	2.20×10^{-5}
	Signal	214	1.60×10^{-3}
3	Palmoplantar keratoderma	10	5.20×10^{-6}
	Keratin	10	2.10×10^{-1}
4	Regulation of cell growth	11	3.60×10^{-3}
	Insulin-like growth factor-binding	4	4.60×10^{-2}
	Negative regulation of cell death	4	4.70×10^{-1}

Table 2. GO analysis.

GO Term	Genes
Hair follicle regulation system gene	INHBA, RUNX1, TGFB2, FST, KRT17
Hair cycle control gene	KRT14

3.5. Effects of Exosomes Derived from Fisetin-Treated HaCaT Cells on HFSCs

In this study, we focused on exosomes as the mediators of the interaction between keratinocytes and HFSCs and hypothesized that the exosomes secreted by keratinocytes activated by fisetin would activate the HFSCs. Therefore, we first prepared exosomes from the supernatant of fisetin-treated HaCaT cells and tested their effect on the proliferation of HFSCs. The results showed that fisetin-treated HaCaT cell-derived exosomes triggered the nuclear translocation of β-catenin (Figure 4A), augmented the expression of AXIN2 (Figure 4B), and increased the number of Ki67$^+$ cells in HFSCs (Figure 4C). These results suggest that fisetin-treated HaCaT cell-derived exosomes activate the proliferation of HFSCs.

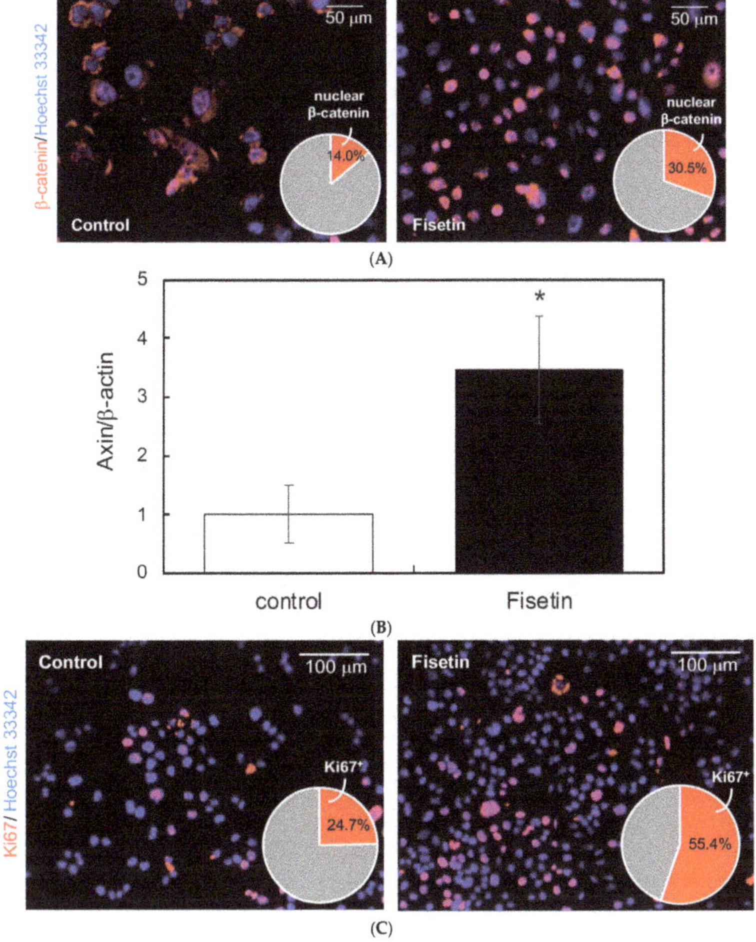

Figure 4. Effects of exosomes derived from fisetin-treated keratinocytes on hair follicle stem cells (HFSCs). (**A**) Localization of β-catenin in HFSCs was evaluated by immunocytochemistry using anti-β-catenin antibody. Nuclear localization of β-catenin (red in pie chart) in HFSCs treated with exosomes was quantitatively determined using IN Cell Analyzer 2200. (**B**) Expression of axin in HFSCs treated with exosomes was quantitatively determined using qRT-PCR. (**C**) Ki67$^+$ cells in HFSCs were detected by immunocytochemistry using anti-Ki67 antibody. The number of Ki67$^+$ cells (red in pie chart) was determined using IN Cell Analyzer 2200. Statistical significance was determined using a two-sided Student's t-test. Statistical significance was defined as $p < 0.05$ (* $p < 0.05$).

We next tested whether fisetin-treated HaCaT cell-derived exosomes could activate the mitochondria in HFSCs. The HFSCs were treated with fisetin-treated HaCaT cell-derived exosomes, and the expression of TOMM20, a mitochondrial marker, was verified by immunostaining. The results showed that the number of TOMM20$^+$ cells was significantly increased in the HFSCs (Figure 5A). Moreover, the number and activity of mitochondria were evaluated using the fluorescence probes MitoTracker Green FM and MitoTracker Red CMXRos, respectively. The results showed that fisetin-treated HaCaT cell-derived exosomes significantly increased the number of cells harboring activated mitochondria relative to the total number of mitochondria (Figure 5B).

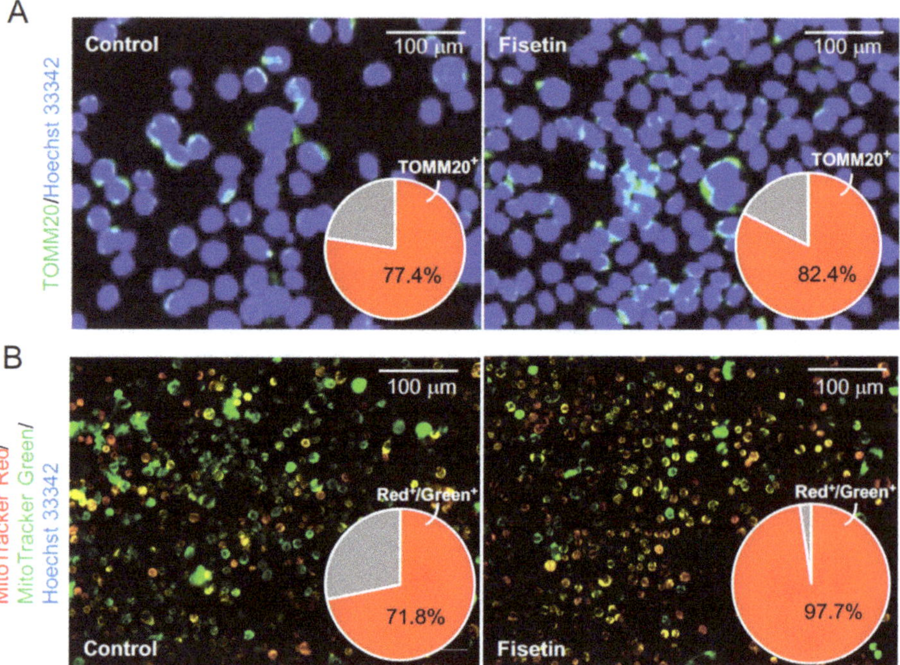

Figure 5. Effects of exosomes derived from fisetin-treated keratinocytes on the mitochondria in HFSCs. (**A**) TOMM20$^+$ cells in HFSCs were detected by immunocytochemistry using anti-TOMM antibody. The number of TOMM20$^+$ cells (red in pie chart) was determined using IN Cell Analyzer 2200. (**B**) Mitochondria in HFSCs were detected by immunocytochemistry using specific probes (MitoTracker Red CMXRos and MitoTracker Green FM). The relative number of cells harboring active mitochondria (yellow cells in photo; red in pie chart) was determined using IN Cell Analyzer 2200.

4. Discussion

In a previous study, we reported that fisetin, which augments the expression of TERT in keratinocytes, induces a shift from telogen to anagen in the hair follicles by inducing the proliferation of HFSCs, thus promoting hair growth [3]. These results suggest that fisetin activates keratinocytes and further strengthens the interaction between keratinocytes and HFSCs. In the present study, we aimed to clarify the molecular basis of fisetin-induced hair growth.

First, after confirming that fisetin is not toxic to keratinocyte at concentrations up to 10 µM, we examined whether keratinocytes are activated by fisetin. We found that fisetin activated the SIRT1-mitochondrial axis as well as keratinocytes by regulating the expression of various genes, such as those encoding secretory factors and those involved in epidermogenesis, cell proliferation, hair follicle regulation, and hair cycle regulation. Since mitochondrial dysfunction in skin cells is believed to cause an increase in reactive oxygen species levels and may be a cause of skin aging [17], activation of the SIRT1–mitochondrial axis by fisetin may lead to keratinocyte improvement and activation [18]. These results suggest that fisetin activates keratinocytes, which could lead to the promotion of hair growth.

Next, we examined the molecular basis for the activation of the keratinocyte–HFSC interaction induced by fisetin. We assumed that exosomes are involved in the activation of cell–cell interactions. Exosomes are an important component of paracrine signaling and can mediate communication between distant cells by directly transferring various biomolecules, including miRNAs, from donor cells to recipient cells [10,19,20]. However, the potential of exosomes as modulators of hair follicle dynamics has not received widespread attention. Previous studies have shown that exosomes derived from dermal papillae [4,5], dermal fibroblasts [6], and mesenchymal stem cells [7–9] can activate the proliferation of HFSCs and consequently induce hair growth. Key findings regarding the relevance of exosomes in skin and hair follicle regeneration have been reported [21], with an emphasis on the signaling pathways that mediate these effects. It is widely known that Wnt is a master regulator of hair follicle morphogenesis and hair growth [22]. Furthermore, mitochondrial aerobic respiration is activated during HFSC differentiation [23]. Thus, activation of the Wnt pathway and mitochondria is essential for hair follicle activation and differentiation.

In the present study, we found that fisetin-treated HaCaT cell-derived exosomes activated β-catenin and promoted HFSC growth, in addition to activating the mitochondria. These results revealed exosomes as the molecular basis of keratinocyte–HFSC interaction and that fisetin, along with its effects on keratinocytes, could cause exosome secretion and the consequent activation of HFSCs. Furthermore, since hair growth is triggered by exosome secretion, the hair growth-promoting effect is expected to continue for some time after fisetin application is discontinued. To the best of our knowledge, this is the first study to show that keratinocyte-derived exosomes can activate HFSCs, which in turn induces hair growth.

In conclusion, based on the obtained results, exosomes can be considered as natural mediators that could be involved in hair cycle regulation as well as serve as promising delivery vehicles for improving skin and hair regeneration because of their potential to target various molecular processes and cells. Further studies are required to clarify the mode of involvement of exosomes in the hair cycle in vivo and the therapeutic effects that can be expected when using exosomes to regulate hair growth in clinical settings.

Author Contributions: Conceptualization, Y.K.; methodology, Y.K.; validation, K.T. and N.U.; investigation, M.O.; data curation, M.U.; writing—original draft preparation, Y.K.; writing—review and editing, Y.K.; project administration, Y.K.; funding acquisition, Y.K. All authors have read and agreed to the published version of the manuscript.

Funding: This research received no external funding.

Institutional Review Board Statement: The study was conducted according to the guidelines of the Declaration of Helsinki, and approved by the Ethics Committee on Animal Experimentation of Kyushu University (approval number: A28-187-0).

Informed Consent Statement: Not applicable.

Data Availability Statement: The data presented in this study are openly available in QIR at [https://catalog.lib.kyushu-u.ac.jp/opac_browse/papers/?lang=0, accessed on 18 June 2021].

Acknowledgments: We would like to thank G. Takada and T. Nakazawa (Cytiva, Tokyo, Japan) for providing expert assistance with IN Cell Analyzer 2200 and K. Yasuda (Cell Innovator, Fukuoka, Japan) for her assistance with the microarray analysis.

Conflicts of Interest: The authors declare no conflict of interest.

References

1. Sarin, K.Y.; Cheung, P.; Gilison, D.; Lee, E.; Tennen, R.I.; Wang, E.; Artandi, M.K.; Oro, A.E.; Artandi, S.E. Conditional telomerase Induction Causes Proliferation of Hair Follicle Stem Cells. *Nature* **2005**, *436*, 1048–1052. [CrossRef] [PubMed]

2. Choi, J.; Southworth, L.K.; Sarin, K.Y.; Venteicher, A.S.; Ma, W.; Chang, W.; Cheung, P.; Jun, S.; Artandi, M.K.; Shah, N.; et al. TERT Promotes Epithelial Proliferation through Transcriptional Control of a Myc- and Wnt-related Developmental Program. *PLoS Genet.* **2008**, *4*, e10. [CrossRef]
3. Kubo, C.; Ogawa, M.; Uehara, N.; Katakura, Y. Fisetin Promotes Hair Growth by Augmenting TERT Expression. *Front. Cell Dev. Biol.* **2020**, *8*, 566617. [CrossRef]
4. Yan, H.; Gao, Y.; Ding, Q.; Liu, J.; Li, Y.; Jin, M.; Xu, H.; Ma, S.; Wang, X.; Zeng, W.; et al. Exosomal Micro RNAs Derived from Dermal Papilla Cells Mediate Hair Follicle Stem Cell Proliferation and Differentiation. *Int. J. Biol. Sci.* **2019**, *15*, 1368–1382. [CrossRef]
5. Zhou, L.; Wang, H.; Jing, J.; Yu, L.; Wu, X.; Lu, Z. Regulation of Hair Follicle Development by Exosomes Derived from Dermal Papilla Cells. *Biochem. Biophys. Res. Commun.* **2018**, *500*, 325–332. [CrossRef] [PubMed]
6. Le Riche, A.; Aberdam, E.; Marchand, L.; Frank, E.; Jahoda, C.; Petit, I.; Bordes, S.; Closs, B.; Aberdam, D. Extracellular Vesicles from Activated Dermal Fibroblasts Stimulate Hair Follicle Growth Through Dermal Papilla-Secreted Norrin. *Stem Cells* **2019**, *37*, 1166–1175. [CrossRef] [PubMed]
7. Rajendran, R.L.; Gangadaran, P.; Bak, S.S.; Oh, J.M.; Kalimuthu, S.; Lee, H.W.; Baek, S.H.; Zhu, L.; Sung, Y.K.; Jeong, S.Y.; et al. Extracellular Vesicles Derived from MSCs Activates Dermal Papilla Cell In Vitro and Promotes Hair Follicle Conversion from Telogen to Anagen in Mice. *Sci. Rep.* **2017**, *7*, 15560. [CrossRef]
8. Kwack, M.H.; Seo, C.H.; Gangadaran, P.; Ahn, B.C.; Kim, M.K.; Kim, J.C.; Sung, Y.K. Exosomes Derived from Human Dermal Papilla Cells Promote Hair Growth in Cultured Human Hair Follicles and Augment the Hair-inductive Capacity of Cultured Dermal Papilla Spheres. *Exp. Dermatol.* **2019**, *28*, 854–857. [CrossRef] [PubMed]
9. Gross, J.C.; Chaudhary, V.; Bartscherer, K.; Boutros, M. Active Wnt Proteins are Secreted on Exosomes. *Nat. Cell. Biol.* **2012**, *14*, 1036–1045. [CrossRef] [PubMed]
10. Sugihara, Y.; Onoue, S.; Tashiro, K.; Sato, M.; Hasegawa, T.; Katakura, Y. Carnosine Induces Intestinal Cells to Secrete Exosomes that Activate Neuronal Cells. *PLoS ONE* **2019**, *14*, e0217394. [CrossRef]
11. Müller-Röver, S.; Foitzik, K.; Paus, R.; Handjiski, B.; van der Veen, C.; Eichmüller, S.; McKay, I.A.; Stenn, K.S. A Comprehensive Guide for the Accurate Classification of Murine Hair Follicles in Distinct Hair Cycle Stages. *J. Investig. Dermatol.* **2001**, *117*, 3–15. [CrossRef]
12. Pi, L.Q.; Lee, W.S.; Min, S.H. Hot Water Extract of Oriental Melon Leaf Promotes Hair Growth and Prolongs Anagen Hair Cycle: In Vivo and In Vitro Evaluation. *Food Sci. Biotechnol.* **2016**, *25*, 575–580. [CrossRef]
13. Kadooka, K.; Fujii, K.; Matsumoto, T.; Sato, M.; Morimatsu, F.; Tashiro, K.; Kuhara, S.; Katakura, Y. Mechanisms and Consequences of Carnosine-induced Activation of Intestinal Epithelial Cells. *J. Funct. Foods* **2015**, *13*, 32–37. [CrossRef]
14. Huang, D.W.; Sherman, B.T.; Lempicki, R.A. Bioinformatics Enrichment Tools: Paths Toward the Comprehensive Functional Analysis of Large Gene Lists. *Nucleic Acids Res.* **2009**, *37*, 1–13. [CrossRef]
15. Huang, D.W.; Sherman, B.T.; Lempicki, R.A. Systematic and Integrative Analysis of Large Gene Lists using DAVID Bioinformatics Resources. *Nat. Protoc.* **2009**, *4*, 44–57. [CrossRef]
16. Nakai, W.; Yoshida, T.; Diez, D.; Miyatake, Y.; Nishibu, T.; Imawaka, N.; Naruse, K.; Sadamura, Y.; Hanayama, R. A novel affinity-based method for the isolation of highly purified extracellular vesicles. *Sci. Rep.* **2016**, *6*, 33935. [CrossRef]
17. Nagashima, S.; Tokuyama, T.; Yanashiro, R.; Inatome, R.; Yanagi, S. Roles of mitochondrial ubiquitin ligase MITOL/MARCH5 in mitochondrial dynamics and diseases. *J. Biochem.* **2014**, *155*, 273–279. [CrossRef]
18. Sreedhar, A.; Aguilera-Aguirre, L.; Singh, K.K. Mitochondria in Skin Health, Aging, and Disease. *Cell Death Dis.* **2020**, *11*, 444. [CrossRef] [PubMed]
19. Sheldon, H.; Heikamp, E.; Turley, H.; Dragovic, R.; Thomas, P.; Oon, C.E.; Leek, R.; Edelmann, M.; Kessler, B.; Sainson, R.C.A.; et al. New Mechanism for Notch Signaling to Endothelium at a Distance by Delta-like 4 Incorporation into Exosomes. *Blood* **2010**, *116*, 2385–2394. [CrossRef] [PubMed]
20. Inotsuka, R.; Uchimura, K.; Yamatsu, A.; Kim, M.; Katakura, Y. γ-aminobutyric acid (GABA) activates neuronal cells though inducing the secretion of exosomes from intestinal cells. *Food Funct.* **2020**, *11*, 9285–9290. [CrossRef] [PubMed]
21. Carrasco, E.; Soto-Heredero, G.; Mittelbrunn, M. The Role of Extracellular Vesicles in Cutaneous Remodeling and Hair Follicle Dynamics. *Int. J. Mol. Sci.* **2019**, *20*, 2758. [CrossRef] [PubMed]
22. Veltri, A.; Lang, C.; Lien, W.H. Concise Review: Wnt signaling pathways in skin development and epidermal stem cells. *Stem Cells* **2018**, *36*, 22–35. [CrossRef] [PubMed]
23. Tang, Y.; Luo, B.; Deng, Z.; Wang, B.; Liu, F.; Li, J.; Shi, W.; Xie, H.; Hu, X.; Li, J. Mitochondrial Aerobic Respiration is Activated During Hair Follicle Stem Cell Differentiation, and its Dysfunction Retards Hair Regeneration. *PeerJ* **2016**, *4*, e1821. [CrossRef] [PubMed]

nutrients

Article

Long-Term Caffeine Intake Exerts Protective Effects on Intestinal Aging by Regulating Vitellogenesis and Mitochondrial Function in an Aged *Caenorhabditis Elegans* Model

Hyemin Min [†], Esther Youn [†] and Yhong-Hee Shim *

Department of Bioscience and Biotechnology, Konkuk University, Seoul 05029, Korea;
mintmin0701@naver.com (H.M.); dbsdptmej02@naver.com (E.Y.)
* Correspondence: yshim@konkuk.ac.kr; Tel.: +82-2-450-4059
† These authors contributed equally to this work.

Abstract: Caffeine, a methylxanthine derived from plants, is the most widely consumed ingredient in daily life. Therefore, it is necessary to investigate the effects of caffeine intake on essential biological activities. In this study, we attempted to determine the possible anti-aging effects of long-term caffeine intake in the intestine of an aged *Caenorhabditis elegans* model. We examined changes in intestinal integrity, production of vitellogenin (VIT), and mitochondrial function after caffeine intake. To evaluate intestinal aging, actin-5 (ACT-5) mislocalization, lumenal expansion, and intestinal colonization were examined after caffeine intake, and the levels of vitellogenesis as well as the mitochondrial activity were measured. We found that the long-term caffeine intake (10 mM) in the L4-stage worms at 25 °C for 3 days suppressed ACT-5 mislocalization. Furthermore, the level of autophagy, which is normally increased in aging animals, was significantly reduced in these animals, and their mitochondrial functions improved after caffeine intake. In addition, the caffeine-ingesting aging animals showed high resistance to oxidative stress and increased the expression of antioxidant proteins. Taken together, these findings reveal that caffeine may be a potential anti-aging agent that can suppress intestinal atrophy during the progression of intestinal aging.

Keywords: caffeine; intestinal aging; anti-aging; vitellogenesis; mitochondrial function; oxidative stress response; *Caenorhabditis elegans*

Citation: Min, H.; Youn, E.; Shim, Y.-H. Long-Term Caffeine Intake Exerts Protective Effects on Intestinal Aging by Regulating Vitellogenesis and Mitochondrial Function in an Aged *Caenorhabditis Elegans* Model. *Nutrients* **2021**, *13*, 2517. https://doi.org/10.3390/nu13082517

Academic Editor: Yoshinori Katakura

Received: 19 June 2021
Accepted: 20 July 2021
Published: 23 July 2021

Publisher's Note: MDPI stays neutral with regard to jurisdictional claims in published maps and institutional affiliations.

Copyright: © 2021 by the authors. Licensee MDPI, Basel, Switzerland. This article is an open access article distributed under the terms and conditions of the Creative Commons Attribution (CC BY) license (https://creativecommons.org/licenses/by/4.0/).

1. Introduction

The physiological effects of caffeine intake have been reported to be beneficial or harmful, depending on the dose of intake [1–4]. In general, caffeine treatment at a lower concentration (<10 mM) shows beneficial effects [3,5,6], while harmful effects are observed at higher concentrations [1,7,8]. However, we previously showed adverse effects of caffeine treatment, even at a lower concentration, on the reproduction and intergenerational effects in a *Caenorhabditis elegans* (*C. elegans*) model during the active reproductive period [4]. These findings suggested that the effects of caffeine intake are dependent not only on the caffeine concentration but also on the developmental stages of the organism when treatment is administered. Therefore, in this study, we examined the effects of caffeine intake (10 mM) during the post-developmental stage and the post-reproductive period in a *C. elegans* model.

The gonad of *C. elegans* hermaphrodites consists of germ lines containing mitotic germ cells, oocytes, and sperm, in which spermatogenesis begins at the L4 larval stage and oogenesis at the adult stage. Reproduction is the most active at day 2 of the adult stage [9,10]. In day-3 adults, fertility begins to decline, and the aging processes become active [10]. At this stage, the cost of reproduction comes at the price of oxidative stress resistance and lifespan extension [11–13]. *C. elegans* is an excellent animal model for studying the effects of caffeine intake on aging processes based on various indicators, including the degeneration of pharynx, presence of tumors in the uterus, atrophy of the intestine and gonads, and accumulation of lipoproteins in the body cavity [14]. Furthermore, the aged intestine of *C. elegans* showed the impaired permeability of the intestinal epithelium, small and fewer microvilli, and accumulation of the indigested food, which have also been reported in the aged intestine in mammals [15,16]. The intestinal atrophy is a prominent aging feature that is closely related to the lifespan of the animal [15,17]. Increased vitellogenesis and accumulation of the pseudocoelomic lipoprotein pool (PLP) are linked to intestinal atrophy at the advanced ages in *C. elegans* [14]. Vitellogenin (VIT) is a yolk protein that supports the development of the progeny in oviparous animals [18,19]. VITs are believed to be similar to low-density lipoproteins in humans based on their sequence similarities [20]. In *C. elegans*, VITs are exclusively synthesized in the intestine of the adult hermaphrodites; therefore, their production shows temporal-, spatial-, and sex-specificity [19,21,22]. VITs are expressed at the adult stage and continue to be produced even after the end of reproduction, and it further accelerates the aging process by increasing the intestinal atrophy [14,19]. In particular, vitellogenesis in aging *C. elegans* has been reported to be related to biomass conversion from intestine to a yolk protein by autophagic activity [14]. Interestingly, we recently reported that caffeine intake reduces VIT production during the adult stage of *C. elegans* [4]. Based on these findings, we hypothesized that caffeine intake could ameliorate intestinal atrophy, prevent intestinal aging, and eventually extend the lifespan of these aging animals by inhibiting the production of VIT.

Aging is an irreversible phenomenon observed across different species. Studies based on the insulin/insulin-like growth factor (IGF-1) signaling in *C. elegans* aiming to reveal the mechanism underlying aging have shown that genetic factors play critical roles in the aging and regulation of the lifespan in these animals [23–25]. Although dietary factors are important to prevent obesity, various adult disorders and aging in addition to genetic backgrounds [26–29], the mechanisms by which the dietary factors affect intestinal aging remain elusive. In addition, our previous results showed that the intestinal atrophy was strongly observed with gliadin intake, which promoted ROS production, and the intestinal atrophy became more severe in the *mev-1* mutant that is hypersensitive to oxidative stress [30]. These findings suggest that intestinal atrophy is induced by oxidative stress. Therefore, it is important to investigate how these factors regulate the aging processes at the organism level to reveal the link among diet, intestinal aging, oxidative stress response, and lifespan of the organism, which is generally difficult to examine unless a proper animal model, such as *C. elegans*, is available.

In this study, we attempted to examine the effects of long-term caffeine intake on the level of intestinal atrophy by evaluating the progression of intestinal aging and changes in the production of VIT, mitochondrial activity, and oxidative stress responses in the aging *C. elegans*. Here, we report that long-term caffeine intake in aging *C. elegans* reduced the VIT production, improved intestinal atrophy, promoted mitochondrial function, induced resistance against oxidative stress, and extended the life span.

2. Materials and Methods

2.1. Caenorhabditis Elegans Strains and Treatment with Caffeine

All strains were maintained at either 15 °C or 20 °C on the nematode growth medium (NGM) agar plates seeded with *Escherichia coli* OP50 as previously described [31]. The following strains were used in this study to analyze aging phenotype: N2 (*C. elegans* wild isolate, Bristol variety), ML2615: *dlg-1(mc103[dlg-1::GFP]) X*, for the junctional morphology [32], ERT60: *jyIs13 [act-5p::GFP::ACT-5+rol-6(su1006)] II*, for the intestinal actin localization [32], DH1033: *bIs1 (vit-2::GFP+rol-6(su10060)) X*, for the vitellogenin expression [19], DA2123: *adIs2122 [lgg-1p::GFP::lgg-1+rol-6(su1006)]*, for the autophagic activity [33], SJ4103: *zcIs14 [myo-3::GFP(mit)]*, for the muscle mitochondria, [34], SJ4143: *zcIs17 [ges-1p::GFP(mit)]*, for the intestinal mitochondria [35], CF1553: *sod-3(muIs84)::GFP*, for the activity of superoxide dismutase [17], CL2166: *dvIs19 [(pAF15)gst-4p::GFP::NLS]II*, for oxidative stress response [3], LD1: *Idls7[skn-1b/c::GFP+rol-6(su1006)]*, for the stress response [36], and GR1352: *xrIs87 [daf-16(alpha)::GFP::daf-16B+rol-6(su1006)]*, for the stress response [17].

For caffeine treatment, 10 mM caffeine (Sigma-Aldrich, St. Louis, MO, USA) was added to NGM before autoclaving, as previously described [4]. Throughout the study, we treated worms with 10 mM caffeine following our previous study showing that 10 mM of caffeine intake reduced vitellogenin production [4]. To investigate the effects of caffeine intake in aging *C. elegans*, synchronized L4 (long-term caffeine intake, caffeine treatment at 25 °C for 72 h), or 72 h post-L4 stage (short-term caffeine intake, caffeine treatment at 20 °C for 24 h) animals were examined. For the detailed experimental scheme, refer to the Supplementary Figure S1.

2.2. Analysis of Intestinal Aging

To analyze intestinal aging, we evaluated the pharyngeal deterioration, intestinal atrophy, localization of Actin-5 (ACT-5)::green fluorescent protein (GFP), intestinal colonization by fluorescent bacteria, and the accumulation of PLP as previously described with minor modifications [14,37]. To observe pharyngeal deterioration, discs large MAGUK scaffold protein 1 (DLG-1)::GFP transgenic animals were observed at 200× magnification under a fluorescence microscope (Zeiss Axioscope, Oberkochen, Germany). Intestinal atrophy was quantified by measuring the intestinal width at a point posterior to either the uterine tumors or vulva region, subtracting the lumenal width, and dividing it by the body width, as previously described with minor modifications [14]. To analyze the ACT-5::GFP localization in aging *C. elegans*, the ACT-5::GFP in the posterior intestine was observed at 200× magnification under a fluorescence microscope (Zeiss Axioscope, Oberkochen, Germany). To quantify the degree of OP50::GFP bacterial colonization in the intestine, synchronized L4-stage animals were fed with OP50::GFP, which is a fluorescent bacteria, on NGM plates containing 0 or 10 mM caffeine at 25 °C for 72 h. The animals were observed at 200× magnification under a fluorescence microscope (Zeiss Axioscope, Oberkochen, Germany). PLP accumulation rate was measured by the presence of the yolk pools in the body cavity at 400× magnification under a microscope (Zeiss Axioscope, Oberkochen, Germany). Between 10 and 20 worms were observed for each set of DLG-1::GFP, OP50::GFP, and PLP accumulation. For intestinal width measure, a total of 24 worms were observed.

2.3. Live Image Observation of Fluorescence-Tagged Transgenic Animals

To observe the gene expression of each transgenic animal, synchronized L4-stage animals were fed 0 or 10 mM caffeine at 25 °C for 72 h. Then, the animals were mounted on a poly-L-lysine (Sigma-Aldrich, St. Louis, MO, USA) coated glass slide using 10 µL M9 containing 0.2 mM tetramisole hydrochloride (Sigma-Aldrich, St. Louis, MO, USA). Live worm images were acquired under a fluorescence microscope (Zeiss Axioscope, Oberkochen, Germany) and processed using the Nikon NIS-Elements Basic Research imaging software v.4.3. The fluorescence intensity was quantified using the ImageJ software. Between 10 and 20 worms were observed with fluorescence-tagged transgenic animals for each set of experiments except for VIT-2::GFP observation. A total of 27 worms were observed with VIT-2::GFP transgenic animals under the condition of 0 mM caffeine treatment.

2.4. Western Blot Analysis

Western blot analysis was performed as described previously [4]. The DH1033: *bIs1 (vit-2::GFP+rol-6(su10060))* transgenic animal protein extract prepared from 30 adult hermaphrodites of each treatment group was subjected to sodium dodecyl sulfate–polyacryl amide gel electrophoresis and transferred to nitrocellulose membrane. Antibodies bound to a nitrocellulose membrane (PROTRAN BA83, Whatman; Sigma-Aldrich, St. Louis, MO, USA) were visualized using an ECL Western blotting detection kit (Amersham, GE Healthcare Life Sciences, Pittsburgh, PA, USA), and the band intensities were measured using a LAS-3000 image analyzer equipped with Multi Gauge v.3.0 (Fuji Film, Tokyo, Japan). The following primary and secondary antibodies were used: rabbit anti-GFP (1:1000; Novus

Biologicals, Centennial, CO, USA), mouse anti-α-tubulin (1:1000; Sigma-Aldrich, St. Louis, MO, USA), horseradish peroxidase (HRP)-conjugated goat anti-rabbit IgG (1:1000; Santa Cruz Biotechnology, Dallas, TX, USA), and HRP-conjugated donkey anti-mouse IgG (1:1000; Jackson ImmunoResearch, West Grove, PA, USA).

2.5. Quantitative Reverse Transcription-Polymerase Chain Reaction (qRT-PCR)

Real-time qRT-PCR was performed as described previously [4]. Briefly, RNA was isolated from adult hermaphrodites treated with 0 or 10 mM caffeine at the L4 stage at 25 °C for 72 h. The animals were placed in the TRIzol reagent (Invitrogen, Waltham, MA, USA), and total RNA was extracted using standard phenol–chloroform extraction and ethanol precipitation method using a phase lock gel (MaXtract High Density; Qiagen, Germantown, MD, USA) with 150 adult hermaphrodites of each treatment group. cDNA was synthesized using oligo-dT primers and the Moloney Murine Leukemia Virus (M-MLV) reverse transcriptase (Invitrogen, Waltham, MA, USA). qRT-PCR assays were performed on ABI 7500 (Applied Biosystems, Waltham, MA, USA) using SYBR Green PCR Master Mix (Applied Biosystems, Waltham, MA, USA). The final PCR volume was 10 μL, with 50 ng of the converted cDNA. The *act-1* mRNA was used as an endogenous control for the normalization of data. The primers used to measure the expression levels of each gene were as follows: *act-1* forward, 5′-CCAGGAATTGCTGATCGTATGCAGAA-′3; *act-1* reverse, 5′-TGGAGAGGGAAGCGAGGATAG-3; *unc-62* forward, 5′-TAAGACATACCCAAGAGAATG CTG-′3; *unc-62* reverse, 5′-TTTGCCTTTCAGACAGACCA-′3; *ceh-60* forward, 5′-AGTTCTA CGGTTGCATCTTCG-′3; *ceh-60* reverse, 5′-AGTGTGGCTGATGGAGAAAC-′3; *pqm-1* forward, 5′-TCTCGAAAATGTCCGCACTG-3′; *pqm-1* reverse, 5′-GAGGTTCTTTCACGAATT GCTTC-3′. The GenBank database accession number for *C. elegans* mRNA and the product size of each qRT-PCR are: *act-1* (NC_003283.11, Chr V: 11081052..11082415): 133 bp, *unc-62* (NC_003283.11, Chr V: 4497463..4511447): 133 bp, *ceh-60* (NC_003284.9, Chr X: 6522596..6530146): 148 bp, *pqm-1* (NC_003280.10, Chr II: 11150391..11152595): 125 bp.

2.6. Analysis of Reactive Oxygen Species (ROS) Production in Mitochondria

To examine the effect of caffeine intake on the mitochondrial ROS, CellROX® Green (Invitrogen, Carlsbad, CA, USA) staining was performed as described previously [4,38]. Briefly, CellROX® Green was freshly prepared as 5 mM stock solutions and diluted in the M9 buffer at a 1:500 dilution before treatment. Then, the animals were transferred to either 0 or 10 mM caffeine plates containing the staining solution and stained at 20 °C for 2 h. The animals were mounted on a poly L-lysine-coated slide and observed under a fluorescence microscope (Zeiss Axioscope, Oberkochen, Germany). The relative quantification of mitochondrial ROS was performed using the ImageJ software. More than 15 worms were observed for each set of the experiments.

2.7. Analysis of the Mitochondrial Membrane Potential (MMP)

To measure MMP, tetramethylrhodamine methyl ester (TMRM; Thermo Fisher Scientific, Waltham, MA, USA) staining was performed as previously described [38]. Briefly, TMRM (final concentration: 30 μM) was added to NGM agar plates containing 0 or 10 mM caffeine. The plates were then seeded with dead *E. coli* OP50, and dried for 24 h in the dark. The synchronized animals were transferred to TMRM plates and incubated at 20 °C for 15 h. Then, the animals were mounted on a poly L-lysine-coated slide and observed under a fluorescence microscope (Zeiss Axioscope, Oberkochen, Germany). Fluorescence intensity was measured using the ImageJ software. More than 10 worms were observed for each set of the experiments except the young adult with 0 mM caffeine treatment in which 12 worms were observed.

2.8. Motility Assay

To analyze the effect of long-term caffeine intake on motility, body bends were measured as previously described [38]. Briefly, animals from each test condition were transferred onto separate NGM plates and scored for the number of body bends at 20 s intervals. One body bend was defined as a complete cycle of terminal bulb motion, starting from the top position of the sinusoidal wave track through to the bottom and back to the top. A total of 21 worms were observed for each treatment.

2.9. Survival Assay

The survival assay was performed as previously described [39]. Synchronized animals were transferred to NGM plates containing 0 or 10 mM caffeine. Following treatment, the plates were observed daily under a dissecting microscope, and the animal viability was scored. The animals were judged to be dead if they did not respond to gentle poking with a platinum wire. Percent survival was calculated as the percentage of surviving animals in the population. A total of 20 worms were observed for each treatment.

2.10. Survival Assay under Paraquat-Induced Oxidative Stress

The paraquat survival assay was performed as previously described, with minor modifications [38]. To analyze the survival rates under oxidative stress conditions in each group, the synchronized 72 h post-L4-stage animals were exposed to 100 mM paraquat solution at 20 °C for 3 h, and subsequently, the number of dead and live animals were counted. The animals were considered dead when they failed to respond to a gentle touch with a platinum wire on their bodies. More than 30 worms were observed for each treatment.

2.11. Statistical Analysis

All experiments were repeated more than three times for statistical evaluation of the data. The p values were calculated using either a two-tailed Student's t-test or one-way analysis of variance (ANOVA) with Tukey's post hoc test. Statistical significance was set at $p < 0.05$. Data are expressed as the mean ± standard deviation (SD). Statistical analyses were performed using the jamovi software (https://www.jamovi.org/, accessed on 13 November 2020).

3. Results

3.1. Long-Term Caffeine Intake Prevents Intestinal Aging in Caenorhabditis elegans

One of the mechanisms underlying intestinal aging was recently identified, demonstrating that vitellogenesis is coupled to intestinal atrophy mediated by autophagy, which facilitates intestinal biomass conversion to sustain vitellogenin synthesis in aging *C. elegans* [14]. In our previous study, caffeine intake was shown to reduce vitellogenin production in *C. elegans* [4]. Here, we investigated whether reduced vitellogenesis caused by caffeine intake affects the progression of intestinal aging in *C. elegans*. To determine the effect of caffeine intake on intestinal aging, we examined intestinal atrophy after short-term and long-term caffeine intake in aging wild-type *C. elegans* (Supplementary Figures S1 and S2). We found that both short-term and long-term caffeine ingestion in these animals significantly improved intestinal atrophy compared to the 0 mM caffeine aging group (Supplementary Figure S2). However, among the caffeine groups, long-term caffeine intake was the most effective in preventing intestinal atrophy during aging (Supplementary Figure S2). These results suggest that caffeine intake exerts a preventive effect on intestinal aging, and long-term caffeine intake is particularly effective in aging *C. elegans*. Therefore, to further understand how caffeine intake effectively protects against intestinal aging, we investigated long-term caffeine intake at 25 °C, which accelerates aging in *C. elegans* (Supplementary Figure S1).

We investigated the effects of long-term caffeine intake on intestinal aging phenotypes, including pharyngeal deterioration and intestinal atrophy, based on a previous report [14], (Figure 1). Pharyngeal deterioration was analyzed by observing a GFP-tagged *dlg-1* transgene to determine the integrity of epithelial junctional localization with age [32,40]. We found that caffeine-ingested animals showed a significantly reduced pharyngeal deterioration with age compared to the animals subjected to the caffeine-free diet (Figure 1A). In particular, severe intestinal atrophy developed with age in caffeine-free diet animals, whereas caffeine-ingested animals showed significantly decreased intestinal atrophy with age (Figure 1B). The mislocalization of ACT-5, indicating the disrupted integrity of the *C. elegans* intestinal barrier and accelerated pathogenesis with age [32], was also confirmed by observing ACT-5::GFP transgenic animals after caffeine intake (Figure 1C). Consistent with the results of ACT-5 mislocalization, a caffeine-free diet in aging animals promoted intestinal colonization of GFP-expressing *E. coli*, as detected by fluorescence microscopy (Figure 1D). However, caffeine-ingested aging animals showed less bacterial colonization in the intestine compared to caffeine-free diet animals (Figure 1D). Moreover, PLP accumulation, which is closely correlated with intestinal atrophy [14], significantly decreased following caffeine intake in aging animals (Figure 1E). These results suggest that long-term caffeine intake prevents intestinal aging by maintaining intestinal integrity in aging *C. elegans*.

3.2. Long-Term Caffeine Intake Reduces Vitellogenesis in Aging Caenorhabditis elegans

How does long-term caffeine intake prevent intestinal aging? It has been shown that *C. elegans* consumes its own intestine via autophagy to produce VIT and biomass conversion during advanced ages [14]. Notably, caffeine intake can reduce the VIT production at the adult stage in *C. elegans* [4]. These observations suggest that caffeine intake modulates intestinal aging by reducing VIT production. To test this hypothesis, we analyzed the level of expression of VIT-2::GFP using transgenic animals by both fluorescence microscopic observation and Western blotting after long-term caffeine treatment (Figure 2A,B). The amount of VIT-2 protein in response to long-term caffeine intake reduced to less than 0.5 fold, as observed by fluorescence microscopy. Quantitative analysis of VIT-2::GFP protein by Western blotting showed that its expression decreased by approximately 80% compared to that in the caffeine-free diet animals (Figure 2A,B). These results suggest that the reduced vitellogenin production in response to caffeine intake prevents intestinal aging in the advanced age of *C. elegans*.

Some transcriptional regulators of vitellogenesis, such as *unc-62*, *ceh-60*, and *pqm-1*, have been identified [41,42]. To explore whether these transcription regulators control vitellogenesis, we measured the expression levels of *unc-62*, *ceh-60*, and *pqm-1* using qRT-PCR analysis after caffeine treatment (Figure 2C). Caffeine-ingested animals showed a 0.38 ($\pm$0.12)-fold decline in the expression of *unc-62* in aging animals as compared to that in the caffeine-free diet animals (Figure 2C), suggesting that the reduction of VIT-2 level in response to caffeine intake is due to the low level of *unc-62*, which is consistent with the fact that *unc-62* is a transcriptional activator of *vit-2* [4]. This finding also suggests that *unc-62* is the major regulator of vitellogenesis in caffeine-treated animals.

Next, we verified whether the previously reported autophagy activity, which plays a major role in promoting intestinal biomass conversion for vitellogenesis, is reduced by caffeine intake in aging animals. We observed autophagy activity using *lgg-1p::GFP::lgg-1* transgenic animals subjected to fluorescence microscopy after long-term caffeine treatment. Caffeine-ingested animals showed a 0.47 ($\pm$0.32)-fold decline in the level of LGG-1 foci in the intestinal cell compared to that in the caffeine-free diet animals (Figure 2D). Taken together, these results suggest that long-term caffeine intake decreases vitellogenesis by reducing *unc-62* expression and intestinal autophagy activity, which prevents intestinal atrophy in advanced age.

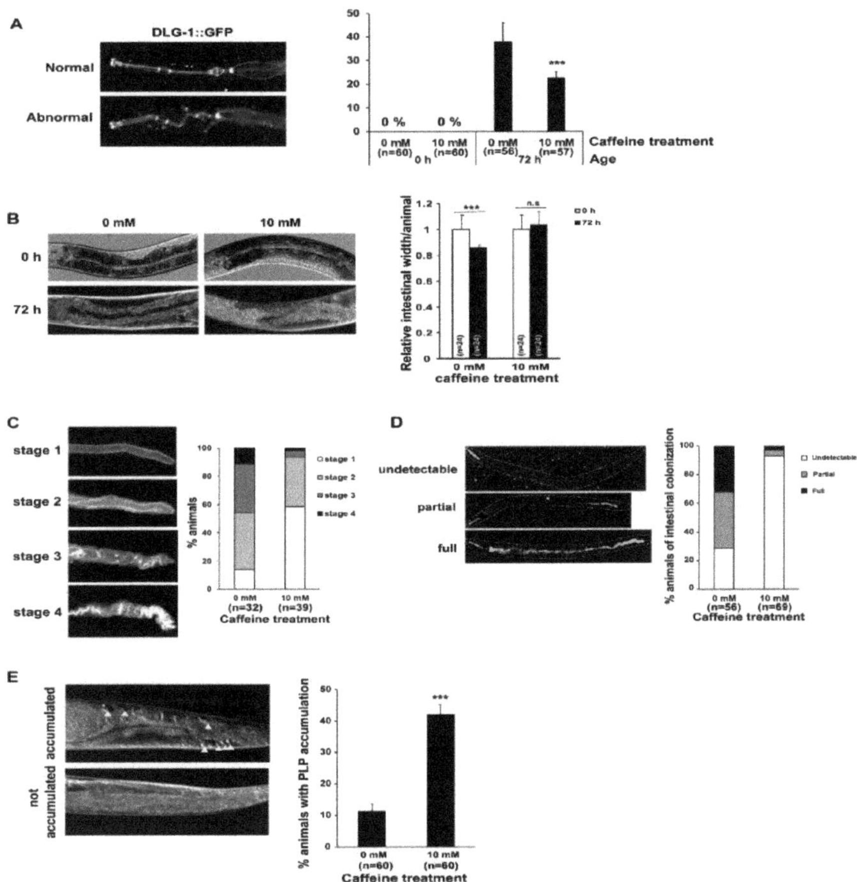

Figure 1. Long-term caffeine intake delays intestinal aging during advanced ages of *Caenorhabditis elegans*: (**A**) Animals expressing discs large MAGUK scaffold protein 1 (DLG-1):: green fluorescent protein (GFP) were treated with 0 or 10 mM caffeine at the L4 stage at 25 °C for 72 h. The location of DLG-1::GFP in the pharynx was observed at 72 h post-L4 stage at 25 °C. Error bars represent standard deviation (SD). *** $p < 0.001$ (one-way analysis of variance (ANOVA) with Tukey's post hoc test); (**B**) The synchronized wild-type L4-stage animals were treated with 0 or 10 mM caffeine at 25 °C for 72 h. The intestinal atrophy was measured at 72 h post-L4 stage at 25 °C. Error bars represent SD. *** $p < 0.001$. n.s., not significant (two-way ANOVA with Tukey's post hoc test); (**C**) Animals expressing actin 5 (ACT-5)::GFP were treated with 0 or 10 mM caffeine at the L4 stage at 25 °C for 72 h. The type of mislocalization was classified into four stages. The percent distributions of the respective stages in animals fed with 0 or 10 mM caffeine are presented; (**D**) The synchronized wild-type L4-stage animals were fed with *E. coli* OP50::GFP, a fluorescent bacteria on 0 or 10 mM caffeine nematode growth medium (NGM) plates at 25 °C for 72 h. The type of bacterial colonization was classified into three categories: (1) undetectable, (2) partial, and (3) full. The percent distributions of the respective categories in animals treated with 0 or 10 mM caffeine are presented; (**E**) The synchronized wild-type L4-stage animals were treated with 0 or 10 mM caffeine at 25 °C for 72 h. Accumulation of the pseudocoelomic lipoprotein pool (PLP) (indicated by yellow arrowheads) was observed at 72 h in post-L4-stage animals at 25 °C. The percentage of animals with PLP accumulation among the total number of animals is shown. Error bars represent SD. *** $p < 0.001$ (*t*-test).

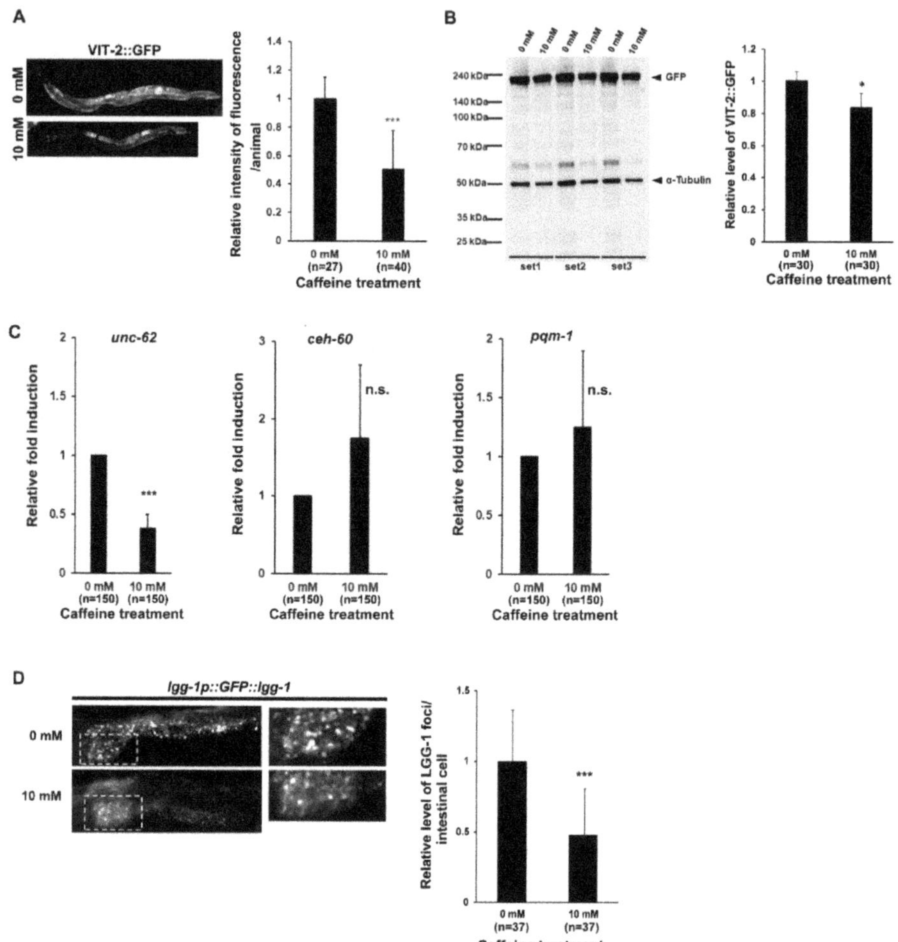

Figure 2. Long-term caffeine intake decreases vitellogenesis in advanced ages of *C. elegans*: (**A**) Animals expressing vitellogenin (VIT)-2::GFP were treated with 0 or 10 mM caffeine at the L4 stage at 25 °C for 72 h. The expression of VIT-2::GFP was observed at 72 h in post-L4-stage animals at 25 °C. Error bars represent SD. *** $p < 0.001$ (*t*-test); (**B**) Western blot analysis of VIT-2::GFP protein levels using an anti-GFP antibody in each test condition. α-tubulin was used as the loading control. The relative expression levels of GFP in each condition are shown. GFP band intensity was normalized to that of α-tubulin on the same lane, and the relative levels of GFP were converted to a relative value against that of the animals fed with 0 mM caffeine as 1. Error bars represent SD. * $p < 0.05$ (*t*-test); (**C**) The synchronized wild-type L4-stage animals were treated with 0 or 10 mM caffeine at 25 °C for 72 h. The mRNA levels of *unc-62*, *ceh-60*, and *pqm-1* in the caffeine-free or caffeine-ingested animals were determined by three independent quantitative reverse transcription-polymerase chain reaction (qRT-PCR) tests using the mRNA level of *act-1* in each sample as an internal control for normalization. Error bars represent SD. *** $p < 0.001$. n.s., not significant (*t*-test); (**D**) Animals expressing *lgg-1p::GFP::lgg-1* were treated with 0 or 10 mM caffeine at the L4 stage at 25 °C for 72 h; then, we imaged GFP::LGG-1 foci in the intestinal cells at 72 h post-L4-stage animals at 25 °C. The white dotted box represents the GFP::LGG-1 foci in the intestinal cells, and the right panel shows an enlarged image. The graph shows the relative levels of GFP::LGG-1 foci in the intestinal cells after caffeine treatment. Error bars represent SD. *** $p < 0.001$. n.s., not significant (*t*-test).

3.3. Long-Term Caffeine Intake Promotes Mitochondrial Function in Aging Caenorhabditis elegans

It has been suggested that immunity and antioxidant defense, which regulate ROS production in response to bacterial colonization during intestinal aging, are major effectors of aging and lifespan in *C. elegans* [43,44]. Given that caffeine intake suppresses bacterial colonization of the intestine in aging animals (Figure 1D), we measured mitochondrial ROS levels using CellROX Green, which is a fluorogenic probe for measuring mitochondrial ROS in live cells [38]. We observed that the fluorescence intensity decreased significantly in the long-term caffeine-ingested wild-type animals (Figure 3A). In addition, long-term caffeine-ingested animals showed normal MMP (Figure 3B). Mitochondrial activity was increased in the intestine of intestinal mitoGFP transgenic animals compared to that in the caffeine-free diet animals of advanced age (Figure 3C). These results suggest that long-term caffeine intake inhibits bacterial colonization and maintains normal mitochondrial function, resulting in a low level of ROS production at advanced ages.

To further investigate the modulations in mitochondria in response to caffeine intake at advanced ages, we examined morphological changes in the mitochondria using *myo-3p::mitoGFP* transgenic animals, which are expressed in the mitochondria of the muscle cells. The caffeine-free diet animals at advanced ages showed increased altered muscle mitochondrial morphology, including 'fused (28.23%)' and 'fragmented (39.53%)' mitochondria, while the caffeine-ingested animals in advanced ages exhibited a decreased 'fragmented (9.93%)' muscle mitochondrial morphology and an increased normal morphology (56.83%) (Figure 3D). This result indicates that long-term caffeine intake improves the integrity of mitochondrial morphology in advanced age groups.

Since abnormal mitochondria in muscle cells are associated with changes in locomotion behavior [45], we tested whether long-term caffeine intake affects motility by measuring the body bending rates at advanced ages. Compared to the caffeine-free diet animals, the body bending rate increased significantly in the caffeine-treated animals at advanced ages (Figure 3E). Consistent with the correlation between motility activity and lifespan [46], we also confirmed that long-term caffeine intake increased the survival rate (Figure 3F). Moreover, *skn-1* activation was evaluated, since it is the primary target to be activated by nuclear localization for regulating the survival rate [47,48]. In our study, SKN-1 was activated in the intestine after caffeine treatment (Figure 3F,G), implying that long-term caffeine intake promotes motility and extends the lifespan mediated by SKN-1 activation.

3.4. Long-Term Caffeine Intake Induces Oxidative Stress Response in Aging Caenorhabditis elegans

Next, we examined whether long-term caffeine intake alters the sensitivity of oxidative stress response in animals at advanced ages. Caffeine-ingested animals showed a significantly higher survival rate compared to the caffeine-free diet animals upon exposure to 100 mM paraquat, indicating that caffeine intake promotes resistance to oxidative stress in advanced ages (Figure 4A).

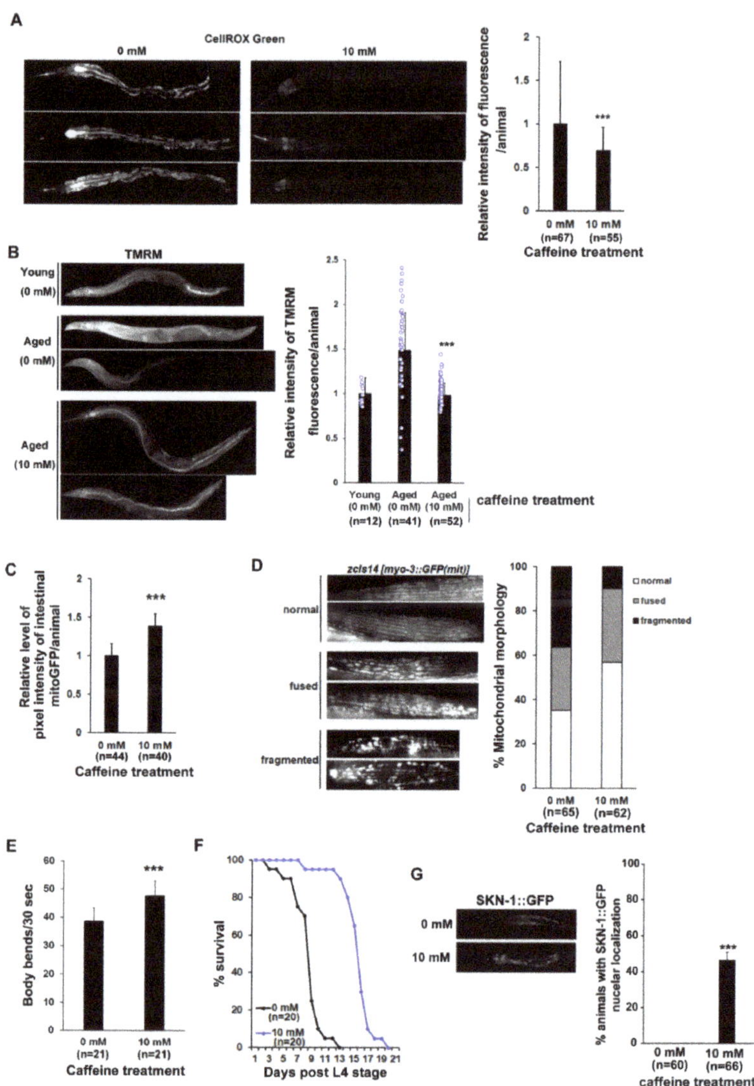

Figure 3. Long-term caffeine intake improves mitochondrial function and motility during advanced ages of *C. elegans*: (**A**) Comparison of the mitochondrial reactive oxygen species (ROS) levels between the caffeine-free and caffeine-ingested animals by CellROX Green staining. Wild-type animals were treated with 0 or 10 mM caffeine at the L4 stage at 25 °C for 72 h. The graph shows the relative levels of mitochondrial ROS analyzed by ImageJ. Error bars represent SD. *** $p < 0.001$ (*t*-test); (**B**) Comparison of the mitochondrial membrane potential (MMP) in young (0 mM) and aged (0 mM;

10 mM) animals using tetramethylrhodamine methyl ester (TMRM) staining. In the aged groups, wild-type animals were treated with 0 or 10 mM caffeine at the L4 stage at 25 °C for 72 h. In the young (0 mM) group, wild-type animals were treated with 0 mM caffeine at the L4 stage at 25 °C for 24 h. The TMRM fluorescence was quantified for each test condition by ImageJ. The graph shows the relative levels of MMP. Error bars represent SD. *** $p < 0.001$ (one-way ANOVA with Tukey's post hoc test); (**C**) Comparison of intestinal mitochondrial activity in *ges-1p::GFP(mit)* transgenic animal expressing *ges-1* promoter driven GFP. The transgenic animals were treated with 0 or 10 mM caffeine at the L4 stage at 25 °C for 72 h. The graph shows the relative levels of fluorescence intensity as analyzed by ImageJ. Error bars represent SD. *** $p < 0.001$ (*t*-test); (**D**) The mitochondrial morphology was analyzed using the SJ4103 transgenic animal expressing a mitochondrial-targeted GFP under the control of the muscle-specific *myo-3* promoter. The transgenic animals were treated with 0 or 10 mM caffeine at the L4 stage at 25 °C for 72 h. The graph indicates the percentage of animals with muscle mitochondria classified into three categories: (1) normal, (2) fused, and (3) fragmented; (**E**) Comparison of body bending in caffeine-free diet animals and caffeine-ingested animals at advanced ages. Wild-type animals were treated with 0 or 10 mM caffeine at the L4 stage at 25 °C for 72 h. Error bars represent SD. *** $p < 0.001$ (*t*-test); (**F**) Comparison of the survival rates between the caffeine-free diet animals and caffeine-ingested animals. Wild-type animals were treated with 0 or 10 mM caffeine at the L4 stage until dead at 25 °C; (**G**) Animals expressing skinhead 1 (SKN-1)::GFP were treated with 0 or 10 mM caffeine at the L4 stage at 25 °C for 72 h. The expression levels of SKN-1::GFP were observed at 72 h post-L4-stage animals at 25 °C. Error bars represent SD. *** $p < 0.001$ (*t*-test).

The SKN-1-mediated oxidative stress response involves the upregulated expression of phase II detoxification enzymes, such as GST-4 [36]. DAF-16 is a forkhead box O (FOXO) transcription factor that responds to various stresses. Superoxide dismutase 3 (SOD-3), an antioxidant enzyme, is one of the transcriptional targets of DAF-16 [17]. Therefore, we hypothesized that the resistance to oxidative stress mediated by caffeine intake may result from the changes in the expression levels of factors that regulate the transcription of essential detoxification genes, such as *gst-4* and *sod-3* [49–51]. Indeed, we found that the activation of GST-4::GFP was significantly increased in caffeine-treated animals, suggesting that caffeine intake induces the detoxification enzyme GST-4 via SKN-1 activation to promote resistance to oxidative stress (Figures 3G and 4B). We also found a significant increase in SOD-3::GFP in caffeine-treated animals at advanced ages (Figure 4C). However, contrary to our expectation, DAF-16 was not activated by long-term caffeine intake in the advanced age groups (Figure 4D), although both SKN-1 and DAF-16 were activated by heat treatment (Supplementary Figure S3). Collectively, these results suggest that long-term caffeine intake increases GST-4 dependent SKN-1 activity and promotes SOD-3 activity in a manner distinct from that of DAF-16.

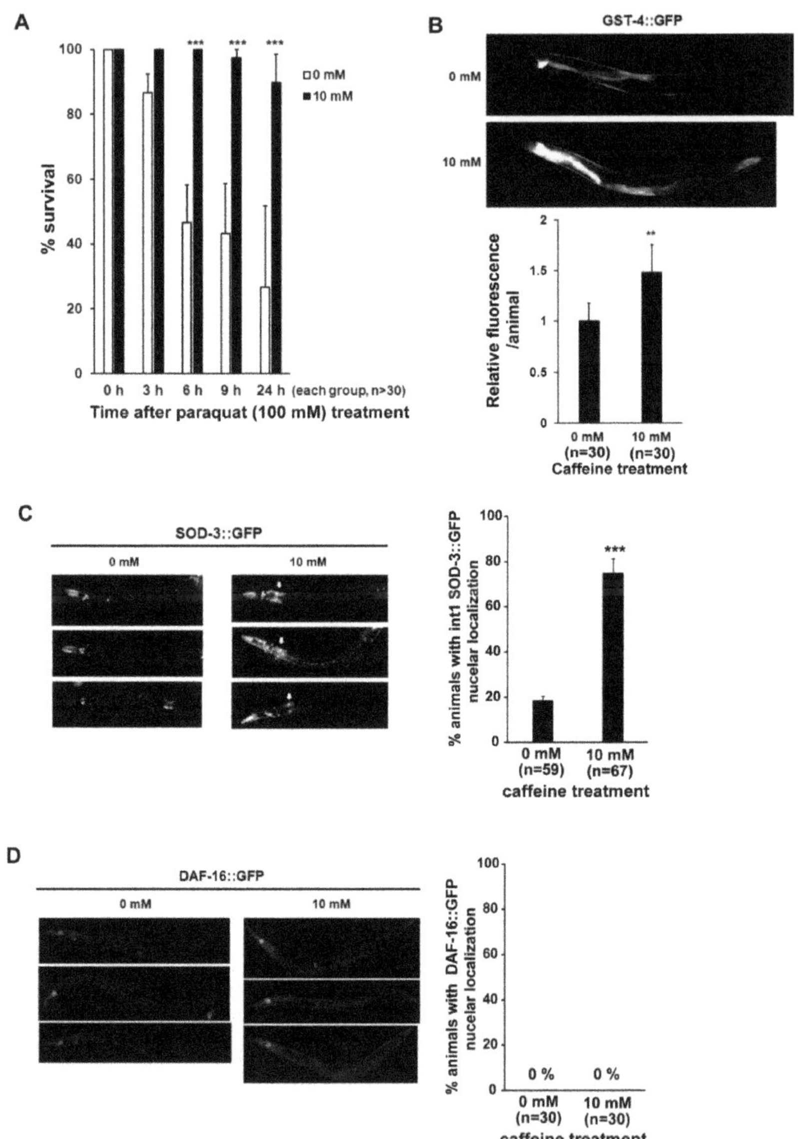

Figure 4. Long-term caffeine intake exerts a protective effect on oxidative stress in advanced ages of *C. elegans*: (**A**) Percentage survival rates of caffeine-free diet and caffeine-ingested animals were analyzed under paraquat (100 mM)-induced oxidative stress condition. Error bars represent SD. *** $p < 0.001$ (two-way ANOVA with Tukey's post hoc test); (**B**) Animals expressing glutathione

S-transferase 4 (GST-4)::GFP were treated with 0 or 10 mM caffeine at the L4 stage at 25 °C for 72 h. The expression of GST-4::GFP was observed at 72 h post-L4-stage animals at 25 °C. The graph shows the relative fluorescence intensity analyzed by ImageJ. Error bars represent SD. ** $p < 0.01$ (*t*-test); (**C**) Animals expressing superoxide dismutase 3 (SOD-3)::GFP were treated with 0 or 10 mM caffeine at the L4 stage at 25 °C for 72 h. The expression levels of SOD-3::GFP were observed at 72 h post-L4-stage animals at 25 °C. The graph indicates the percentage of animals with SOD-3::GFP nuclear localization in the intestinal cells. Error bars represent SD. *** $p < 0.001$ (*t*-test); (**D**) Animals expressing DAF-16::GFP were treated with 0 or 10 mM caffeine at the L4 stage at 25 °C for 72 h. The expression levels of DAF-16::GFP were observed at 72 h post-L4-stage animals at 25 °C. The graph indicates the percentage of animals with DAF-16::GFP nuclear localization in the intestinal cells.

4. Discussion

To examine the effects of caffeine intake on aging in somatic tissues, we first investigated the intestinal aging of *C. elegans*, since the most significant changes related to aging were documented in the intestine. In addition, *C. elegans* is an excellent animal model for evaluating the effects of long-term nutrient intake due to its short life span, which is further accelerated by growth at higher temperatures (25 °C). Therefore, in this study, we explored the effects of long-term caffeine intake on intestinal aging during adulthood. Here, we showed that long-term caffeine intake during adulthood prevented intestinal atrophy and dysfunction in aging *C. elegans*. We demonstrated that the intestinal aging phenotypes, including pharyngeal deterioration, intestinal atrophy, ACT-5 mislocalization, bacterial colonization, VIT production, and PLP accumulation were significantly suppressed by long-term caffeine intake. Our study also provides evidence that long-term caffeine intake contributes to the maintenance of redox homeostasis and mitochondrial function, including the membrane potential, activity, and morphology of mitochondria in the aging animals. Furthermore, long-term caffeine intake significantly improves the motility and lifespan of these organisms. These findings indicate that long-term caffeine intake at a low dose (10 mM) during adulthood contributes to improving the integrity of the intestine as well as the mitochondrial function in advanced ages. Although the effects of long-term use of caffeine for the elderly are still controversial with both positive and adverse effects in human [52–54], our results obtained from the *C. elegans* model showed positive effects on the intestinal aging.

Our results revealed that long-term caffeine intake prevents intestinal aging via regulating vitellogenesis in aging *C. elegans*, suggesting that long-term dietary habits affect the regulation of lipoprotein production, which is linked to aging. Based on our results, we propose a mechanism by which the suppression of vitellogenesis in long-term caffeine intake is regulated by *unc-62*, which is an essential transcriptional factor for the expression of VIT in *C. elegans* [55]. It was previously reported that *unc-62* RNAi increased the expression of SOD-3::GFP in aged animals and extended the lifespan of *C. elegans* [55]. It was also shown that autophagy promotes intestinal atrophy and yolk steatosis [14]. Therefore, we speculate that the decrease in *unc-62* expression due to long-term caffeine intake may suppress vitellogenesis and contribute to intestinal integrity by inhibiting the autophagy, improving the mitochondrial function and redox homeostasis, and extending the lifespan of the organism. It is necessary to further investigate the effects of inhibiting the expression of genes involved in autophagy or *unc-62* activity on intestinal aging, mitochondrial function, and oxidative stress responses in aging organisms. The *C. elegans* VIT proteins contain domains homologous to apoB-100, which is the apoprotein of the low-density lipoprotein (LDL) in humans [20]. VIT proteins bind to and transport lipids, such as triglycerides and cholesterol, to oocytes, thereby showing a similar function to the LDL in mammals [56]. In *C. elegans*, a mechanism causing intestinal atrophy-mediated bioconversion (via intestinal autophagy) for yolk synthesis has been identified [14]. Apart from autophagy, this intestine-to-yolk biomass conversion is also mediated by insulin/insulin-like growth factor (IGF-1) signaling [14]. Loss of function of genes activating autophagy suppresses intestinal atrophy, indicating that autophagy facilitates intestine-to-yolk biomass conversion, and vitellogenesis plays a crucial role in intestinal atrophy in aging *C. elegans*. In humans, hyperlipidemia occurs with advanced age; in particular, LDL hypercholesterolemia plays a causative role in the pathogenesis of cardiovascular diseases, indicating that the levels of cholesterol-rich LDL and other apolipoprotein B (apoB)-containing lipoproteins are directly implicated in the development of cardiovascular diseases [57,58]. In terms of hyperlipidemia, enhanced vitellogenesis in advanced ages of *C. elegans* imitates this condition. Therefore, identifying the mechanism underlying the regulation of lipoprotein production during aging is imperative in understanding the basis of the aging. In particular, the discovery of dietary lipoprotein regulators related to aging, such as vitellogenin, is important because of its close relevance to humans.

Mitochondrial alterations lead to a decline in energy production at the cellular level, which is associated with aging, as shown in animal models as well as human tissues [45,59–62]. The human colon is markedly affected by the progression of mitochondrial aging, which has emerged as an important player in intestinal tissue homeostasis and pathogenesis [63]. An association between aging human colonic cells and defective complexes of the respiratory chain has also been described [59]. Our findings showed a strong association between intestinal integrity and mitochondrial function in response to long-term caffeine intake in aging *C. elegans*. This speculation is supported by a recent report, which showed that mitochondrial dysfunction with increased reactive oxygen species production is a potential cause of intestinal aging [64]. Furthermore, inflammation along with mitochondrial dysfunction is a major pathological factor for the intestinal atrophy [65]. Therefore, anti-inflammatory effects of caffeine intake remain to be determined. In our study, long-term caffeine intake increased the expression of essential detoxification genes, including *gst-4* and *sod-3*, which were accompanied by SKN-1 activity, but not DAF-16, suggesting that the effects of long-term caffeine intake in aging animals are associated with SKN-1 activity but are independent of DAF-16 activity. However, the possibility of short-term induction of DAF-16, which was not sustained at the point of observation in the long-term caffeine intake condition, cannot be ruled out. This notion is supported by previous observations that DAF-16 is activated by short-term caffeine intake and is also temporally activated by probiotic microorganisms in *C. elegans* [4,66]. DAF-16 is a FOXO transcription factor that responds to various stresses, and one of the transcriptional targets of DAF-16 is the *sod-3* encoding the antioxidant enzyme SOD-3 [17].

Based on our study outcomes, we propose a model for the regulation of vitellogenesis via long-term caffeine intake (Figure 5), thereby providing a possible molecular mechanism that links intestinal aging, mitochondrial function, and health in the context of diet-induced regulation of aging. We also propose that intestinal integrity and mitochondrial functions are closely interconnected, and that "what to eat on a long-term basis" and "when to eat" are important factors regulating the aging process. Furthermore, caffeine and related purine alkaloids, such as theophylline and theobromine, have been reported to exhibit pro-oxidant and lifespan extension effects in *C. elegans* at low concentrations (5 mM) [3]. It will be interesting to study whether other caffeine-analogs also exhibit similar effects on the intestinal aging and mitochondrial function in *C. elegans*.

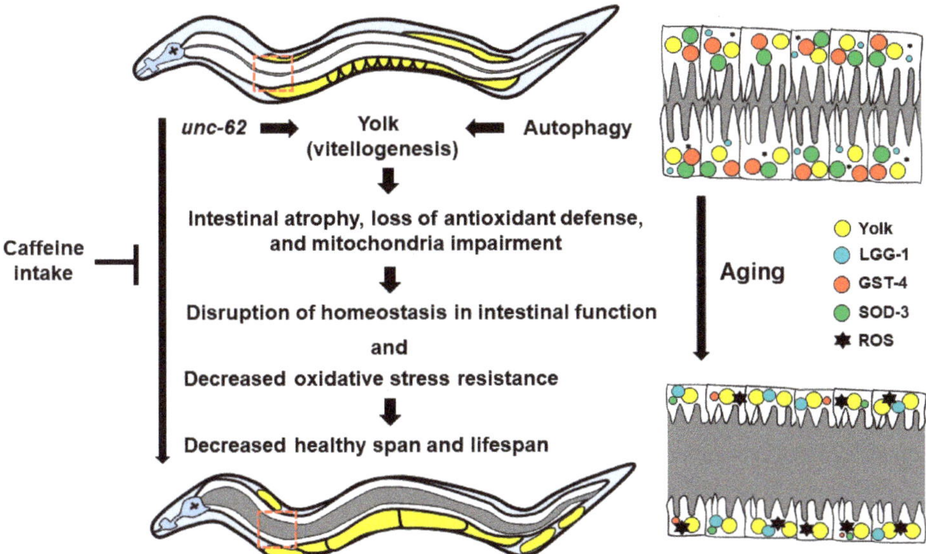

Figure 5. A working model to explain the protective effects of long-term caffeine intake on intestinal aging *C. elegans* at advanced ages. Long-term caffeine intake reduces vitellogenesis via regulating the expression of *unc-62* and autophagy. The decrease in vitellogenesis in response to caffeine intake delays the intestinal atrophy along with improving the mitochondrial function and antioxidant defense. Maintaining homeostasis of intestinal function via caffeine intake supports the health and lifespan in *C. elegans* at advanced ages.

Supplementary Materials: The following are available online at https://www.mdpi.com/article/10.3390/nu13082517/s1, Figure S1: Scheme of the assay used to determine the effects of caffeine intake on intestinal aging, Figure S2: Effects of the caffeine exposure duration on intestinal atrophy, Figure S3: Effects of heat shock on the localization of skinhead 1 (SKN-1)::green fluorescent protein (GFP) and DAF-16::GFP in *Caenorhabditis elegans*.

Author Contributions: Conceptualization, H.M., E.Y. and Y.-H.S.; Methodology, H.M. and E.Y.; Formal Analysis, H.M., E.Y. and Y.-H.S.; Investigation, H.M., E.Y. and Y.-H.S.; Resources, Y.-H.S.; Data Curation, H.M., E.Y. and Y.-H.S.; Writing—Original Draft Preparation, H.M., E.Y., and Y.-H.S.; Writing—Review and Editing, H.M., E.Y. and Y.-H.S.; Visualization, H.M. and E.Y.; Supervision, Y.-H.S.; Project Administration, Y.-H.S.; Funding Acquisition, Y.-H.S. All authors have read and agreed to the published version of the manuscript.

Funding: This study was supported by a grant from the National Research Foundation of Korea (NRF) funded by the Korean Ministry of Science and ICT (grant number: NRF-2021R1A2C1011658 to Y.-H.S.) and was written as a part of Konkuk University's research support program for its faculty on sabbatical leave in 2020.

Acknowledgments: The *Caenorhabditis elegans* strains were provided by the *Caenorhabditis* Genetics Center, which is funded by the National Institutes of Health (NIH) Office of Research Infrastructure Programs (P40 OD010440).

Conflicts of Interest: The authors declare no conflict of interest.

References

1. Min, H.; Kawasaki, I.; Gong, J.; Shim, Y.H. Caffeine induces high expression of *cyp-35A* family genes and inhibits the early larval development in *Caenorhabditis elegans*. *Mol. Cells.* **2015**, *38*, 236–242. [CrossRef]
2. Du, X.; Guan, Y.; Huang, Q.; Lv, M.; He, X.; Yan, Y.; Hayashi, S.; Fang, C.; Wang, X.; Sheng, J. Low concentrations of caffeine and its analogs extend the lifespan of *Caenorhabditis elegans* by modulating IGF-1-Like pathway. *Front. Aging Neurosci.* **2018**, *10*, 211. [CrossRef]

3. Li, H.; Roxo, M.; Cheng, X.; Zhang, S.; Cheng, H.; Wink, M. Pro-oxidant and lifespan extension effects of caffeine and related methylxanthines in *Caenorhabditis elegans*. *Food Chem. X* **2019**, *1*, 100005. [CrossRef] [PubMed]
4. Min, H.; Youn, E.; Shim, Y.H. Maternal caffeine intake disrupts eggshell integrity and retards larval development by reducing yolk production in a *Caenorhabditis elegans* model. *Nutrients* **2020**, *12*, 1334. [CrossRef] [PubMed]
5. Sutphin, G.L.; Bishop, E.; Yanos, M.E.; Moller, R.M.; Kaeberlein, M. Caffeine extends life span, improves healthspan, and delays age-associated pathology in *Caenorhabditis elegans*. *Logev. Healthspan.* **2012**, *1*, 9. [CrossRef]
6. Bridi, J.C.; Barros, A.G.A.; Sampaio, L.R.; Ferreira, J.C.D.; Antunes Soares, F.A.; Romano-Silva, M.A. Lifespan extension induced by caffeine in *Caenorhabditis elegans* is partially dependent on adenosine signaling. *Front. Aging Neurosci.* **2015**, *7*, 220. [CrossRef] [PubMed]
7. Al-Amin, M.; Kawasaki, I.; Gong, J.; Shim, Y.H. Caffeine induces the stress response and up-regulates heat shock proteins in *Caenorhabditis elegans*. *Mol. Cells.* **2016**, *39*, 163–168. [CrossRef]
8. Min, H.; Youn, E.; Kawasaki, I.; Shim, Y.H. Caffeine-induced food-avoidance behavior is mediated by neuroendocrine signals in *Caenorhabditis elegans*. *BMB Rep.* **2017**, *50*, 31–36. [CrossRef]
9. Hubbard, E.J.; Greenstein, D. Introduction to the germ line. *Wormbook* **2005**, 1–4. [CrossRef]
10. Luo, S.; Murphy, C.T. *Caenorhabditis elegans* reproductive aging: Regulation and underlying mechanisms. *Genesis* **2011**, *49*, 53–65. [CrossRef]
11. Hsin, H.; Kenyon, C. Signals from the reproductive system regulate the lifespan of *C. elegans*. *Nature* **1999**, *399*, 362–366. [CrossRef]
12. Berman, J.R.; Kenyon, C. Germ-cell loss extends *C. elegans* life span through regulation of DAF-16 by kri-1 and lipophilic-hormone signaling. *Cell* **2006**, *124*, 1055–1068. [CrossRef]
13. Zhou, K.I.; Pincus, Z.; Slack, F.J. Longevity and stress in *Caenorhabditis elegans*. *Aging* **2011**, *3*, 733–753. [CrossRef] [PubMed]
14. Ezcurra, M.; Benedetto, A.; Sornda, T.; Gilliat, A.F.; Au, C.; Zhang, Q.; van Schelt, S.; Petrache, A.L.; Wang, H.; de la Guardia, Y.; et al. *C. elegans* eats its own intestine to make yolk leading to multiple senescent pathologies. *Curr. Biol.* **2018**, *28*, 2544–2556. [CrossRef] [PubMed]
15. McGee, M.D.; Weber, D.; Day, N.; Vitelli, C.; Crippen, D.; Herndon, L.A.; Hall, D.H.; Melov, S. Loss of intestinal nuclei and intestinal integrity in aging *C. elegans*. *Aging Cell* **2011**, *10*, 699–710. [CrossRef] [PubMed]
16. Drozdowski, L.; Thomson, A.B. Aging and the intestine. *World J. Gastroenterol.* **2006**, *12*, 7578–7584. [CrossRef]
17. Libina, N.; Berman, J.R.; Kenyon, C. Tissue-specific activities of *C. elegans* DAF-16 in the regulation of lifespan. *Cell* **2003**, *115*, 489–502. [CrossRef]
18. Tata, J.R.; Smith, D.F. Vitellogenesis: A versatile model for hormonal regulation of gene expression. *Recent Prog. Horm. Res.* **1979**, *35*, 47–95. [CrossRef]
19. Perez, M.F.; Lehner, B. Vitellogenins—Yolk gene function and regulation in *Caenorhabditis elegans*. *Front. Physiol.* **2019**, *10*, 1067. [CrossRef] [PubMed]
20. Baker, M.E. Is vitellogenin an ancestor of apolipoprotein B-100 of human low-density lipoprotein and human lipoprotein lipase? *Biochem. J.* **1988**, *255*, 1057–1060. [CrossRef]
21. Kimble, J.; Sharrock, W.J. Tissue-specific synthesis of yolk proteins in *Caenorhabditis elegans*. *Dev. Biol.* **1983**, *96*, 189–196. [CrossRef]
22. Spieth, J.; MacMorris, M.; Broverman, S.; Greenspoon, S.; Blumenthal, T. Regulated expression of a vitellogenin fusion gene in transgenic nematodes. *Dev. Biol.* **1988**, *130*, 285–293. [CrossRef]
23. Barbieri, M.; Bonafè, M.; Franceschi, C.; Paolisso, G. Insulin/IGF-I-signaling pathway: An evolutionarily conserved mechanism of longevity from yeast to humans. *Am. J. Physiol. Endocrinol. Metab.* **2003**, *285*, E1064–E1071. [CrossRef]
24. Kenyon, C. A pathway that links reproductive status to lifespan in *Caenorhabditis elegans*. *Ann. N. Y. Acad. Sci.* **2010**, *1204*, 156–162. [CrossRef]
25. Mao, K.; Quipildor, G.A.; Tabrizian, T.; Novaj, A.; Guan, F.; Walters, R.O.; Delahaye, F.; Hubbard, G.B.; Ikeno, Y.; Ejima, K. Late-life targeting of the IGF-1 receptor improves healthspan and lifespan in female mice. *Nat. Commun.* **2018**, *9*, 2394. [CrossRef] [PubMed]
26. Garcidueñas-Fimbres, T.E.; Paz-Graniel, I.; Nishi, S.K.; Salas-Salvadó, J.; Babio, N. Eating speed, eating frequency, and their relationships with diet quality, adiposity, and metabolic syndrome, or its components. *Nutrients* **2021**, *13*, 1687. [CrossRef] [PubMed]
27. Blekkenhorst, L.C.; Sim, M.; Bondonno, C.P.; Bondonno, N.P.; Ward, N.C.; Prince, R.L.; Devine, A.; Lewis, J.R.; Hodgson, J.M. Cardiovascular health benefits of specific vegetable types: A narrative review. *Nutrients* **2018**, *10*, 595. [CrossRef] [PubMed]
28. Kaźmierczak-Barańska, J.; Boguszewska, K.; Karwowski, B.T. Nutrition can help DNA repair in the case of aging. *Nutrients* **2020**, *12*, 3364. [CrossRef] [PubMed]
29. Wood-Bradley, R.J.; Barrand, S.; Giot, A.; Armitage, J.A. Understanding the role of maternal diet on kidney development; an opportunity to improve cardiovascular and renal health for future generations. *Nutrients* **2015**, *7*, 1881–1905. [CrossRef]
30. Min, H.; Kim, J.S.; Ahn, J.; Shim, Y.H. Gliadin intake causes disruption of the intestinal barrier and an increase in germ cell apoptosis in a *Caenorhabditis elegans* model. *Nutrients* **2019**, *11*, 2587. [CrossRef] [PubMed]
31. Brenner, S. The genetics of *Caenorhabditis elegans*. *Genetics* **1974**, *77*, 71–94. [CrossRef]
32. Egge, N.; Arneaud, S.L.B.; Wales, P.; Mihelakis, M.; McClendon, J.; Fonseca, R.S.; Savelle, C.; Gonzalez, I.; Ghorashi, A.; Yadavalli, S.; et al. Age-onset phosphorylation of a minor actin variant promotes intestinal barrier dysfunction. *Dev. Cell.* **2019**, *51*, 587–601. [CrossRef] [PubMed]

33. Palmisano, N.J.; Meléndez, A. Detection of autophagy in *Caenorhabditis elegans* using *GFP::LGG-1* as an autophagy marker. *Cold Spring Harb. Protoc.* **2016**, *2016*. [CrossRef]
34. Regmi, S.G.; Rolland, S.G.; Conradt, B. Age-dependent changes in mitochondrial morphology and volume are not predictors of lifespan. *Aging* **2014**, *2*, 118–130. [CrossRef]
35. Gatsi, R.; Schulze, B.; Rodríguez-Palero, M.J.; Hernando-Rodríguez, B.; Baumeister, R.; Artal-Sanz, M. Prohibitin-mediated lifespan and mitochondrial stress implicate SGK-1, insulin/IGF and mTORC2 in *C. elegans*. *PLoS ONE* **2014**, *9*, e107671. [CrossRef] [PubMed]
36. Tullet., J.M.A.; Hertweck, M.; An, J.H.; Baker, J.; Hwang, J.Y.; Liu, S.; Oliveira, R.P.; Baumeister, R.; Blackwell, T.K. Direct inhibition of the longevity promoting factor SKN-1 by Insulin-like signaling in *C. elegans*. *Cell* **2008**, *132*, 1025–1038. [CrossRef] [PubMed]
37. Palominos, M.F.; Calixto, A. Quantification of bacteria residing in *Caenorhabditis elegans* intestine. *Bio Protoc.* **2020**, *10*, e3605. [CrossRef]
38. Min, H.; Lee, M.; Cho, K.S.; Lim, H.J.; Shim, Y.H. Nicotinamide supplementation improves oocyte quality and offspring development by modulating mitochondrial function in an aged *Caenorhabditis elegans* model. *Antioxidants* **2021**, *10*, 519. [CrossRef] [PubMed]
39. Lim, S.D.; Min, H.; Youn, E.; Kawasaki, I.; Shim, Y.H. Gliadin intake induces oxidative-stress responses in *Caenorhabditis elegans*. *Biochem. Biophys. Res. Commun.* **2018**, *503*, 2139–2145. [CrossRef]
40. Lockwood, C.A.; Lynch, A.M.; Hardin, J. Dynamic analysis identifies novel roles for DLG-1 subdomains in AJM-1 recruitment and LET-413-dependent apical focusing. *J. Cell Sci.* **2008**, *121*, 1477–1487. [CrossRef]
41. Dowen, R.H.; Breen, P.C.; Tullius, T.; Conery, A.L.; Ruvkun, G. A microRNA program in the *C. elegans* hypodermis couples to intestinal mTORC2/PQM-1 signaling to modulate fat transport. *Genes. Dev.* **2016**, *30*, 1515–1528. [CrossRef] [PubMed]
42. Dowen, R.H. CEH-60/PBX and UNC-62/MEIS coordinate a metabolic switch that supports reproduction in *C. elegans*. *Dev. Cell* **2019**, *49*, 235–250. [CrossRef] [PubMed]
43. Chávez, V.; Mohri-Shiomi, A.; Maadani, A.; Vega, L.A.; Garsin, D.A. Oxidative stress enzymes are required for DAF-16-mediated immunity due to generation of reactive oxygen species by *Caenorhabditis elegans*. *Genetics* **2007**, *176*, 1567–1577. [CrossRef]
44. Van Raamsdonk, J.M.; Hekimi, S. Reactive oxygen species and aging in *Caenorhabditis elegans*: Causal or causal relationship? *Antioxid. Redox Signal* **2010**, *13*, 1911–1953. [CrossRef] [PubMed]
45. Gaffney, C.J.; Pollard, A.; Barratt, T.F.; Constantin-Teodosiu, D.; Greenhaff, P.L.; Szewczyk, N.J. Greater loss of mitochondrial function with ageing is associated with earlier onset of sarcopenia in *C. elegans*. *Aging* **2018**, *10*, 3382–3396. [CrossRef]
46. Hsu, A.L.; Feng, Z.; Hsieh, M.Y.; Xu, X.Z.S. Identification by machine vision of the rate of motor activity decline as a lifespan predictor in *C. elegans*. *Neurobiol. Aging.* **2009**, *30*, 1498–1503. [CrossRef]
47. Blackwell, T.K.; Steinbaugh, M.J.; Hourihan, J.M.; Ewald, C.Y.; Isik, M. SKN-1/Nrf, stress responses, and aging in *Caenorhabditis elegans*. *Free Radic. Biol. Med.* **2015**, *88*, 290–301. [CrossRef]
48. Inoue, H.; Hisamoto, N.; An, J.H.; Riva, P. Oliveira, R.P.; Nishida, E.; Blackwell, T.K.; Matsumoto, K. The *C. elegans* p38 MAPK pathway regulates nuclear localization of the transcription factor SKN-1 in oxidative stress response. *Genes Dev.* **2005**, *19*, 2278–2283. [CrossRef]
49. Oliveira, R.P.; Abate, J.P.; Dilks, K.; Landis, J.; Ashraf, J.; Murphy, C.T.; Blackwell, T.K. Condition-adapted stress and longevity gene regulation by *Caenorhabditis elegans* SKN-1/Nrf. *Aging Cell* **2009**, *8*, 524–541. [CrossRef]
50. Wang, J.; Robida-Stubbs, S.; Tullet, J.M.A.; Rual, J.F.; Vidal, M.; Blackwell, T.K. RNAi screening implicates a SKN-1-dependent transcriptional response in stress resistance and longevity deriving from translation inhibition. *PLoS Genet.* **2010**, *6*, e1001048. [CrossRef]
51. Shore, D.E.; Ruvkun, G. A cytoprotective perspective on longevity regulation. *Trends Cell Biol.* **2013**, *23*, 409–420. [CrossRef]
52. De Pooter-Stijnman, L.M.M.; Vrijkotte, S.; Smalbrugge, M. Effect of caffeine on sleep and behaviour in nursing home residents with dementia. *Eur. Geriatr. Med.* **2018**, *9*, 829–835. [CrossRef]
53. Vercambre, M.N.; Berr, C.; Ritchie, K.; Kang, J.H. Caffeine and cognitive decline in elderly women at high vascular risk. *J. Alzheimer's Dis.* **2013**, *35*, 413–421. [CrossRef]
54. Gunter, M.J.; Murphy, N.; Cross, A.J.; Dossus, L.; Dartois, L.; Fagherazzi, G.; Kaaks, R.; Kühn, T.; Boeing, H.; Aleksandrova, K.; et al. Coffee drinking and mortality in 10 european countries: A multinational cohort study. *Ann. Intern. Med.* **2017**, *167*, 236–247. [CrossRef]
55. Van Nostrand, E.L.; Sánchez-Blanco, A.; Wu, B.; Nguyen, A.; Kim, S.K. Roles of the developmental regulator unc-62/Homothorax in limiting longevity in *Caenorhabditis elegans*. *PLoS Genet.* **2013**, *9*, e1003325. [CrossRef] [PubMed]
56. Matyash, V.; Geier, C.; Henske, A.; Mukherjee, S.; Hirsh, D.; Thiele, C.; Grant, B.; Maxfield, F.R.; Kurzchalia, T.V. Distribution and transport of cholesterol in *Caenorhabditis elegans*. *Mol. Biol. Cell.* **2001**, *12*, 1725–1736. [CrossRef] [PubMed]
57. Ference, B.A.; Ginsberg, H.N.; Graham, I.; Ray, K.K.; Packard, C.J.; Bruckert, E.; Hegele, R.A.; Krauss, R.M.; Raal, F.J.; Schunkert, H. Low-density lipoproteins cause atherosclerotic cardiovascular disease. 1. Evidence from genetic, epidemiologic, and clinical studies. A consensus statement from the European Atherosclerosis Society Consensus Panel. *Eur. Heart J.* **2017**, *38*, 2459–2472. [CrossRef] [PubMed]
58. Rosada, A.; Kassner, U.; Weidemann, F.; König, M.; Buchmann, N.; Steinhagen-Thiessen, E.; Spira, D. Hyperlipidemias in elderly patients: Results from the Berlin Aging Study II (BASEII), a cross-sectional study. *Lipids Health Dis.* **2020**, *19*, 92. [CrossRef]

59. Greaves, L.C.; Barron, M.J.; Plusa, S.; Kirkwood, T.B.; Mathers, J.C.; Taylor, R.W.; Turnbull, D.M. Defects in multiple complexes of the respiratory chain are present in ageing human colonic crypts. *Exp. Gerontol.* **2010**, *45*, 573–579. [CrossRef] [PubMed]
60. Trifunovic, A.; Wredenberg, A.; Falkenberg, M.; Spelbrink, J.N.; Rovio, A.T.; Bruder, C.E.; Bohlooly, -Y.M.; Gidlöf, S.; Oldfors, A.; Wibom, R.; et al. Premature ageing in mice expressing defective mitochondrial DNA polymerase. *Nature* **2004**, *429*, 417–423. [CrossRef]
61. Vermulst, M.; Wanagat, J.; Kujoth, G.C.; Bielas, J.H.; Rabinovitch, P.S.; Prolla, T.A.; Loeb, L.A. DNA deletions and clonal mutations drive premature aging in mitochondrial mutator mice. *Nat. Genet.* **2008**, *40*, 392–394. [CrossRef]
62. Yen, T.C.; Chen, Y.S.; King, K.L.; Yeh, S.H.; Wei, Y.H. Liver mitochondrial respiratory functions decline with age. *Biochem. Biophys. Res. Commun.* **1989**, *165*, 944–1003. [CrossRef]
63. Urbauer, E.; Rath, E.; Haller, D. Mitochondrial metabolism in the intestinal stem cell niche-sensing and signaling in health and disease. *Front. Cell Dev. Biol.* **2021**, *8*, 602814. [CrossRef]
64. Schneider, A.M.; Özsoy, M.; Zimmermann, F.A.; Feichtinger, R.G.; Mayr, J.A.; Kofler, B.; Sperl, W.; Weghuber, D.; Mörwald, K. Age-related deterioration of mitochondrial function in the intestine. *Oxidative Med. Cell Longev.* **2020**, *2020*, 4898217. [CrossRef] [PubMed]
65. Novak, E.A.; Mollen, K.P. Mitochondrial dysfunction in inflammatory bowel disease. *Front. Cell Dev. Biol.* **2015**, *3*, 62. [CrossRef] [PubMed]
66. Poupet, C.; Saraoui, T.; Veisseire, P.; Bonnet, M.; Dausset, C.; Gachinat, M.; Camarès, O.; Chassard, C.; Nivoliez, A.; Bornes, S. *Lactobacillus rhamnosus* Lcr35 as an effective treatment for preventing *Candida albicans* infection in the invertebrate model *Caenorhabditis elegans*: First mechanistic insights. *PLoS ONE* **2019**, *14*, e0216184. [CrossRef] [PubMed]

Article

Exosome-Mediated Activation of Neuronal Cells Triggered by γ-Aminobutyric Acid (GABA)

Ryo Inotsuka [1], Miyako Udono [2], Atsushi Yamatsu [3,4], Mujo Kim [4] and Yoshinori Katakura [1,2,*]

[1] Graduate School of Bioresources and Bioenvironmental Sciences, Kyushu University, Fukuoka 819-0395, Japan; inotsuka.ryo.840@s.kyushu-u.ac.jp

[2] Faculty of Agriculture, Kyushu University, Fukuoka 819-0395, Japan; mudono@grt.kyushu-u.ac.jp

[3] International GABA Research Center, Kyoto 615-8245, Japan; a-yamatsu@pharmafoods.co.jp

[4] Pharma Foods International Co., Ltd., Kyoto 615-8245, Japan; mujokim@pharmafoods.co.jp

* Correspondence: katakura@grt.kyushu-u.ac.jp

Abstract: γ-Aminobutyric acid (GABA) is a potent bioactive amino acid, and several studies have shown that oral administration of GABA induces relaxation, improves sleep, and reduces psychological stress and fatigue. In a recent study, we reported that exosomes derived from GABA-treated intestinal cells serve as signal transducers that mediate brain–gut interactions. Therefore, the purpose of this study was to verify the functionality of GABA-derived exosomes and to examine the possibility of improving memory function following GABA administration. The results showed that exosomes derived from GABA-treated intestinal cells (Caco-2) activated neuronal cells (SH-SY5Y) by regulating genes related to neuronal cell functions. Furthermore, we found that exosomes derived from the serum of GABA-treated mice also activated SH-SY5Y cells, indicating that exosomes, which are capable of activating neuronal cells, circulate in the blood of mice orally administered GABA. Finally, we performed a microarray analysis of mRNA isolated from the hippocampus of mice that were orally administered GABA. The results revealed changes in the expression of genes related to brain function. Gene Set Enrichment Analysis (GSEA) showed that oral administration of GABA affected the expression of genes related to memory function in the hippocampus.

Keywords: GABA; exosome; gut-brain interaction; Caco-2; SH-SY5Y

Citation: Inotsuka, R.; Udono, M.; Yamatsu, A.; Kim, M.; Katakura, Y. Exosome-Mediated Activation of Neuronal Cells Triggered by γ-Aminobutyric Acid (GABA). *Nutrients* **2021**, *13*, 2544. https://doi.org/10.3390/nu13082544

Academic Editor: Paolo Brambilla

Received: 1 June 2021
Accepted: 23 July 2021
Published: 25 July 2021

Publisher's Note: MDPI stays neutral with regard to jurisdictional claims in published maps and institutional affiliations.

Copyright: © 2021 by the authors. Licensee MDPI, Basel, Switzerland. This article is an open access article distributed under the terms and conditions of the Creative Commons Attribution (CC BY) license (https://creativecommons.org/licenses/by/4.0/).

1. Introduction

γ-Aminobutyric acid (GABA) is a naturally occurring nonprotein amino acid that is one of the principal inhibitory neurotransmitters of the central nervous system (CNS) [1]. The human GABA$_B$ receptor-a member of the class C family of G-protein-coupled receptors (GPCRs)-mediates inhibitory neurotransmission [2]. It has been reported that GABAergic synapses constitute 20–50% of all synapses present in the CNS [3] and play important roles in information encoding and behavioral control, regulation of motor learning, and motor functions [4,5]. Generally, genes characterized by GABAergic function are down-regulated during normal aging in humans [6], and in certain studies involving animals, it has been suggested that GABA function declines with age. For example, the number of GABA-immunoreactive neurons declines with age in the hippocampus, inferior colliculus, and striate visual cortex of animals [7–9]. There are also reported reductions in GABA levels, GABA release, and GABA receptor binding from the baseline levels with aging [8]. When the concentration of GABA in the brain diminishes below a certain threshold, various neurological disorders including epilepsy, seizures, convulsions, and Alzheimer disease have also been reported [10–12]. Furthermore, it has been recently reported that serum GABA levels are decreased in various diseases including stroke [13]. These studies suggest that GABA has health-promoting effects that may also be related to the enhancement of brain function.

The development of functional foods containing GABA has been actively studied. In a recent study, a randomized, double-blind, placebo-controlled clinical trial involving healthy Japanese adults was reported, where administration of 100 mg of GABA resulted in increased participant memory and spatial cognitive function [14]. Indeed, several studies have shown that oral administration of GABA induces relaxation, improves sleep, and reduces psychological stress and fatigue [15–17]. Additionally, healthy adults who consumed 50 mg of GABA dissolved in a beverage reported less occupational fatigue after completion of required tasks [18]. However, in view of the lack of evidence regarding the blood–brain barrier permeability of GABA, the mechanisms through which GABA might exert these beneficial effects in humans remain unclear. It is considered that the oral intake of GABA exerts such effects through an indirect pathway [19].

Furthermore, GABA has been reported to have a variety of health promoting effects including immunomodulatory, anti-diabetes, anti-cancer, anti-oxidant, and so on [20–23]. Future research is expected to focus not only on neurological and psychological disorders linked to a decrease in amount of GABA, but also on the systemic health-promoting effects of GABA and its molecular basis.

In our previous study, we clarified the molecular basis of GABA-induced gut–brain interactions and reported that exosomes derived from GABA-treated intestinal cells serve as signal transducers that mediate brain–gut interactions [24]. Exosomes are a family of particles released from the cell that are delimited by a lipid bilayer, and attention has recently been focused on the role of exosomes as biomarker candidates for diagnosis, prognosis and even therapeutic tools of various diseases [25]. In the present study, we clarified that exosomes derived from serum of mice administered GABA as well as from GABA-treated Caco-2 cells activated neuronal cells, and that GABA administration changes the expression of memory-related genes in hippocampus.

2. Materials and Methods

2.1. Cell Culture and Reagents

The human colorectal cancer cell line Caco-2 (ATCC, Manassas, VA, USA) and the human neuroblastoma cell line SH-SY5Y (ATCC) were cultured in Dulbecco's modified Eagle's medium (DMEM; Nissui, Tokyo, Japan) supplemented with 10% heat-inactivated fetal bovine serum (FBS, Life Technologies, Gaithersburg, MD, USA) at 37 °C in an atmosphere containing 5% CO_2. γ-Aminobutyric acid (GABA) was purchased from Abcam (Cambridge, UK). 5-Aminoimidazole-4-carboxamide 1-β-D-ribofuranoside (AICAR) and retinoic acid (RA) were purchased from FUJIFILM Wako Pure Chemical Corp. (Osaka, Japan).

2.2. Exosome Isolation and Treatment

Firstly, Caco-2 cells (1.4×10^5 cells/mL) were cultured in DMEM containing 10% Exosome-depleted FBS media supplement heat inactivated (System Bioscience, Mountain View, CA, USA) and 500 or 1000 μM GABA. After 24 h of culture, the MagCapture Exosome Isolation Kit PS (FUJIFILM Wako Pure Chemical Corp.) was used to isolate exosomes from the media of Caco-2 cells, according to the manufacturer's instructions. Exosomes were isolated from mouse serum using ExoQuick Exosome Precipitation Solution (System Biosciences, Palo Alto, CA, USA), according to the manufacturer's instructions. SH-SY5Y cells (2.0×10^5 cell/mL) were cultured for 24 h, and treated with exosomes (equivalent to 90 ng protein) derived from GABA-treated Caco-2 cells for 24 h.

2.3. Quantitative Evaluation of Neurite Growth

SH-SY5Y cells were seeded onto a μClear fluorescence black plate (Greiner-Bio One, Tokyo, Japan), fixed with 4% paraformaldehyde for 15 min, and blocked with blocking buffer (1 × PBS, 5% goat serum, and 0.3% Triton X-100) for 1 h. The cells were subsequently incubated with Milli-Mark Pan Neuronal Marker (Merck Millipore, Billerica, MA, USA) at 25 °C overnight. After washing with PBS, the cells were stained with Alexa Fluor 555 goat anti-rabbit IgG antibody (Thermo Fisher Scientific, Inc., Waltham, MA, USA) for 1 h at 25 °C. After washing with PBS, cells were further stained with Hoechst 33342 (Dojindo, Kumamoto, Japan) for 15 min, and neurite length was measured using the IN Cell Analyzer 2200 (GE Healthcare, Amersham Place, UK), as previously described [24]. The total neurite length for each cell is shown in the figure.

2.4. Mitochondria

Cells were stained with 250 nM MitoTracker Red CMXRos (Thermo Fischer Scientific) at 37 °C for 30 min, and subsequently with 200 nM MitoTracker Green FM (Thermo Fischer Scientific) at 37 °C for 30 min. Finally, the cells were stained with Hoechst 33342 at 37 °C for 30 min. Stained cells were analyzed using IN Cell Analyzer 2200 (GE Healthcare, Amersham Place, UK) to quantitatively determine the number, area, and activity of mitochondria using IN Cell Investigator high-content image analysis software (GE Healthcare).

2.5. Quantitative Reverse Transcriptase-Polymerase Chain Reaction (RT-qPCR)

RNA was prepared from cells using the High Pure RNA Isolation kit (Roche Diagnostics GmbH, Mannheim, Germany) according to the manufacturer's protocols. RT-qPCR was performed using the GoTaq 1-Step RT-PCR System (Promega, WI, USA) and Thermal Cycler Dice Real Time System TP-800 (Takara). Samples were analyzed in triplicate. The PCR primer sequences used were as follows: human β-actin (*ACTB*) forward primer 5′-TGGCACCCAGCACAATGAA-3′ and reverse primer 5′-CTAAGTCATAGTCCGCCTAGAA GCA-3′: human brain-derived growth factor (*BDNF*) forward primer 5′-GTCAAGTTGGGA GCCTGAAATAGTG-3′ and reverse primer 5′-AGGATGCTGGTCCAAGTGGTG-3′: peroxisome proliferator-activated receptor γ coactivator 1-α (*PGC-1α*) forward primer 5′-GCTGACAGATGGAGACGTGA-3′ and reverse primer 5′-TAGCTGAGTGTTGGCTGGTG-3′: human *NESTIN* forward primer 5′-ACTGGGAAGGAGGAGGTGGT-3′ and reverse primer 5′-CACACTGGCTCCCTCAACCA-3′: human neurofilament medium chain (*NEFM*) forward primer 5′-AGACATCCACCGGCTCAAGG-3′ and reverse primer 5′-CGACGCCTC CTCGATGTCT-3′. β-actin was used as a housekeeping gene. Samples were normalized and analyzed by the ΔΔCT method [26].

2.6. miRNA Microarray Assay

The expression profile of miRNA in exosomes prepared using the MagCapure Exosome Isolation Kit PS was evaluated using microarray analysis with an Affimetrix GeneChip miRNA 4.0 Array (Affymetrix, Santa Clara, CA, USA). Total RNA was isolated from exosomes using TRIzol Reagent (Thermo Fisher Scientific) and purified using the miRNeasy Mini Kit (Qiagen, Valencia, CA, USA). Subsequent operations were outsourced to Cell Innovator (Fukuoka, Japan), a commercial contract analysis provider. To identify up- or down-regulated genes, we calculated ratios (non-log scaled fold-change) from the normalized intensities of each gene for comparisons between control and experimental samples. Then, we established criteria for regulated genes: (up-regulated genes) ratio $\geq$ 1.3-fold; (down-regulated genes) ratio $\leq$ 0.77 [27]. miRNA target genes were predicted using miRWalk (http://mirwalk.umm.uni-heidelberg.de, access on 20 February 2021). To determine significantly over-represented gene ontology (GO) categories and significantly enriched pathways, we used tools and data provided by the Database for Annotation, Visualization, and Integrated Discovery (DAVID, http://david.abcc.ncifcrf.gov, access on 20 February 2021) [28,29].

2.7. mRNA Microarray Assay

The mRNA expression profile was evaluated using a DNA microarray (SurePrint G3 Human Gene Expression 8 × 60 K v.3, Agilent). Total RNA was isolated from SH-SY5Y cells and mouse brains using Isogen II (Nippon Gene, Tokyo, Japan). The subsequent operations were outsourced to Cell Innovator. We then established criteria for significantly up- or down-regulated genes: up-regulated genes, Z-score $\geq$ 2.0 and ratio $\geq$ 1.5-fold; down-regulated genes: Z-score ≤ -2.0 and ratio ≤ 0.66-fold. Significantly over-represented GO categories and enriched pathways were determined as described above. Gene set enrichment analysis (GSEA) was performed to determine the enrichment score (ES), which indicates the degree to which each gene set is overrepresented at the top or bottom of a ranked list of genes.

2.8. Integrated Analysis

We then performed integrated analysis of miRNAs with altered expression in Exo-GABA and of mRNAs with altered expression in SH-SY5Y cells in response to Exo-GABA treatment, and selected genes.

2.9. Animal Experiments

Six-week-old female C57BL/6 mice were obtained from KBT Oriental Co. Ltd. (Saga, Japan) and allowed to adapt for 2 weeks, with food and water provided ad libitum. All mouse experiments and protocols were in accordance with the Guide for the Care and Use of Laboratory Animals and were approved by the Ethics Committee on Animal Experimentation (Kyushu University; approval number: A28-187-0). One group was composed of five mice. GABA solution was orally administered to mice once a day in 100 µL (200 mg/kg) doses using a sonde. The control group received 100 µL of sterile water orally. After 7 days of treatment, cardiac blood samples were collected from the mice, and serum samples were prepared. Serum was used for exosome isolation by using the MagCapture Exosome Isolation Kit PS. Simultaneously, hippocampal tissue was collected from mouse brains.

2.10. Statistical Analysis

All experiments were performed at least three times, and the corresponding data are shown. The results are presented as means $\pm$ standard deviation. Multiple comparisons between groups were carried out by one-way ANOVA with Tukey's post-hoc test. Statistical significance was defined as $p < 0.05$ (* $p < 0.05$; ** $p < 0.01$; *** $p < 0.001$).

3. Results

3.1. Exosomes Derived from GABA-Treated Caco-2 Cells Activate SH-SY5Y Cells

Caco-2 cells are used in this study because they are known to exhibit differentiation functions such as digestion and absorption. First, we tested whether exosomes derived from GABA-treated Caco-2 cells activated SH-SY5Y cells. Exo-ctrl shows the exosomes derived from non-treated Caco-2 cells. As shown in Figure 1A, although the treatment of Caco-2 cells with 500 or 1000 µM GABA did not change the amount of exosomes that could be isolated (data not shown), exosomes derived from GABA-treated Caco-2 cells (Exo-GABA) exhibited elongated neurites in SH-SY5Y cells as compared to Exo-ctrl. However, the neurite outgrowth effect of Exo-GABA was lower than that of the positive control retinoic acid (RA). Furthermore, Exo-GABA was found to augment the expression of *PGC-1α*, a master gene of mitochondrial biogenesis, and increased the number, area, and activity of mitochondria in SH-SY5Y cells (Figure 1B–E), as compared to Exo-ctrl. 5-Aminoimidazole-4-carboxamide 1-β-D-ribofuranoside (AICAR) is known to induce mitochondrial biogenesis [30], and then used as a positive control. As can be seen from some of the results, exosomes derived from Caco-2 cells treated with 1000 µM GABA were found to be more effective.

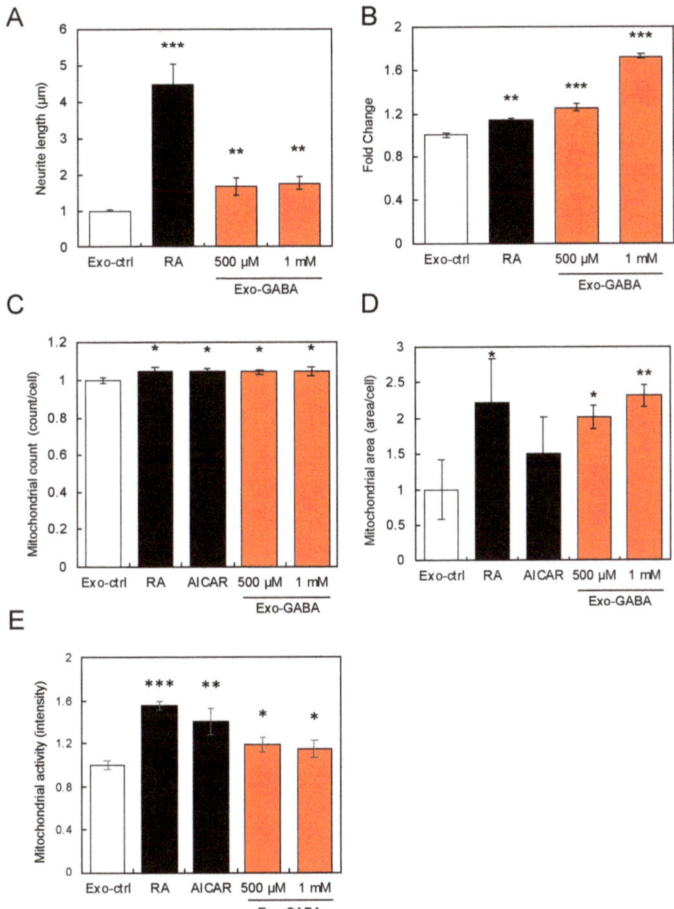

Figure 1. Exosomes derived from GABA-treated Caco-2 cells activate SH-SY5Y cells. Neurite length (**A**) and *PGC-1α* expression (**B**) in SH-SY5Y cells treated with exosomes (equivalent to 90 ng protein) derived from GABA-treated Caco-2 cells. Exo-ctrl shows the exosomes derived from non-treated Caco-2 cells. Count (**C**), area (**D**), and activity (**E**) of mitochondria in SH-ST5Y cells treated with exosomes derived from GABA-treated Caco-2 cells. Retinoic acid (RA) was used as positive control. Multiple comparisons between groups were carried out by one-way ANOVA with Tukey's post-hoc test. Statistical difference was evaluated by comparing to Exo-ctrl. Statistical significance was defined as $p < 0.05$ (* $p < 0.05$; ** $p < 0.01$; *** $p < 0.001$) (value means ± SEM, $n = 3$).

Next, we tested the effects of Exo-GABA on the expression of marker genes (brain-derived growth factor, *BDNF*; Nestin; and neurofilament medium chain, *NEFM*) of neuronal cells in SH-SY5Y cells. The results showed that Exo-GABA augmented the expression of these marker genes (Figure 2A–C). These results indicated that Exo-GABA activated SH-SY5Y cells.

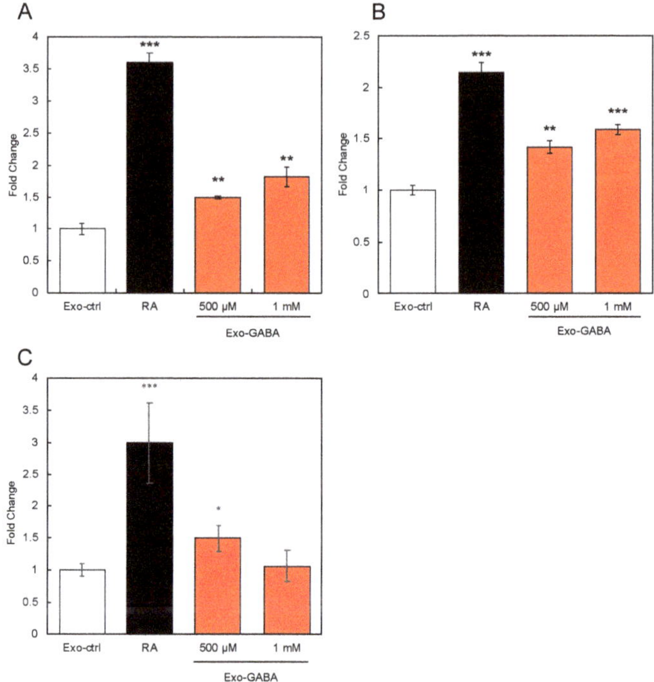

Figure 2. Effects of exosomes derived from GABA-treated Caco-2 cells on gene expression. Expression of brain-derived growth factor (*BDNF*) (**A**), Nestin (**B**), and neurofilament medium chain (*NEFM*) (**C**) in SH-SY5Y cells treated with exosomes derived from GABA-treated Caco-2 cells was evaluated by RT-qPCR. Retinoic acid (RA) was used as positive control. Multiple comparisons between groups were carried out by one-way ANOVA with Tukey's post-hoc test. Statistical difference was evaluated by comparing to Exo-ctrl. Statistical significance was defined as $p < 0.05$ (* $p < 0.05$; ** $p < 0.01$; *** $p < 0.001$) (value means $\pm$ SEM, $n = 3$).

3.2. Molecular Basis for the Exo-GABA-Induced Activation of SH-SY5Y

We then tried to identify miRNAs contained in Exo-GABA and their target genes that activate SH-SY5Y cells using integrated analysis. Exo-ctrl, Exo-GABA$_{500}$, and Exo-GABA$_{1000}$ show the exosomes derived from non-treated, 500 µM GABA-treated and 1000 µM GABA-treated Caco-2 cells, respectively. First, microarray analysis of miRNAs in Exo-GABA was used to search for miRNAs in exosomes that varied with GABA treatment. The results showed that GABA treatment decreased the expression levels of the four miRNAs in Caco-2 cells (Table 1). The target gene of these four miRNAs can be found in the Supplementary Materials Table S1. KEGG pathway analysis showed that all four miRNAs were involved in neuronal cell regulation and activation (Table 1).

Table 1. Functional classification of miRNAs differentially expressed in GABA-treated Caco-2 cells.

miRNA	KEGG Pathway
miR-6732-5p	Calcium signaling pathway, Axon guidance, Neurotrophin signaling pathway
miR-8075	Calcium signaling pathway, Regulation of actin cytoskeleton, Long-term potentiation, Neurotrophin signaling pathway, Axon guidance
miR-3665	Axon guidance, Long-term potentiation, Calcium signaling pathway, Neurotrophin signaling pathway
miR-5787	Axon guidance, Long-term potentiation, Calcium signaling pathway, Neurotrophin signaling pathway, Long-term depression

Changes in mRNA expression in Exo-GABA-treated SH-SY5Y cells were examined using microarray analysis (Figure 3), and we identified 641 mRNAs whose expression was commonly altered in SH-SY5Y treated with Exo-GABA$_{500}$ and Exo-GABA$_{1000}$. These genes can be directly or indirectly regulated by miRNAs and include both up-regulated and down-regulated genes. Similarly, using DAVID, KEGG pathway analysis estimated the involvement of pathways implicated in neuronal activity (Table 2).

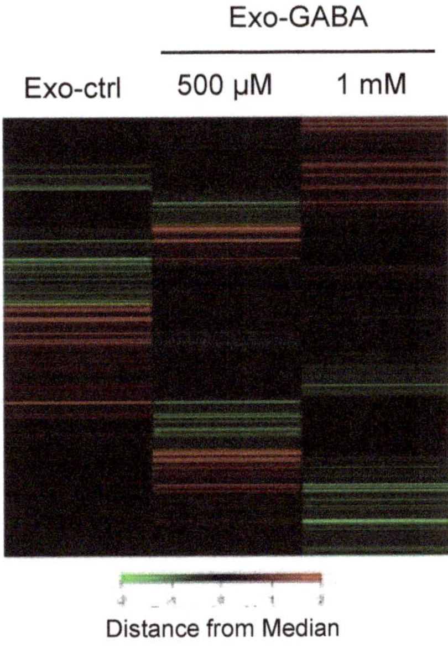

Figure 3. Microarray analysis of genes in exosome-treated SH-SY5Y cells. Heatmap was used to visualize differential expressed genes.

Table 2. Functional classification of genes differentially expressed in SH-SY5Y cells treated with exosomes derived from GABA-treated Caco-2 cells.

KEGG Pathway	*p*-Value
Cytokine–cytokine receptor interaction	0.0037
Neuroactive ligand–receptor interaction	0.025
Inflammatory bowel disease	0.029

We then performed integrated analysis of miRNAs with altered expression in Exo-GABA$_{500}$ and Exo-GABA$_{1000}$ and of mRNAs with altered expression in SH-SY5Y cells in response to Exo-GABA$_{500}$ and Exo-GABA$_{1000}$ treatment. After comparison of miRNA target genes whose expression was altered both in Exo-GABA$_{500}$ and Exo-GABA$_{1000}$ and those whose expression was altered in SH-SY5Y cells in response to Exo-GABA$_{500}$ and Exo-GABA$_{1000}$, 185 common genes were found, 12 of which were involved in the regulation of brain function (Table 3).

Table 3. Neuronal gene and its function selected by integration analysis.

Gene	Function
CIT	Development of the central nervous system
SLC6A17	Neurotransmitter uptake

Table 3. *Cont.*

Gene	Function
MPPED2	Brain development
NPAS3	Neurogenesis
SHANK2	Scaffolding at the synapse
NEUROG1	Neuronal differentiation
RIT1	Neuron development and regeneration
SLC5A7	Depression
GCSAM	Signaling pathway
KCNN2	Regulation of neuronal excitability
KCNK13	Neurotransmitter release
KAT6B	Brain development

3.3. Effects of Exosomes Derived from the Serum of GABA-Treated Mice in SH-SY5Y Cells

First, we tested whether serum-derived exosomes from mice orally administered GABA (seExo-GABA) could induce neurite outgrowth in SH-SY5Y cells. After isolating exosomes from serum of mice administered GABA, SH-SY5Y cells were similarly treated with seExo-GABA (equivalent to 90 ng protein) for 24 h, as mentioned above. The results clearly showed that seExo-GABA induced neurite outgrowth in SH-SY5Y cells (Figure 4A). Furthermore, seExo-GABA augmented the expression of *PGC-1α* and increased the number and area of mitochondria in SH-SY5Y cells, but not activity (Figure 4B–E).

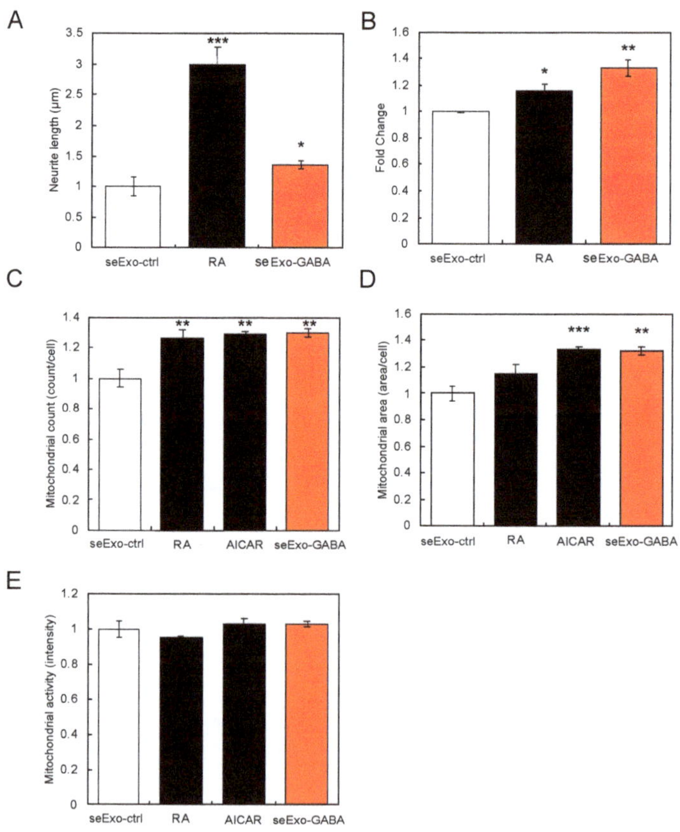

Figure 4. Effects of exosomes derived from serum of GABA-treated mice (seExo-GABA) on SH-SY5Y

cells. Exosomes derived from serum of non-treated mice (seExo-ctrl) was used as control. Neurite length (**A**) and *PGC-1α* expression (**B**) in SH-SY5Y cells treated with seExo-GABA. Count (**C**), area (**D**), and activity (**E**) of mitochondria in SH-ST5Y cells treated with seExo-GABA. RA was used as positive control. AICAR is used as positive control for mitochondrial biogenesis. Multiple comparisons between groups were carried out by one-way ANOVA with Tukey's post-hoc test. Statistical difference was evaluated by comparing to seExo-ctrl. Statistical significance was defined as $p < 0.05$ (* $p < 0.05$; ** $p < 0.01$; *** $p < 0.001$) (value means $\pm$ SEM, $n = 3$).

Next, we tested the effects of seExo-GABA on the expression of neuronal cell marker genes in SH-SY5Y cells. The results showed that seExo-GABA augmented the expression of some of marker genes except Nestin (Figure 5A–C). These results indicated that seExo-GABA activated SH-SY5Y cells, and revealed that exosomes, which are capable of activating neuronal cells, circulate in the blood of mice orally administered GABA.

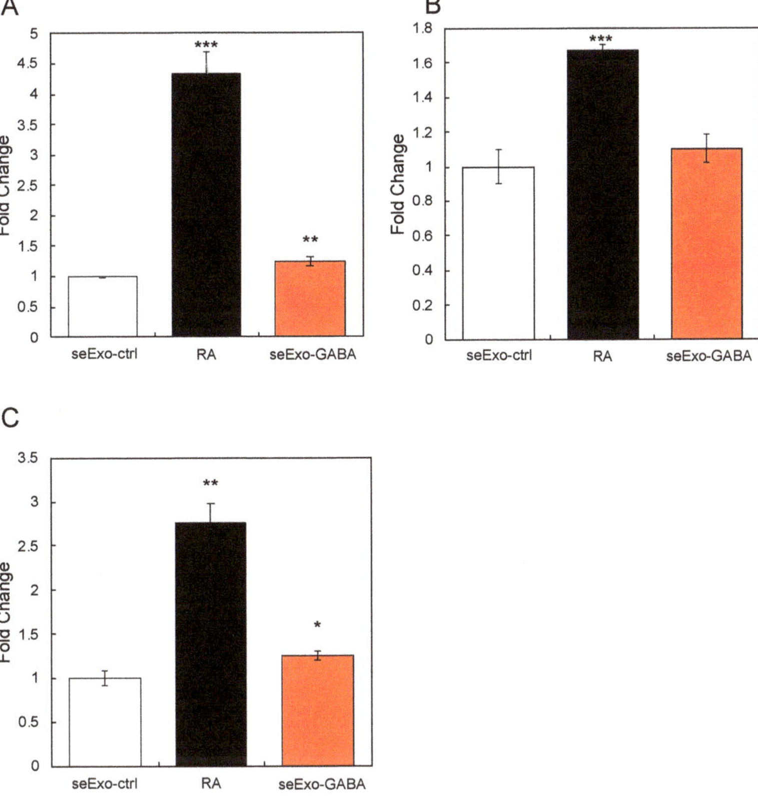

Figure 5. Effects of exosomes derived from serum of GABA-treated mice (seExo-GABA) on gene expression. Exosomes derived from serum of non-treated mice (seExo-ctrl) was used as control. Expression of *BDNF* (**A**), Nestin (**B**), and *NEFM* (**C**) in SH-SY5Y cells treated with seExo-GABA was evaluated by RT-qPCR. RA was used as positive control. Multiple comparisons between groups were carried out by one-way ANOVA with Tukey's post-hoc test. Statistical difference was evaluated by comparing to seExo-ctrl. Statistical significance was defined as $p < 0.05$ (* $p < 0.05$; ** $p < 0.01$; *** $p < 0.001$) (value means $\pm$ SEM, $n = 3$).

3.4. Microarray Analysis of Hippocampal Tissue of Mice Orally Administered GABA

Here, we performed microarray analysis of hippocampal mRNA of mice orally administered GABA. Results showed that in the hippocampus of mice orally administered GABA, 1127 and 249 genes were significantly up- and down-regulated, respectively. Among these genes, many genes were observed to be related to brain function (Tables 4 and 5). Furthermore, GSEA analysis of these genes showed that oral administration of GABA up-regulated the expression of genes related to memory function in the hippocampus (Figure 6).

Table 4. Functional categories of genes differentially expressed in the hippocampus of GABA-administered mice.

Functional Categories	*p*-Value
Synapse	3.1×10^{-7}
Zinc-finger	5.6×10^{-5}
Postsynaptic membrane	7.3×10^{-3}
Actin nucleation	4.5×10^{-3}

Table 5. KEGG pathway of genes differentially expressed in hippocampus of GABA-administered mice.

KEGG Pathway	*p*-Value
Long-term potentiation	2.0×10^{-4}
Regulation of actin cytoskeleton	2.2×10^{-4}
Alzheimer's disease	5.9×10^{-2}
Long-term depression	1.5×10^{-3}

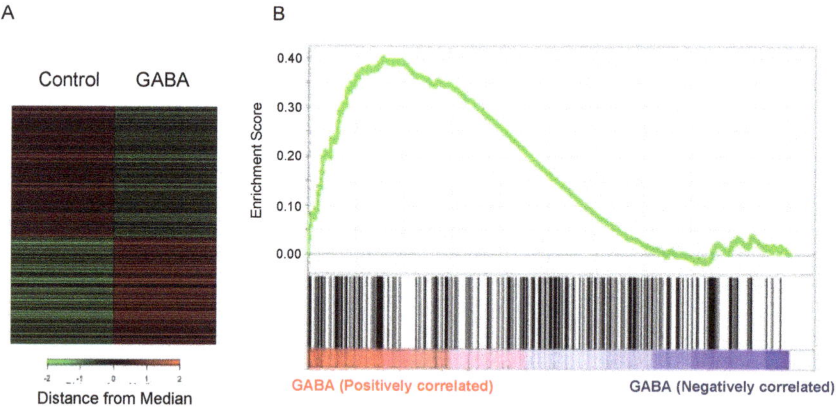

Figure 6. Microarray analysis of genes in the hippocampus of GABA-treated mice. (**A**) Heatmap was used to visualize differential expressed genes. (**B**) Gene set enrichment analysis (GSEA) analysis of gene sets relating to memory in the hippocampus of GABA-treated mice. Enrichment plots (green curve) show the running sum of enrichment score (ES) for memory-related gene set. The score at the peak of the plot is the ES for the gene set. The black bars show where the member of the gene set appear in the ranked list of genes. Each black bar represents a memory-related gene. A predominance of black bars to the left or right side indicates that these genes are up-regulated or down-regulated in the hippocampus of GABA-treated mice.

4. Discussion

Exosomes derived from various types of cells have been reported to induce neuroprotection and neural recovery by modulating the expression of genes, proteins, and miRs in target cells and tissues [31,32]. Emerging evidence shows that exosome-mediated multiple communication axes between various organs such as the brain, heart, kidneys, and intestines, as well as systemic immune responses, can influence health status. In the present study, we showed that GABA activated intestinal cells to secrete exosomes that activate neurons, which may shed light on the exosome-mediated activation of gut–brain interactions caused by GABA. Since it has been reported that $GABA_A$ receptor is expressed in Caco-2 cells [33], it will be necessary to verify whether GABA activates Caco-2 cells via its receptors to change the content of exosomes, or whether GABA itself is taken up by cells and incorporated into exosomes to function in the target tissue. Recent research has shown that GABA, a type of amino acid, is found in many fermented foods such as yogurt and pickled vegetables. We have previously shown that carnosine, as well as GABA, activates gut–brain interactions via exosomes, but this is a rare example of exosome-mediated activation of gut–brain interactions by a food component [24,34]. In this study, we clarified that serum exosomes of mice administered GABA (seExo-GABA) as well as exosomes derived from GABA-treated Caco-2 cells (Exo-GABA) activated neuronal cells to induce neurite growth and mitochondrial activation. These results suggest that these exosomes function as mediators that carry signals to activate neurons, and that certain foods including carnosine and GABA can produce these neuron-activating exosomes. This study could lead to the creation of a new research area: the development of foods that control brain function through the secretion of functional exosomes from the gut.

Microarray analysis of miRNA in Exo-GABA and of mRNA in SH-SY5Y cells treated with Exo-GABA identified signaling pathways which might be activated by Exo-GABA and be involved in GABA-induced phenotypes. Integrated analysis of miRNAs with altered expression in Exo-GABA and of mRNAs with altered expression in SH-SY5Y cells in response to Exo-GABA treatment identified 12 genes which were involved in the GABA-induced activation of brain function. In the future study, we would like to clarify the functionality of these 12 genes in the regulation of brain function.

There are many possible mechanisms by which exosomes regulate neuronal function. In a previous study, we reported that carnosine augmented the expression of miR-6769-5p in exosomes derived from carnosine-treated Caco-2 cells, which led to repression of target gene (ATXN1) expression, thereby activating neurons [34]. With further analysis, we would like to clarify the molecular mechanism of GABA-mediated activation of neuronal cells through exosomes. In particular, it is necessary to verify whether exosomes contain GABA or other bioactive parts in the future.

The present study also showed that serum exosomes of mice administered GABA activated neuronal cells (Figure 7). This result indicated that exosomes, which can activate neurons, circulate in the blood of mice after GABA administration. Indeed, exosomes derived from mesenchymal stromal cells may be useful for remodeling and functional recovery of neurons after stroke by transferring miR-133b to astrocytes and neurons and regulating gene expression [35]. Recently, various functions of milk exosomes derived from bovine milk have been reported. The milk exosome has been shown to carry specific miRNAs (miR-148a) to various target cells and to exert various functions on them. A characteristic example of milk exosome functionalities include effects against α-synuclein pathology in Parkinson's disease and Type 2 diabetes mellitus [36]. In other words, both milk exosomes and GABA-derived exosomes are similar in the sense that they are carried in the bloodstream to the target tissue. These reports indicate that serum exosomes are involved in the regulation of brain function, and suggest that by analyzing serum exosomes after ingestion, the functionality of foodstuffs can be verified. Furthermore, it will be possible to develop diagnostic methods for brain function based on serum exosomes.

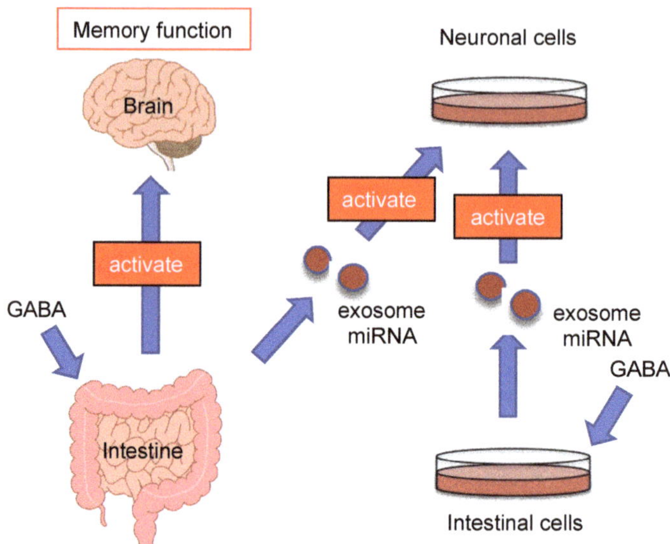

Figure 7. Schematic diagram of GABA function.

To verify the effect of GABA administration on brain function, the effects of GABA on sleep, relaxation, and psychological and physical fatigue have been investigated in humans [15,18,37]. These results indicate that GABA strongly affects the early stages of sleep, reduces both psychological and physical fatigue, and acts as a natural relaxant inducer. These results also suggest that GABA may also contribute to improved brain function, especially memory function. GSEA analysis of genes in the hippocampus of GABA-treated mice revealed that GABA administration could have some effect on brain function through the regulation of genes related to memory function. The detailed molecular mechanisms by which GABA regulate gene expression in the hippocampus need to be clarified in the future.

5. Conclusions

In this study, we clarified that serum exosomes of mice administered GABA (seExo-GABA) as well as exosomes derived from GABA-treated Caco-2 cells (Exo-GABA) activated neuronal cells to induce neurite growth and mitochondrial activation. These results suggest that these exosomes function as mediators that carry signals to activate neurons, and that certain foods including GABA can produce these neuron-activating exosomes. This study could lead to the creation of a new research area: the development of foods that control brain function through the secretion of functional exosomes from the gut.

Supplementary Materials: The following are available online at https://www.mdpi.com/article/10.3390/nu13082544/s1, Table S1: Target genes of four miRNA.

Author Contributions: Conceptualization, A.Y., M.K. and Y.K.; methodology, Y.K.; validation, M.U.; investigation, R.I.; data curation, M.U.; writing—original draft preparation, Y.K.; writing—review and editing, Y.K.; project administration, Y.K.; funding acquisition, Y.K. All authors have read and agreed to the published version of the manuscript.

Funding: Y.K. was supported by JSPS KAKENHI (Grant Number 16K14929, 21H02141). Y.K. received the collaborative research cost from Pharma Foods International Co., Ltd.

Institutional Review Board Statement: The study was conducted according to the guidelines of the Declaration of Helsinki, and approved by the Ethics Committee on Animal Experimentation of Kyushu University (approval number: A28-187-0).

Informed Consent Statement: Not applicable.

Data Availability Statement: The data that support the findings of this study are available from the corresponding author, Y.K., upon reasonable request.

Acknowledgments: The authors would like to thank G. Takada and N. Oshima (GE Healthcare) for their expert assistance with the IN Cell Analyzer 2200, and K. Yasuda (Cell Innovator, Fukuoka, Japan) for her assistance with microarray analysis.

Conflicts of Interest: Authors from Pharma Foods International Co., Ltd. had no role in the design of the study; in the collection, analyses, or interpretation of data; in the writing of the manuscript, or in the decision to publish the results.

References

1. Obata, K. Synaptic Inhibition and γ-Aminobutyric Acid in the Mammalian Central Nervous System. *Proc. Jpn. Acad. Ser. B Phys. Biol. Sci.* **2013**, *89*, 139–156. [CrossRef] [PubMed]
2. Park, J.; Fu, Z.; Frangaj, A.; Liu, J.; Mosyak, L.; Shen, T.; Slavkovich, V.N.; Ray, K.M.; Taura, J.; Cao, B.; et al. Structure of human GABA B receptor in an inactive state. *Nature* **2020**, *584*, 304–309. [CrossRef] [PubMed]
3. Scott, S.; Aricescu, A.R. A structural perspective on GABAA receptor pharmacology. *Curr. Opin. Struct. Biol.* **2019**, *54*, 189–198. [CrossRef]
4. Stagg, C.J.; Bachtiar, V.; Johansen-Berg, H. The Role of GABA in Human Motor Learning. *Curr. Biol.* **2011**, *21*, 480–484. [CrossRef]
5. Floyer-Lea, A.; Wylezinska, M.; Kincses, T.; Matthews, P.M. Rapid Modulation of GABA Concentration in Human Sensorimotor Cortex During Motor Learning. *J. Neurophysiol.* **2006**, *95*, 1639–1644. [CrossRef]
6. Loerch, P.M.; Lu, T.; Dakin, K.A.; Vann, J.M.; Isaacs, A.; Geula, C.; Wang, J.; Pan, Y.; Gabuzda, D.H.; Li, C.; et al. Evolution of the Aging Brain Transcriptome and Synaptic Regulation. *PLoS ONE* **2008**, *3*, e3329. [CrossRef] [PubMed]
7. Stanley, D.P.; Shetty, A.K. Aging in the Rat Hippocampus Is Associated With Widespread Reductions in the Number of Glutamate decarboxylase-67 Positive Interneurons but Not Interneuron Degeneration. *J. Neurochem.* **2004**, *89*, 204–216. [CrossRef] [PubMed]
8. Robinson, L.C.; Barat, O.; Mellott, J.G. GABAergic and glutamatergic cells in the inferior colliculus dynamically express the GABA$_A$R γ$_1$ subunit during aging. *Neurobiol. Aging* **2019**, *80*, 99–110. [CrossRef]
9. Hua, T.; Kao, C.; Sun, Q.; Li, X.; Zhou, Y. Decreased Proportion of GABA Neurons Accompanies Age-Related Degradation of Neuronal Function in Cat Striate Cortex. *Brain Res. Bull.* **2008**, *75*, 119–125. [CrossRef] [PubMed]
10. Han, S.; Tai, C.; Westenbroek, R.E.; Yu, F.H.; Cheah, C.S.; Potter, G.B.; Rubenstein, J.L.; Scheuer, T.; de la Iglesia, H.O.; Catterall, W.A. Autistic-Like Behaviour in Scn1a+/− Mice and Rescue by Enhanced GABA-Mediated Neurotransmission. *Nature* **2012**, *489*, 385–390. [CrossRef]
11. Verret, L.; Mann, E.O.; Hang, G.B.; Barth, A.M.I.; Cobos, I.; Ho, K.; Devidze, N.; Masliah, E.; Kreitzer, A.C.; Mody, I.; et al. Inhibitory Interneuron Deficit Links Altered Network Activity and Cognitive Dysfunction in Alzheimer Model. *Cell* **2012**, *149*, 708–721. [CrossRef]
12. Xu, Y.; Zhao, M.; Han, Y.; Zhang, H. GABAergic Inhibitory Interneuron Deficits in Alzheimer's Disease: Implications for Treatment. *Front. Neurosci.* **2020**, *14*, 660. [CrossRef] [PubMed]
13. Hosinian, M.; Qujeq, D.; Ahangar, A.A. The Relation Between GABA and L-Arginine Levels With Some Stroke Risk Factors in Acute Ischemic Stroke Patients. *Int. J. Mol. Cell. Med.* **2016**, *5*, 100–105.
14. Yamatsu, A.; Nakamura, U.; Saddam, H.M.; Horie, N.; Kaneko, T.; Kim, M. Improvement of Memory and Spatial Cognitive Function by Continuous Ingestion of 100mg/Day of γ-Aminobutyric Acid (GABA)-A Randomized, Double-Blind, Placebo-Controlled Parallel-Group Clinical Trial-. *Jpn. Pharmacol. Ther.* **2020**, *48*, 475–486.
15. Abdou, A.M.; Higashiguchi, S.; Horie, K.; Kim, M.; Hatta, H.; Yokogoshi, H. Relaxation and Immunity Enhancement Effects of γ-Aminobutyric Acid (GABA) Administration in Humans. *BioFactors* **2006**, *26*, 201–208. [CrossRef] [PubMed]
16. Nakamura, H.; Takishima, T.; Kometani, T.; Yokogoshi, H. Psychological Stress-Reducing Effect of Chocolate Enriched With γ-Aminobutyric Acid (GABA) in Humans: Assessment of Stress Using Heart Rate Variability and Salivary Chromogranin A. *Int. J. Food Sci. Nutr.* **2009**, *60* (Suppl. 5), 106–113. [CrossRef] [PubMed]
17. Yamatsu, A.; Yamashita, Y.; Maru, I.; Yang, J.; Tatsuzaki, J.; Kim, M. The Improvement of Sleep by Oral Intake of GABA and Apocynum venetum Leaf Extract. *J. Nutr. Sci. Vitaminol. (Tokyo)* **2015**, *61*, 182–187. [CrossRef]
18. Kanehira, T.; Nakamura, Y.; Nakamura, K.; Horie, K.; Horie, N.; Furugori, K.; Sauchi, Y.; Yokogoshi, H. Relieving Occupational Fatigue by Consumption of a Beverage Containing γ-Amino Butyric Acid. *J. Nutr. Sci. Vitaminol. (Tokyo)* **2011**, *57*, 9–15. [CrossRef]
19. Boonstra, E.; de Klejin, R.; Colzato, L.S.; Alkemade, A.; Forstmann, B.U.; Nieuwenhuis, S. Neurotransmitters as food supplements: The effects of GABA on brain and behavior. *Front. Psychol.* **2015**, *6*, 1520. [CrossRef]
20. Cataldo, P.G.; Villena, J.; Elean, M.; de Giori, G.S.; Saavedra, L.; Hebert, E.M. Immunomodulatory Properties of a γ-Aminobutyric Acid-Enriched Strawberry Juice Produced by *Levilactobacillus brevis* CRL 2013. *Front. Microbiol.* **2020**, *11*, 610016. [CrossRef]
21. Al-Kuraishy, H.M.; Hussian, N.R.; Al-Naimi, M.S.; Al-Gareeb, A.I.; Al-Mamorri, F.; Al-Buhadily, A.K. The Potential Role of Pancreatic γ-Aminobutyric Acid (GABA) in Diabetes Mellitus: A Critical Reappraisal. *Int. J. Prev. Med.* **2021**, *12*, 19.

22. An, J.; Seok, H.; Ha, E.M. GABA-producing Lactobacillus plantarum inhibits metastatic properties and induces apoptosis of 5-FU-resistant colorectal cancer cells via GABAB receptor signaling. *J. Microbiol.* **2021**, *59*, 202–216. [CrossRef]
23. Zareian, M.; Oskoueian, E.; Majdinasab, M.; Forghani, B. Production of GABA-enriched idli with ACE inhibitory and antioxidant properties using Aspergillus oryzae: The antihypertensive effects in spontaneously hypertensive rats. *Food Funct.* **2020**, *11*, 4304–4313. [CrossRef]
24. Inotsuka, R.; Uchimura, K.; Yamatsu, A.; Kim, M.; Katakura, Y. γ-Aminobutyric Acid (GABA) Activates Neuronal Cells by Inducing the Secretion of Exosomes From Intestinal Cells. *Food Funct.* **2020**, *11*, 9285–9290. [CrossRef]
25. Bazzan, E.; Tinè, M.; Casara, A.; Biondini, D.; Semenzato, U.; Cocconceli, E.; Baestro, E.; Damin, M.; Radu, C.M.; Taruto, G.; et al. Critical Review of the Evolution of Extracellular Vesicles' Knowledge: From 1946 to Today. *Int. J. Mol. Sci.* **2021**, *22*, 6417. [CrossRef]
26. Livak, K.J.; Schmittgen, T.D. Analysis of relative gene expression data using real-time quantitative PCR and the $2^{-\Delta\Delta CT}$ method. *Methods* **2001**, *25*, 402–408. [CrossRef] [PubMed]
27. Bolstad, B.M.; Irizarry, R.A.; Åstrand, M.; Speed, T.P. A Comparison of Normalization Methods for High Density Oligonucleotide Array Data Based on Variance and Bias. *Bioinformatics* **2003**, *19*, 185–193. [CrossRef] [PubMed]
28. Huang, D.W.; Sherman, B.T.; Lempicki, R.A. Bioinformatics Enrichment Tools: Paths Toward the Comprehensive Functional Analysis of Large Gene Lists. *Nucl. Acids Res.* **2009**, *37*, 1–13. [CrossRef] [PubMed]
29. Huang, D.W.; Sherman, B.T.; Lempicki, R.A. Systematic and Integrative Analysis of Large Gene Lists Using DAVID Bioinformatics Resources. *Nat. Protoc.* **2009**, *4*, 44–57. [CrossRef]
30. Fu, X.; Wan, S.; Lyu, Y.L.; Liu, L.F.; Qi, H. Etoposide induces ATM-dependent mitochondrial biogenesis through AMPK activation. *PLoS ONE* **2008**, *6*, e2009. [CrossRef]
31. Venkat, P.; Chen, J.; Chopp, M. Exosome-Mediated Amplification of Endogenous Brain Repair Mechanisms and Brain and Systemic Organ Interaction in Modulating Neurological Outcome After Stroke. *J. Cereb. Blood Flow Metab.* **2018**, *38*, 2165–2178. [CrossRef]
32. Xin, H.; Li, Y.; Chopp, M. Exosomes/miRNAs as Mediating Cell-Based Therapy of Stroke. *Front. Cell. Neurosci.* **2014**, *8*, 377. [CrossRef] [PubMed]
33. Ma, X.; Sun, Q.; Sun, X.; Chen, D.; Wei, C.; Yu, X.; Liu, C.; Li, Y.; Li, J. Activation of GABA$_A$ receptors in colon epithelium exacerbates acute colitis. *Front. Immunol.* **2018**, *9*, 987. [CrossRef] [PubMed]
34. Sugihara, Y.; Onoue, S.; Tashiro, K.; Sato, M.; Hasegawa, T.; Katakura, Y. Carnosine Induces Intestinal Cells to Secrete Exosomes That Activate Neuronal Cells. *PLoS ONE* **2019**, *14*, e0217394. [CrossRef]
35. Xin, H.; Li, Y.; Liu, Z.; Wang, X.; Shang, X.; Cui, Y.; Zhang, Z.G.; Chopp, M. MiR-133b Promotes Neural Plasticity and Functional Recovery After Treatment of Stroke With Multipotent Mesenchymal Stromal Cells in Rats via Transfer of Exosome-Enriched Extracellular Particles. *Stem Cells* **2013**, *31*, 2737–2746. [CrossRef]
36. Melnik, B.C. Synergistic Effects of Milk-Derived Exosomes and Galactose on α-Synuclein Pathology in Parkinson's Disease and Type 2 Diabetes Mellitus. *Int. J. Mol. Sci.* **2021**, *22*, 1059. [CrossRef] [PubMed]
37. Yamatsu, A.; Yamashita, Y.; Pandharipande, T.; Maru, I.; Kim, M. Effect of Oral γ-Aminobutyric Acid (GABA) Administration on Sleep and Its Absorption in Humans. *Food Sci. Biotechnol.* **2016**, *25*, 547–551. [CrossRef]

nutrients

MDPI

Article

Elucidation of Melanogenesis-Associated Signaling Pathways Regulated by Argan Press Cake in B16 Melanoma Cells

Thouria Bourhim [1,‡], **Myra O. Villareal** [2,3,‡], **Chemseddoha Gadhi** [1,2,*] and **Hiroko Isoda** [2,3,*]

[1] Faculty of Sciences Semlalia, Cadi Ayyad University, Avenue Prince Moulay Abdellah, B.P. 2390, Marrakesh 40000, Morocco; t.bourhim@gmail.com
[2] Alliance for Research on the Mediterranean and North Africa (ARENA), University of Tsukuba, Tennodai 1-1-1, Tsukuba 305-8572, Japan; villareal.myra.o.gn@u.tsukuba.ac.jp
[3] Faculty of Life and Environmental Sciences, University of Tsukuba, Tennodai 1-1-1, Tsukuba 305-8572, Japan
[*] Correspondence: dgadhi@uca.ac.ma (C.G.); isoda.hiroko.ga@u.tsukuba.ac.jp (H.I.)
[‡] These authors contributed equally to this work.

Abstract: The beneficial effect on health of argan oil is recognized worldwide. We have previously reported that the cake that remains after argan oil extraction (argan press-cake or APC) inhibits melanogenesis in B16 melanoma cells in a time-dependent manner without cytotoxicity. In this study, the global gene expression profile of B16 melanoma cells treated with APC extract was determined in order to gain an understanding of the possible mechanisms of action of APC. The results suggest that APC extract inhibits melanin biosynthesis by down-regulating microphthalmia-associated transcription factor (*Mitf*) and its downstream signaling pathway through JNK signaling activation, and the inhibition of Wnt/β-catenin and cAMP/PKA signaling pathways. APC extract also prevented the transport of melanosomes by down-regulating *Rab27a* expression. These results suggest that APC may be an important natural skin whitening product and pharmacological agent used for clinical treatment of pigmentary disorders.

Keywords: argan press-cake; MITF; JNK; cAMP/PKA; Wnt/β-catenin; microarray analysis

Citation: Bourhim, T.; Villareal, M.O.; Gadhi, C.; Isoda, H. Elucidation of Melanogenesis-Associated Signaling Pathways Regulated by Argan Press Cake in B16 Melanoma Cells. *Nutrients* **2021**, *13*, 2697. https://doi.org/10.3390/nu13082697

Academic Editor: Yoshinori Katakura

Received: 19 June 2021
Accepted: 2 August 2021
Published: 4 August 2021

Publisher's Note: MDPI stays neutral with regard to jurisdictional claims in published maps and institutional affiliations.

Copyright: © 2021 by the authors. Licensee MDPI, Basel, Switzerland. This article is an open access article distributed under the terms and conditions of the Creative Commons Attribution (CC BY) license (https://creativecommons.org/licenses/by/4.0/).

1. Introduction

Argania spinosa L. (family Sapotaceae) is a Moroccan endemic tree. It covers more than 800,000 hectares in the south-western region of the country [1]. In 1998, it was recognized as a biosphere reserve by the UNESCO and is known worldwide for its edible oil, which is now one of the world's priciest oils [2]. Argan oil is extremely rich in unsaturated fatty acids and bioactive phytochemicals including tocopherols, phenolic compounds, and carotenoids [3] contributing to its pharmacological and cosmetic properties [4]. Traditionally, this oil is used as an anti-aging skin care [5]. Consumption of argan oil and/or topical application decrease trans-epidermal water loss and improve skin elasticity [6,7].

At agronomic level, this species has led to great financial returns, significantly reducing the poverty of the local population. In addition to its considerable socio-economic importance, argan trees also contribute to preventing soil erosion and desertification in the Southern part of Morocco [1]. However, destructive logging, climate change, and overexploitation have negatively affected the argan tree ecosystem and about 44% of the argan forest was lost between 1970 and 2007 [8]. The increase in demand for argan oil now requires a more efficient use of this valuable resource.

One of the by-products of argan oil production is argan press-cake (APC). While it was used previously as an animal feed, the discovery that it is abundant in functional secondary metabolites that give argan oil its functional properties made it a possible source of health benefits that argan oil users are looking for. APC contains a high phenolic compounds content that includes procyanidin B1 and B2, catechin, epicatechin, epigallocatechin gallate, phloridzin, myricetin, and quercitrin [9]. It also has a large amount of saponins, steroids, and triterpenoids [10], suggesting that APC could have the same biological effects as argan oil.

Skin exposure to ultraviolet (UV) radiation generates reactive oxygen species (ROS) that may cause skin aging and increased pigmentation [11]. Skin pigmentation is a result of melanin synthesis in the melanocytes, and its subsequent distribution to keratinocytes [12]. Melanin protects the skin against harmful UV radiation and stress resulting from exposure to various environmental pollutants [13]. However, increased production of melanin could cause several unwanted localized skin hyperpigmentation issues such as freckles and age spots, and these could be due to melasma and other post-inflammatory-associated hyperpigmentation [14]. The unwanted side effects of existing treatments for hyperpigmentation increase the demand for safe melanogenesis regulators of plant origin to treat skin hyperpigmentation diseases.

Melanin biosynthesis is regulated by a variety of signal transduction pathways that include cyclic adenosine monophosphate (cAMP) and mitogen activated protein kinase (MAPK), as well as the Wnt signaling pathway [15,16]. cAMP, via PKA, induces cAMP-response element-binding protein (CREB) family phosphorylation and activation, which then leads to the expression of microphthalmia transcription factor (MITF) [17], with CREB being one of the transcription factors that regulate MITF expression [18]. The transcription factor MITF is responsible for the expression of tyrosinase (TYR), tyrosinase related protein (TRP1), and dopachrome tautomerase (DCT), the major enzymes that catalyze relevant reaction in the melanogenesis process [17]. The Wnt pathway regulates MITF expression through the ß-catenin, the pivotal component of the Wnt pathway. When Wnt proteins bind to their receptors, it stabilizes the cytoplasmic ß-catenin, which then leads to its localization in the nucleus, where it regulates the expression of MITF [19]. The effect of the Wnt signaling pathway on MITF makes it an important pathway for regulating melanocyte differentiation [20].

Extracellular signal-regulated protein kinase (ERK), p38, and c-jun N-terminal kinase (JNK) have essential roles in melanogenesis regulation [21–23]. Interfering with p38, MAPK expression has been reported to promote melanogenesis and tyrosinase expression [24], while the active form of ERK, on the other hand, phosphorylates MITF at serine-73 during the posttranslational process, leading to its ubiquitination and, subsequently, its degradation [25]. JNK can interfere with CREB-regulated transcription co-activator 3 (CRTC3)-dependent MITF expression leading to melanogenesis inhibition [26].

Another key factor in melanogenesis regulation is the melanosome pH, which controls the maturation of melanosome in melanocytes and the rate of melanogenesis, as well as the ratio of eumelanin to phaeomelanin [27].

We have demonstrated the melanogenesis inhibitory effect of APC extract via *Mitf* expression down-regulation in B16 murine melanoma cells [28]. However, the mechanism by which APC regulates melanogenesis is not yet understood. In this study, the global gene expression analysis was done to identify the signaling involved in the melanogenesis inhibitory effect of APC extract.

2. Materials and Methods

2.1. Chemicals and Reagents

The Dulbecco's modified eagle's medium (DMEM), fetal bovine serum (FBS), and L-glutamine were from Sigma-Aldrich (Burlington, MA, USA). Penicillin/streptomycin solution was from Lonza, Walkersville Inc., (Walkersville, MD, USA). All the other chemicals were from Wako (Saitama, Japan).

2.2. Extraction of APC

Mature *Argania spinosa* fruits of the spherical type were harvested in July 2012 from the Sidi Ifni region (southwest of Morocco). Voucher samples (MARK10888-1) kept at the regional herbarium of Marrakech (Marrakesh, Morocco) were verified by Prof. Ahmed Ouhammou. The samples were dried at 25 °C, after which the argan fruits were manually peeled, and the nut's shells cracked. The press-cake was obtained by extracting the oil from argan kernels using the mechanical press Komet DD 85 G press (IBG Monforts Oekotec GmbH & Co. KG, Mönchengladbach, Germany).

APC (10 g) was extracted with 100 mL ethanol 70% for 2 weeks at room temperature, after which it was centrifuged ($1000 \times g$, 15 min). The supernatant was then filter-sterilized using a 0.45 μm pore size filter (Millipore, Billerica, MA, USA) and kept at $-80\ ^\circ$C in a freezer until use.

2.3. Cells and Cell Culture

The B16 murine melanoma cells used in this study were purchased from Riken Cell Bank (Tsukuba, Japan). Cells were cultured in DMEM supplemented with 10% FBS, 4 mmol/L L-glutamine, 50 units/ml penicillin, and 50 μg/mL streptomycin, and they were maintained in a humidified incubator at $37\ ^\circ$C with 5% CO_2.

2.4. Total RNA Extraction

B16 cells were seeded at a density of 3×10^6 cells per 100-mm Petri-dish and were allowed to attach before treatment. The growth medium was replaced with a fresh one containing 0, 50 μg/mL of APC extract or 100 μmol/L of arbutin. The extraction of total RNA was done as described previously [28]. After the specified treatment time, the cells were washed twice with cold PBS before RNA extraction using the ISOGEN kit (Nippon Gene, Tokyo, Japan). The quality and quantity of the RNA was examined using a Nanodrop 2000 spectrophotometer (Nanodrop Technologies, Wilmington, DE, USA). The RNAs used were reverse transcribed using the SuperScript III Reverse Transcriptase Kit (Invitrogen, Carlsbad, CA, USA).

2.5. Quantitative Real-Time PCR Analysis

The effect of APC on gene expression in B16 cells was determined by real-time PCR (Applied Biosystems, Carlsbad, CA, USA) using TaqMan master mix, and performed using the TaqMan 7500 Fast Real-time PCR System (Applied Biosystems, Carlsbad, CA, USA). Cycling conditions were as follows: 2 min at $50\ ^\circ$C and 10 min at $95\ ^\circ$C, followed by 40 cycles of $95\ ^\circ$C for 15 s and $60\ ^\circ$C for 1 min. The assay IDs of the TaqMan primers used were: *Rab27a*—Mm00469997_m1, *Ctnnb1*—Mm00483039_m1, *Map3k12*—Mm00437378_m1, and *Gapdh*–Mm99999915_g1. *Gapdh* was used as the internal control. All the reactions were run in triplicates.

2.6. DNA Microarrays

The microarray analysis was performed to determine the global transcriptional response of B16 melanoma cells to argan press-cake or arbutin (positive control) treatment. Total RNA (100 ng) was reverse transcribed to synthesize the first-strand cDNA, which was then converted into double-stranded cDNA (ds-cDNA). The ds-cDNA template and biotin-labeled aRNA was generated using the 30 IVT Express Labeling Kit (Affymetrix, Santa Clara, CA, USA). The ds-cDNA was used to synthesize biotin-modified aRNA, 10 μg of which was fragmented using the GeneAtlas 3 IVT Express Kit, before hybridization to the Affymetrix Mouse 430 PM Array strips (Affymetrix) for 16 h at $45\ ^\circ$C. Following hybridization, the microarray DNA array was washed and stained in the GeneAtlas Fluidics Station 400 (Affymetrix), and then scanned using the GeneAtlas Imaging Station (Affymetrix). Data analysis was carried out using Affymetrix Expression Console Software and Affymetrix Transcriptome Analysis Console (TAC) 2.0 Software (Affymetrix). Hierarchical clustering was performed using Euclidean distance by TIGR's MultiExperiment Viewer v4.9.0 software. The rows represent genes while columns represent the experimental samples. The heat map represents the gene expression ratios with the green and red color of cells indicating gene down- and up-regulation. In addition, the DNA microarray data have been deposited in the ArrayExpress database at EMBI-EBI (Available online:

www.ebi.ac.uk/arrayexpress (accessed on 27 May 2020)), under the reference number E-MTAB-9089.

2.7. Statistical Analysis

The results were expressed as mean ± standard deviation (SD) of three independent experiments. The differences between means were analyzed for significance using one-way analysis of variance (ANOVA) with a Fisher's Least Significant Difference (LSD) post hoc test. A value of $p \leq 0.05$ was considered significant.

3. Results

3.1. Gene Expression Profile of APC Extract-Treated Cells

The expression level of genes in APC extract- or arbutin-treated cells was determined. Furthermore, genes expressed in APC-treated cells that were 1.5-fold different from the untreated controls were subjected to GO Enrichment Analysis. GO Enrichment Analysis identifies the genes that are relevant to biological processes and that were affected by APC treatment. The genes that were significantly changed in expression are presented in Table 1. The processes down-regulated by APC extract included those that are associated with pigmentation, melanosome transport, keratinocyte differentiation, melanocyte differentiation, cell differentiation, and apoptosis. Moreover, nervous system development, Wnt receptor signaling pathway through beta-catenin, the glutathione metabolic process, and MAP kinase kinase kinase activity were differentially regulated by APC extract.

Table 1. List of genes that were differentially expressed (1.5-fold change in expression) in B16 melanoma cells treated with argan press-cake (APC) or arbutin (ARB) as determined by DNA microarray ($p < 0.05$) [1].

Gene Symbol	Gene Name	Function	Fold Change	
			ARB	APC
Brca1	Breast cancer 1	Double-strand break repair via homologous recombination, DNA repair, lipid metabolic process	−1.5	−2.5
Slc24a4	Solute carrier family 24 (sodium/potassium/calcium exchanger), member 4	Calcium, potassium: Sodium transporter activity	1.0	−2.1
Ctnnb1	Catenin (cadherin associated protein), beta 1	Wnt receptor signaling pathway, positive regulation of I-kappaB kinase/NF-kappaB cascade, positive regulation of MAPK cascade, skin development	−1.1	−1.8
Trpm1	Transient receptor potential cation channel, subfamily M, member 1	G-protein coupled glutamate receptor signaling pathway, calcium ion transport into cytosol	−1.0	−1.7
Oca2	Oculocutaneous albinism II	Transport, spermatid development, cell proliferation, melanocyte differentiation, melanin biosynthetic process, pigmentation, developmental pigmentation, transmembrane transport	−1.0	−1.7
Tyr	Tyrosinase	Melanin biosynthetic process	−1.0	−1.6
Rab27a	RAB27A, member RAS oncogene family	Protein transport, melanocyte differentiation, melanosome localization, melanosome transport, pigmentation	−1.2	−1.6
Map3k12	Mitogen-activated protein kinase kinase kinase 12	Activation of MAPKK activity, protein phosphorylation, JNK cascade	−2.3	−1.6
Slc6a17	Solute carrier family 6 (neurotransmitter transporter), member 17	Neurotransmitter: sodium symporter activity, neurotransmitter transport	1.2	−1.6
Vat1	Vesicle amine transport protein 1 homolog (T californica)	Zinc ion binding, oxidoreductase activity, negative regulation of mitochondrial fusion	−1.1	−1.6
Atp6v0b	Atpase, H+ transporting, lysosomal V0 subunit B	Hydrogen-exporting ATPase activity, phosphorylative mechanism, hydrogen ion transmembrane transporter activity ATP catabolic process, ion transport, ATP hydrolysis coupled proton transport	−1.1	−1.5
Rbm39	RNA binding motif protein 39	Nucleotide binding, transcription coactivator activity, poly(A) RNA binding, regulation of transcription, DNA-templated	−1.6	−1.5
Usp9x	Ubiquitin specific peptidase 9, X chromosome	Cysteine-type peptidase activity, hydrolase activity, transforming growth factor beta receptor signaling pathway, BMP signaling pathway, hippocampus development	−1.5	−1.5
Ccs	Copper chaperone for superoxide dismutase	ROS catabolism, superoxide dismutase copper chaperone activity	1.5	1.4

Table 1. *Cont.*

Gene Symbol	Gene Name	Function	Fold Change	
			ARB	APC
Hbegf	Heparin-binding EGF-like growth factor	Positive regulation of keratinocyte migration, positive regulation of protein kinase B signaling cascade, positive regulation of wound healing	−1.1	1.5
Plcb1	Phospholipase C, beta 1	Enzyme binding, positive regulation of JNK cascade,	−1.1	1.5
Prkar1b	Protein kinase, camp dependent regulatory, type I beta	Camp-dependent protein kinase inhibitor activity, camp-dependent protein kinase regulator activity, regulation of protein phosphorylation	1.3	1.5
Slc7a11	Solute carrier family 7 (cationic amino acid transporter, y+ system), member 11	Amino acid transmembrane transporter activity, response to toxic substance, platelet aggregation	1.1	1.5
Maoa	Monoamine oxidase A	Primary amine oxidase activity, oxidoreductase activity, dopamine catabolic process	−1.2	1.5
Mcm3	Minichromosome maintenance deficient 3	DNA replication initiation	1.5	1.6
Erbb3	V-erb-b2 erythroblastic leukemia viral oncogene homolog 3 (avian)	Protein tyrosine kinase activity, receptor signaling protein tyrosine kinase activity, protein phosphorylation	1.1	1.6
Rasa1	RAS p21 protein activator 1	Positive regulation of Ras GTPase activity, negative regulation of Ras protein signal transduction	1.1	1.6
Slc4a4	Solute carrier family 4 (anion exchanger), member 4	Transporter activity, inorganic anion exchanger activity, sodium/bicarbonate symporter activity, regulation of pH	−1.2	1.6
Dzip3	DAZ interacting protein 3, zinc finger	Zinc ion binding	−1.1	1.6
Slc35a5	Solute carrier family 35, member A5	Nucleotide-sugar transmembrane transporter activity, carbohydrate transport	1.2	1.6
Slc12a6	Solute carrier family 12, member 6	Potassium/chloride symporter activity, ion transport, cation chloride transport	1.4	1.6
Cbl	Casitas B-lineage lymphoma	Phosphotyrosine binding, positive regulation of phosphatidylinositol 3-kinase cascade, protein binding, calcium ion binding, zinc ion binding	1.2	1.6
Tcf7l2	Transcription factor 7 like 2, T cell specific, HMG box	Protein kinase binding, skin development, canonical Wnt receptor signaling pathway involved in positive regulation of epithelial to mesenchymal transition	1.0	1.6
Figf	C-fos induced growth factor	Protein binding, vascular endothelial growth factor receptor binding	1.2	1.6
Ccrl1	Atypical chemokine receptor 4	G-protein coupled receptor signaling pathway, scavenger receptor activity	1.5	1.7
Slc6a6	Solute carrier family 6 (neurotransmitter transporter, taurine), member 6	Neurotransmitter/sodium symporter activity, beta-alanine transport	1.1	1.7
Slc4a4	Solute carrier family 4 (anion exchanger), member 4	Anion transmembrane transporter activity, sodium/bicarbonate symporter activity, regulation of pH, bicarbonate transport	1.1	1.7
Braf	Braf transforming gene	MAP kinase kinase kinase activity, activation of MAPKK activity, positive regulation of ERK1 and ERK2 cascade	1.1	1.7
Akap12	A kinase (PRKA) anchor protein 13	Regulation of protein kinase activity, phosphorylation	1.2	1.7
Akap13	A kinase (PRKA) anchor protein (gravin) 12	Positive regulation of protein kinase A signaling cascade	1.1	1.7
Tpr	Translocated promoter region	MAPK import into nucleus	1.1	1.8
Crebbp	CREB binding protein	Negative regulation of transcription from RNA polymerase II promoter, p53 binding	−1.0	1.8
Adh7	Alcohol dehydrogenase 7 (class IV), mu or sigma polypeptide	Aldehyde oxidase activity, oxidoreductase activity	1.1	1.9
Cxcl10	Chemokine (C-X-C motif) ligand 10	Protein secretion	1.4	1.9
Taok1	TAO kinase 1	Protein kinase activator activity, protein phosphorylation	1.2	1.9
Atp2b1	Atpase, Ca++ transporting, plasma membrane 1	Nucleotide binding, hydrolase activity, calcium ion transport	1.1	2.2
Slc7a11	Solute carrier family 7 (cationic amino acid transporter, y+ system), member 11	Amino acid transmembrane transporter activity, response to toxic substance, platelet aggregation	−1.2	2.2
Gclm	Glutamate-cysteine ligase, modifier subunit	Glutamate-cysteine ligase activity, glutamate-cysteine ligase catalytic subunit, protein heterodimerization activity, cysteine, glutamate and glutathione metabolic process, response to oxidative stress, apoptotic mitochondrial changes, negative regulation of neuron apoptotic process, negative regulation of extrinsic apoptotic signaling pathway	−1.1	2.3

Table 1. *Cont.*

Gene Symbol	Gene Name	Function	Fold Change	
			ARB	APC
Spry4	Sprouty homolog 4 (Drosophila)	Protein binding, multicellular organismal development, regulation of signal transduction, negative regulation of MAP kinase activity	1.5	2.4
Gsta3	Glutathione S-transferase, alpha 3	Glutathione transferase activity, metabolic process	1.1	2.7

[1] Based on gene ontology annotations in Mouse Genome Informatics (MGI).

DNA microarray analysis also revealed that several Wnt signaling pathway associated genes were significantly changed in response to APC extract treatment (Table 1). Among those genes, several genes were down-regulated (*Ctnnb1* and *Cxxc4*), while some were up-regulated (*Tcf7L2*). Interestingly, a number of intermediate genes of the MAPK pathway were up-regulated (*Akap9, Rasa1, Spry4*), while *Map3k12* was down-regulated by APC extract treatment. Moreover, the microarray dataset revealed an up-regulation by the APC extract of the *Prkar1b* gene, a regulatory subunit of PKA inactivating its catalytic domains. Additionally, several genes under the transcriptional regulation of MITF, such as *Tyr, Trpm1, Vat1, Atp6v0b, Rbm39*, and *Usp9x*, were significantly down-regulated, including *Rab27a*, which is a melanosome transport protein. The obtained data also showed that many of the solute carrier genes (SLCs) (*Slc24a4, Slc35a5, Slc12a6, Slc6a6, Slc4a4, and Slc7a11*) were differentially expressed following APC extract treatment (Table 1).

Genes that were significantly modulated by APC extract or arbutin were subjected to hierarchical clustering and the results grouped the genes into two main groups: APC-down-regulated and APC-up-regulated genes (Figure 1). The down-regulated genes formed five subgroups, while up-regulated genes formed three subgroups. The first cluster is composed of genes relevant to transporter activity represented by two solute carrier genes *Slc24a4* and *Slc6a17* ($p = 0.52$). The second cluster are genes significant for DNA repair (*Brca1*, $p = 1.04$). The next cluster constituted of genes that are relevant in melanin synthesis, melanosome transport, ATP binding, ion transport, and metal binding (*Tyr, Rab27a, Trpm1, Vat1, Atp6v0b, Cxxc4, Adam10, Epb41l2* ($p = 0.30$)). In addition, a regulation of *Ctnnb1, Map3k12* genes that play a role in the Wnt signaling pathway and MAP kinase activity was observed in cluster four ($p = 0.34$). The fifth cluster are genes in the Wnt signaling pathway and c-AMP signaling pathway, or those that function in metal binding, protein binding ($p = 0.33$). The last three clusters represent the up-regulated genes by APC extract treatment. Those genes play roles in the c-AMP signaling pathway, Wnt signaling pathway, transporter activity, protein binding, and kinase activity (Figure 1, Table 2).

Table 2. Gene clusters obtained by hierarchical clustering of significantly expressed genes in APC-treated B16 cells [a,b].

Cluster No. (*p*-Value)	Signaling Pathway	Genes
1 (0.52)	Transporter activity	*Slc24a4, Slc6a17*
2 (1.04)	DNA repair	*Brca1*
3 (0.30)	Melanogenesis regulation, melanosome transport, ATP binding, ion transport, metal binding	*Tyr, Rab27a, Trpm1, Vat1, Atp6v0b, Cxxc4, Adam10, Epb41l2*
4 (0.34)	Wnt signaling pathway, MAP kinase activity	*Ctnnb1, Map3k12*
5 (0.33)	Wnt signaling pathway, metal binding, protein binding c-AMP signaling pathway	*Rbm39, Usp9x, Ghrhr, Pkia, Nfkbia, Ctnna1*
6 (0.68)	Positive regulation of keratinocyte migration, positive regulation of JNK cascade, transporter activity, glutamate-cysteine ligase activity	*Hbegf, Plcb, Slc4a4, Slc7a11, Gclm*
7 (0.37)	Ros catabolism, transporter activity, cAMP dependent regulatory, protein binding, regulation of Ras GTPase activity, Wnt signaling	*Ccs, Slc12a6, Prkar1b, Slc35a5, Figf, Rasa1, Tcf7l2, Slc6a6*
8 (0.50)	Protein binding, MAP kinase activity, glutathione transferase activity	*Spry4, Gsta3*

[a] Euclidean distance by TIGR's MultiExperiment Viewer v4.9.0 software. [b] Based on gene ontology annotations in Mouse Genome Informatics (MGI).

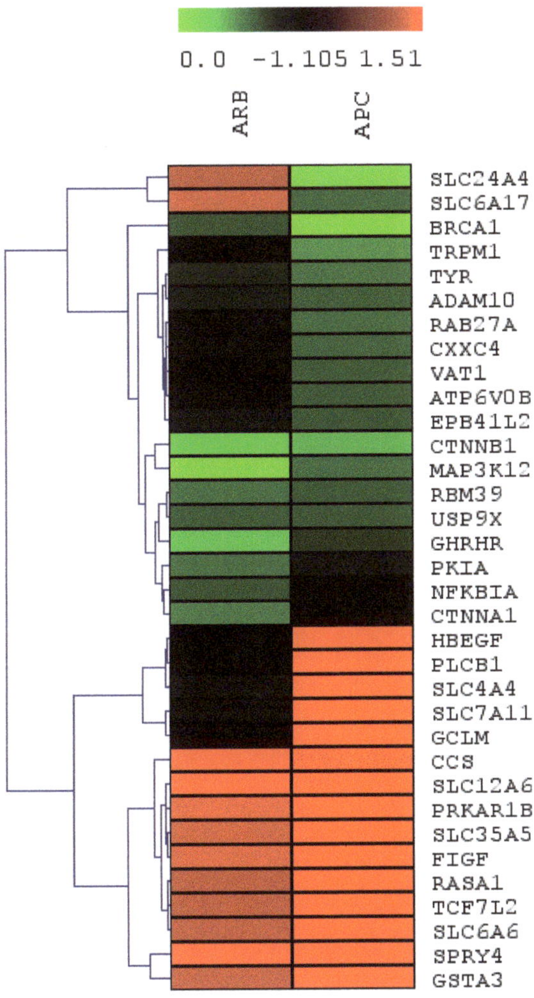

Figure 1. Heat map and hierarchical clustering of 34 genes in response to treatment with argan press-cake (APC) or by arbutin (ARB) in B16 melanoma cells. Clustering was calculated using Euclidian distance in the TIGR's MultiExperiment Viewer v4.9.0 software. Rows and columns represent genes and experimental samples, respectively. Gene expression ratios are presented in the heat map with green and red color indicating down and up-regulation, respectively.

3.2. Validation of Global Gene Expression Results

The DNA microarray results were validated using rt-PCR. *Rab27a* was significantly down-regulated by 51.7%. *Ctnnb1* (catenin cadherin associated protein beta1) and *Map3k12* (mitogen-activated protein kinase kinase kinase 12), the genes that code for proteins that regulate *Mitf* expression, were decreased significantly (Figure 2). The changes in genes expression were consistent with the microarray data.

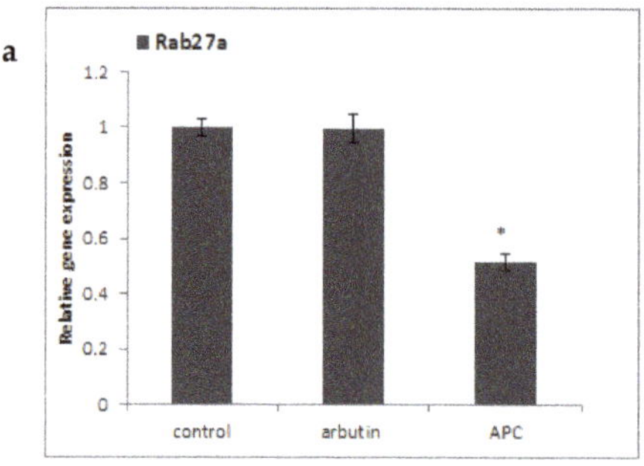
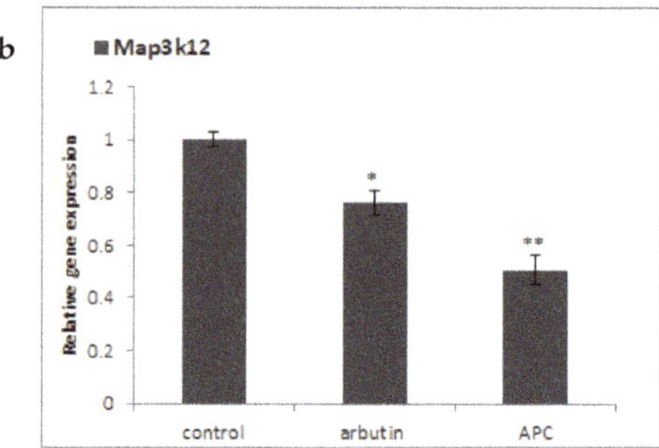
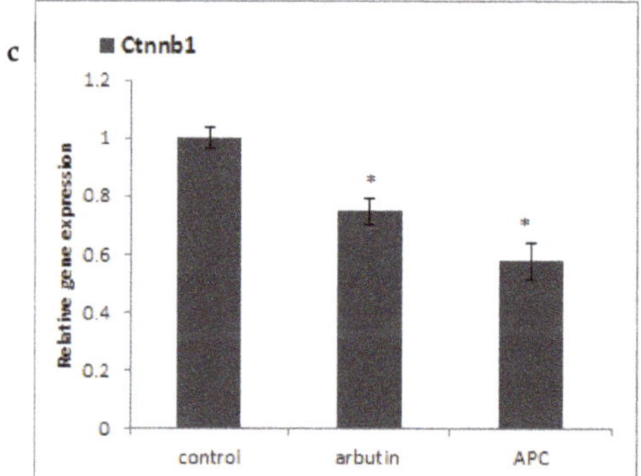

Figure 2. Effect of argan press-cake (APC) on the expression of (**a**) *Rab27a* (member RAS oncogene family), (**b**) mitogen-activated protein kinase kinase kinase 12 (*Map3k12*), and (**c**) catenin (cadherin associated protein) beta1 (*Ctnnb1*) genes in B16 cells. Cells were seeded onto a 100-mm dish at a density of 3×10^6. B16 cells were treated with or without arbutin (100 μM) or argan press-cake extract (50 μg/mL) for 24 h. Rt-PCR was used to determine the select genes expression level. Data are expressed as mean $\pm$ SD (n = 3), * $p < 0.05$, ** $p < 0.01$.

4. Discussion

Previously, we demonstrated that APC extract has a melanogenesis inhibitory effect in B16 cells. APC inhibits the melanogenic enzymes expression through *Mitf* down-regulation [28]. In this study, the global gene expression in the B16 cells in response to APC treatment was analyzed in order to fully understand the effect of APC on B16 cells and to identify APC targets that may have directly or indirectly contributed to its inhibitory effect, specifically on MITF and on melanogenesis in general.

As presented in Table 1, APC up-regulated the *Prkar1b* gene, which codes for the protein that is a regulatory subunit of cyclic AMP-dependent protein kinase A (PKA) [29]. As previously reported, APC treatment inhibits the *Mitf* expression [28], which could be due to the inhibition of signals that are upstream of *Mitf*. In short, an inactivation of PKA and the down-regulation of cAMP due to APC treatment could explain the decreased *Mitf* expression. Down-regulation of *Mitf* also led to inhibition of melanosome transport by its direct regulatory effect on *Rab27a* expression [30]. In this study, *Rab27a* expression was down-regulated by 51.7%. In addition, it has also been observed that *β-catenin* was down-regulated. β-catenin, as a transcriptional regulator, can be redirected by MITF away from genes that are under the regulation of the Wnt signaling toward *Mitf*-specific targets [31]. Additionally, β-catenin is also known to regulate *Mitf* expression [32]. β-catenin may be degraded via a ubiquitin-dependent, PKA-attenuated GSK3β (glycogen synthase kinase-3β) action [24]. MITF-regulated gene *Rab27a*, in addition to its function in melanosome transport, also has a significant role in various cell activities that include cell growth, invasion, and metastasis [33,34]. Several studies have associated increased expression of *Rab27a* with carcinogenesis, and it has been reported to promote the stemness of colon cancer cells [35], leading to poor survival in pancreatic cancer [36]. The inhibition of *Rab27a* by APC therefore suggests an anti-cancer effect of APC by targeting *Rab27a*.

Cheli et al. (2009) showed that cAMP controls the melanosome pH through PKA-independent mechanism. It has also been suggested that cAMP modulates vacuolar ATPases and ion transporters expression [37]. The melanosome pH of light-colored human skin melanocytes is more acidic, which explains the observed low tyrosinase activity observed in Caucasian skin. In contrast, melanosomes in dark human skin are less acidic and have higher tyrosinase activity [38]. The results of this study revealed that many of the solute carrier genes (*Slc24a4, Slc35a5, Slc12a6, Slc6a6, Slc4a4, and Slc7a11*), several vacuolar ATPases, and ion transporters were differentially expressed following APC extract treatment. In addition to its action at the mRNA level, the results of this study further suggest that APC extract may decrease the activity of TYR by reducing the pH of melanosome. This may also explain why despite the fact that there was no change in the TYR expression, a decrease in B16 cells melanin content at 48 h was observed [28]. Moreover, a growing number of solute carrier genes have been reported to play pivotal roles in melanogenesis regulation and in ethnic skin color determination [39,40], and this includes *Slc24a4*, which was down-regulated by 2-fold in APC extract-treated cells.

The effect of APC extract on the JNK signaling most likely contributed to the decreased melanogenesis. JNK is regulated by several molecules, but in this study, a down-regulation of the expression of *Map3k12* and up-regulation of *Plcb1* could induce the activation of the JNK pathway (Figure 3). The activation of JNK causes phosphorylation of MITF at serine 73, which could then result to subsequent ubiquitin-dependent proteasomal degradation of MITF [21]. Although MITF was observed to be down-regulated at the transcriptional level, it is also possible that post-transcriptional modifications occurred in response to APC extract treatment. That effect could be through the several pathways mentioned earlier, which include the MAPK (JNK) pathway.

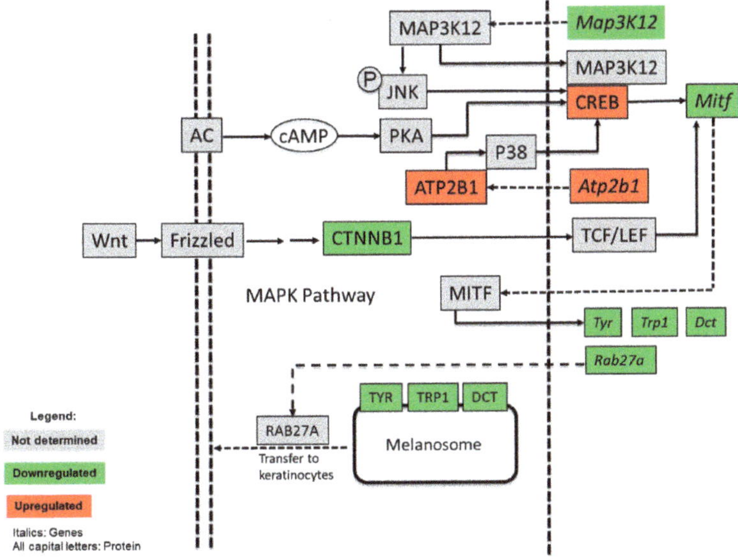

Figure 3. Argan press-cake (APC) down-regulated genes relevant to melanogenesis. This figure is adapted from Kyoto Encyclopedia of Genes and Genomes (KEGG) Pathway for melanogenesis (Available online: https://www.genome.jp/pathway/map04916 (accessed on 15 February 2021)).

DNA microarray results also showed that the Wnt/β-catenin and the cAMP/PKA signaling pathways were inactivated (Figure 3). The cAMP's effect on melanogenesis is mainly of its effect on the process' rate limiting enzyme tyrosinase, and cAMP stimulating tyrosinase activity [41]. However, what has been well reported is its effect on PKA activation that will then lead to CREB activation. CREB activates MITF expression. Compounds that inhibit the cAMP pathway are expected to inhibit melanogenesis [42].

This complex effect of APC extract on B16 cells could be attributed to the presence of different bioactive compounds, polyphenols, saponins, steroids, and triterpenoids.

In this study, the APC-treated cells were not just compared to untreated cells, but also to arbutin, the positive control. Arbutin is a D-glucopyranoside derivative of hydroquinone, which is a widely used skin lightning agent in cosmetic and healthcare industry [43,44]. Our microarray data revealed that *Nfkbia*, which is involved in nuclear factor kappa B (NF-κB) binding, and cytoplasmic sequestering were down-regulated by arbutin. NF-κB signaling has been reported to play a key role in melanogenesis regulation [45–47]. In addition, Ahn et al. (2003) reported that the treatment of human keratinocytes with arbutin effectively down-regulate NF-κB activation, which is consistent with our results [48]. It is well-known that the MAP3K-related kinase is associated with NF-κB stimulation by TNF, CD95, and IL-1 [49,50]. Therefore, *Map3k12*, which was significantly decreased by arbutin, appears to be a plausible target in regulating melanogenesis through NF-κB inhibition.

5. Conclusions

Argan press-cake extract significantly decreases melanin synthesis in B16 cells [28]. In this study, microarray analysis sheds light on the exact mechanism by which the anti-melanogenesis effect occurs. The inhibitory effect on melanogenesis was a result of the regulation of several signals at once, including (i) the inactivation of the cAMP/PKA signaling pathway through an up-regulation of *Prkar1b* gene that lead to PKA down-regulation and subsequently inhibition of *Mitf* expression; (ii) the down-regulation of the Wnt/β-catenin signaling pathway; and (iii) the activation of the JNK MAP kinase pathway. JNK activation causes a subsequent ubiquitin-dependent proteasomal degradation of MITF. In addition, APC may inhibit not only the melanin production but also melanosome transport by *Rab27a* down-regulation.

By inhibiting the overproduction and accumulation of melanin in the skin that cause numerous related pigment disorders like melasma, freckles or lentigines, dermatitis, and geriatric skin pigmentation [14], APC may become an alternative natural and noncytotoxic therapeutic agent against hyperpigmentation disorders. It also has a potential use in cosmetics like argan oil [51] and other argan oil extraction by-products, argan fruit shells [52], argan leaves [53], etc.

Author Contributions: Conceptualization: H.I. and C.G.; methodology: H.I. and M.O.V.; funding acquisition: H.I. and C.G; investigation: T.B. and M.O.V.; supervision: C.G. and H.I.; writing—original draft: T.B.; writing—review and editing: T.B., M.O.V. and C.G. All authors have read and agreed to the published version of the manuscript.

Funding: This research was funded by the Japan International Cooperation Agency (JICA)-Japan Science and Technology Agency (JST)'s Science and Technology Research Partnership for Sustainable Development (SATREPS) project entitled, "Valorization of Bioresources Based on Scientific Evidence in Semi- and Arid Land for Creation of New Industry"; and by Ministry of Higher Education, Scientific Research and Executive Training (MHESRET) of Kingdom of Morocco.

Institutional Review Board Statement: Not applicable.

Informed Consent Statement: Not applicable.

Data Availability Statement: The DNA microarray data have been deposited in the ArrayExpress database at EMBI-EBI (Available online: www.ebi.ac.uk/arrayexpress (accessed on 27 May 2020)), under the reference number E-MTAB-9089.

Acknowledgments: Authors are grateful to the following organizations for their support by providing grants to the first author (TB): Japan Student Services Organization (JASSO); and Centre National pour la Recherche Scientifique et Technique (CNRST) Morocco.

Conflicts of Interest: The authors declare no conflict of interest. The funders had no role in the design of the study; in the collection, analyses, or interpretation of data; in the writing of the manuscript, or in the decision to publish the results.

References

1. Msanda, F.; El Aboudi, A.; Peltier, J.P. Biodiversité et biogéographie de l'arganeraie marocaine. *Cah. Agric.* **2005**, *14*, 357–364.
2. Lybbert, T.J.; Aboudrare, A.; Chaloud, D.; Magnan, N.; Nash, M. Booming markets for Moroccan argan oil appear to benefit some rural households while threatening the endemic argan forest. *Proc. Natl. Acad. Sci. USA* **2011**, *108*, 13963–13968. [CrossRef] [PubMed]
3. Cabrera-Vique, C.; Marfil, R.; Giménez, R.; Martínez-Augustin, O. Bioactive compounds and nutritional significance of virgin argan oil–an edible oil with potential as a functional food. *Nutr. Rev.* **2012**, *70*, 266–279. [CrossRef] [PubMed]
4. Lizard, G.; Filali-Zegzouti, Y.; Midaoui, A.E. Benefits of Argan Oil on Human Health. *Int. J. Mol. Sci.* **2017**, *18*, 1383. [CrossRef] [PubMed]
5. Charrouf, Z.; Guillaume, D.; Driouich, A. Argan tree: An asset for Morocco. *Biofutur* **2002**, *220*, 54–57. (In French)
6. Boucetta, K.Q.; Charrouf, Z.; Aguenaou, H.; Derouiche, A.; Bensouda, Y. The effect of dietary and/or cosmetic argan oil on postmenopausal skin elasticity. *Clin. Interv. Aging* **2015**, *10*, 339–349. [PubMed]
7. Boucetta, K.Q.; Charrouf, Z.; Derouiche, A.; Rahali, Y.; Bensouda, Y. Skin hydration in postmenopausal women: Argan oil benefit with oral and/or topical use. *Prz. Menopauzalny* **2014**, *13*, 280–288. [CrossRef]
8. De Waroux, Y.L.P.; Lambin, E.F. Monitoring degradation in arid and semi-arid forests and woodlands: The case of the argan woodlands (Morocco). *Appl. Geogr.* **2012**, *32*, 777–786. [CrossRef]
9. El Monfalouti, H.; Charrouf, Z.; Belviso, S.; Ghirardello, D.; Scursatone, B.; Guillaume, G.; Denhez, C.; Zeppa, G. Analysis and antioxidant capacity of the phenolic compounds from argan fruit (*Argania spinosa* (L.) Skeels). *Eur. J. Lipid Sci. Technol.* **2012**, *114*, 446–452. [CrossRef]
10. Henry, M.; Kowalczyk, M.; Maldini, M.; Piacente, S.; Stochma, A.; Oleszek, W. Saponin inventory from *Argania spinosa* kernel cakes by liquid chromatography and mass spectrometry. *Phytochem. Anal.* **2013**, *24*, 616–622. [CrossRef]
11. Perez Davo, A.; Truchuelo, M.T.; Vitale, M.; Gonzalez-Castro, J. Efficacy of an antiaging treatment against environmental factors: Deschampsia antarctica extract and high-tolerance retinoids combination. *J. Clin. Aesthet. Dermatol.* **2019**, *12*, E65–E70. [PubMed]
12. Bonaventure, J.; Domingues, M.J.; Larue, L. Cellular and molecular mechanisms controlling the migration of melanocytes and melanoma cells. *Pigment. Cell Melanoma Res.* **2013**, *26*, 316–325. [CrossRef] [PubMed]
13. Brenner, M.; Hearing, V.J. The protective role of melanin against UV damage in human skin. *Photochem. Photobiol.* **2008**, *84*, 539–549. [CrossRef] [PubMed]
14. Briganti, S.; Camera, E.; Picardo, M. Chemical and instrumental approaches to treat hyperpigmentation. *Pigment. Cell Melanoma Res.* **2003**, *16*, 101–110. [CrossRef]
15. Yamaguchi, Y.; Hearing, V.J. Physiological factors that regulate skin pigmentation. *Biofactors* **2009**, *35*, 193–199. [CrossRef]
16. Drira, R.; Sakamoto, K. Sakuranetin induces melanogenesis in B16 melanoma cells through inhibition of ERK and PI3K/AKT signaling pathways. *Phytother. Res.* **2016**, *30*, 997–1002. [CrossRef]
17. D'Mello, S.A.; Finlay, G.J.; Baguley, B.C.; Askarian-Amri, M.E. Signaling Pathways in Melanogenesis. *Int. J. Mol. Sci.* **2016**, *17*, 1144. [CrossRef]
18. Goding, C.R. Mitf from neural crest to melanoma: Signal transduction and transcription in the melanocyte lineage. *Genes Dev.* **2000**, *14*, 1712–1728. [PubMed]
19. Steingrimsson, E.; Copeland, N.G.; Jenkins, N.A. Melanocytes and the microphthalmia transcription factor network. *Annu. Rev. Genet.* **2004**, *38*, 365–411. [CrossRef] [PubMed]
20. Dunn, K.J.; Brady, M.; Ochsenbauer-Jambor, C.; Snyder, S.; Incao, A.; Pavan, W.J. WNT1 and WNT3a promote expansion of melanocytes through distinct modes of action. *Pigment. Cell Res.* **2005**, *18*, 167–180. [CrossRef]
21. Wu, M.; Hemesath, T.J.; Takemoto, C.M.; Horstmann, M.A.; Wells, A.G.; Price, E.R.; Fisher, D.Z.; Fisher, D.E. c-Kit triggers dual phosphorylations, which couple activation and degradation of the essential melanocyte factor Mi. *Genes Dev.* **2000**, *14*, 301–312.
22. Gu, W.J.; Ma, H.J.; Zhao, G.; Yuan, X.Y.; Zhang, P.; Liu, W.; Ma, L.J.; Lei, X.B. Additive effect of heat on the UVB-induced tyrosinase activation and melanogenesis via ERK/p38/MITF pathway in human epidermal melanocytes. *Arch. Dermatol. Res.* **2014**, *306*, 583–590. [CrossRef]
23. Karunarathne, W.A.H.M.; Molagoda, I.M.N.; Kim, M.S.; Choi, Y.H.; Oren, M.; Park, E.K.; Kim, G.Y. Flumequine-mediated upregulation of p38 MAPK and JNK results in melanogenesis in B16F10 cells and zebrafish larvae. *Biomolecules* **2019**, *9*, 596. [CrossRef] [PubMed]
24. Bellei, B.; Pitisci, A.; Catricalà, C.; Larue, L.; Picardo, M. Wnt/β-catenin signaling is stimulated by α-melanocyte-stimulating hormone in melanoma and melanocyte cells: Implication in cell differentiation. *Pigment. Cell Melanoma Res.* **2011**, *24*, 309–325. [CrossRef] [PubMed]

25. Marshall, C.J. Specificity of receptor tyrosine kinase signaling: Transient versus sustained extracellular signal-regulated kinase activation. *Cell* **1995**, *80*, 179–185. [CrossRef]
26. Kim, J.H.; Hong, A.R.; Kim, Y.H.; Yoo, H.; Kang, S.W.; Chang, S.E.; Song, Y. JNK suppresses melanogenesis by interfering with CREB-regulated transcription coactivator 3-dependent MITF expression. *Theranostics* **2020**, *10*, 4017. [CrossRef]
27. Ancans, J.; Tobin, D.J.; Hoogduijn, M.J.; Smit, N.P.; Wakamatsu, K.; Thody, A.J. Melanosomal pH controls rate of melanogenesis, eumelanin/phaeomelanin ratio and melanosome maturation in melanocytes and melanoma cells. *Exp. Cell Res.* **2001**, *268*, 26–35. [CrossRef]
28. Bourhim, T.; Villareal, M.O.; Gadhi, C.; Hafidi, A.; Isoda, H. Depigmenting effect of argan press-cake extract through the down-regulation of Mitf and melanogenic enzymes expression in B16 murine melanoma cells. *Cytotechnology* **2018**, *70*, 1389–1397. [CrossRef] [PubMed]
29. Roesler, W.J.; Park, E.A.; McFie, P.J. Characterization of CCAAT: Enhancer-binding protein alpha as a cyclic AMP-responsive nuclear regulator. *J. Biol. Chem.* **1998**, *273*, 14950–14957. [CrossRef]
30. Chiaverini, C.; Beuret, L.; Flori, E.; Busca, R.; Abbe, P.; Bille, K.; Bahadoran, P.; Ortonne, J.P.; Bertolotto, C.; Ballotti, R. Microphthalmia-associated transcription factor regulates RAB27A gene expression and controls melanosome transport. *J. Biol. Chem.* **2008**, *283*, 12635–12642. [CrossRef]
31. Schepsky, A.; Bruser, K.; Gunnarsson, G.J.; Goodall, J.; Hallsson, J.H.; Goding, C.R.; Steingrimsson, E.; Hecht, A. The microphthalmia-associated transcription factor Mitf interacts with beta-catenin to determine target gene expression. *Mol. Cell. Biol.* **2006**, *26*, 8914–8927. [CrossRef]
32. Vachtenheim, J.; Borovanský, J. "Transcription physiology" of pigment formation in melanocytes: Central role of MITF. *Exp. Dermatol.* **2010**, *19*, 617–627. [CrossRef]
33. Wang, J.S.; Wang, F.B.; Zhang, Q.G.; Shen, Z.Z.; Shao, Z.M. Enhanced expression of Rab27A gene by breast cancer cells promoting invasiveness and the metastasis potential by secretion of insulin-like growth factor-II. *Mol. Cancer Res.* **2008**, *6*, 372–382. [CrossRef]
34. Bobrie, A.; Krumeich, S.; Reyal, F.; Recchi, C.; Moita, L.F.; Seabra, M.C.; Ostrowski, M.; Théry, C. Rab27a supports exosome-dependent and -independent mechanisms that modify the tumor microenvironment and can promote tumor progression. *Cancer Res.* **2012**, *72*, 4920–4930. [CrossRef] [PubMed]
35. Feng, F.; Jiang, Y.; Lu, H.; Lu, X.; Wang, S.; Wang, L.; Wei, M.; Lu, W.; Du, Z.; Ye, Z.; et al. Rab27A mediated by NF-κB promotes the stemness of colon cancer cells via up-regulation of cytokine secretion. *Oncotarget* **2016**, *7*, 63342–63351. [CrossRef]
36. Wang, Q.; Ni, Q.; Wang, X.; Zhu, H.; Wang, Z.; Huang, J. High expression of RAB27A and TP53 in pancreatic cancer predicts poor survival. *Med. Oncol.* **2015**, *32*, 372. [CrossRef] [PubMed]
37. Cheli, Y.; Luciani, F.; Khaled, M.; Beuret, L.; Bille, K.; Gounon, P.; Ortonne, J.P.; Bertolotto, C.; Ballotti, R. α-MSH and Cyclic AMP elevating agents control melanosome pH through a protein kinase A-independent mechanism. *J. Biol. Chem.* **2009**, *284*, 18699–18706. [CrossRef]
38. Fuller, B.B.; Spaulding, D.T.; Smith, D.R. Regulation of the catalytic activity of preexisting tyrosinase in black and caucasian human melanocyte cell cultures. *Exp. Cell. Res.* **2001**, *262*, 197–208. [CrossRef]
39. Chintala, S.; Li, W.; Lamoreux, M.L.; Ito, S.; Wakamatsu, K.; Sviderskaya, E.V.; Bennett, D.C.; Park, Y.M.; Gahl, W.A.; Huizing, M.; et al. Slc7a11 gene controls production of pheomelanin pigment and proliferation of cultured cells. *Proc. Natl. Acad. Sci. USA* **2005**, *102*, 10964–10969. [CrossRef] [PubMed]
40. Han, J.; Kraft, P.; Nan, H.; Guo, Q.; Chen, C.; Qureshi, A.; Hankinson, S.E.; Hu, F.B.; Duffy, D.L.; Zhao, Z.Z.; et al. Genome-wide association study identifies novel alleles associated with hair color and skin pigmentation. *PLoS Genet.* **2008**, *4*, 1000074. [CrossRef]
41. Halaban, R.; Pomerantz, S.H.; Marshall, S.; Lambert, D.T.; Lerner, A.B. Regulation of tyrosinase in human melanocytes grown in culture. *J. Cell Biol.* **1983**, *97*, 480–488. [CrossRef] [PubMed]
42. Shallreuter, K.U.; Kothari, S.; Chavan, B.; Spencer, J.D. Regulation of melanogenesis—controversies and new concepts. *Exp. Dermatol.* **2008**, *17*, 395–404. [CrossRef]
43. Migas, P.; Krauze-Baranowska, M. The significance of arbutin and its derivatives in therapy and cosmetics. *Phytochem. Lett.* **2015**, *13*, 35–40. [CrossRef]
44. Zhou, H.; Zhao, J.; Li, A.; Reetz, M.T. Chemical and Biocatalytic Routes to Arbutin. *Molecules* **2019**, *24*, 3303. [CrossRef]
45. Englaro, W.; Bertolotto, C.; Buscà, R.; Brunet, A.; Pagès, G.; Ortonne, J.P.; Ballotti, R. Inhibition of the mitogen-activated protein kinase pathway triggers B16 melanoma cell differentiation. *J. Biol. Chem.* **1998**, *273*, 9966–9970. [CrossRef] [PubMed]
46. Sun, L.; Pan, S.; Yang, Y.; Sun, J.; Liang, D.; Wang, X.; Xie, X.; Hu, J. Toll-like receptor 9 regulates melanogenesis through NF-κB activation. *Exp. Biol. Med.* **2016**, *241*, 1497–1504. [CrossRef]
47. Kim, K.H.; Choi, H.; Kim, H.J.; Lee, T.R. TNFSF14 inhibits melanogenesis via NF-kB signaling in melanocytes. *Cytokine* **2018**, *110*, 126–130. [CrossRef]
48. Ahn, K.S.; Moon, K.Y.; Lee, J.; Kim, Y.S. Downregulation of NF-κB activation in human keratinocytes by melanogenic inhibitors. *J. Dermatol. Sci.* **2003**, *31*, 193–201. [CrossRef]
49. Malinin, N.L.; Boldin, M.P.; Kovalenko, A.V.; Wallach, D. MAP3K-related kinase involved in NFkappaB induction by TNF, CD95 and IL-1. *Nature* **1997**, *385*, 540–544. [CrossRef] [PubMed]
50. Shi, J.H.; Sun, S.C. Tumor necrosis factor receptor-associated factor regulation of nuclear factor κB and mitogen-activated protein kinase pathways. *Front. Immunol.* **2018**, *9*, 1849. [CrossRef]

51. Villareal, M.O.; Kume, S.; Bourhim, T.; Bakhtaoui, F.Z.; Kashiwagi, K.; Han, J.; Gadhi, C.; Isoda, H. Activation of MITF by argan oil leads to the inhibition of the tyrosinase and dopachrome tautomerase expressions in B16 murine melanoma cells. *Evid. Based Complement. Alternat. Med.* **2013**, *2013*, 340107. [CrossRef] [PubMed]

52. Makbal, R.; Villareal, O.M.; Gadhi, C.; Hafidi, A.; Isoda, H. *Argania spinosa* fruit shell extract-induced melanogenesis via cAMP signaling pathway activation. *Int. J. Mol. Sci.* **2020**, *21*, 2539. [CrossRef] [PubMed]

53. Bourhim, T.; Villareal, M.O.; Couderc, F.; Hafidi, A.; Isoda, H.; Gadhi, C. Melanogenesis promoting effect, antioxidant activity, and UPLC-ESI-HRMS characterization of phenolic compounds of argan leaves extract. *Molecules* **2021**, *26*, 371. [CrossRef] [PubMed]

nutrients

Article

Maslinic Acid Attenuates Denervation-Induced Loss of Skeletal Muscle Mass and Strength

Yuki Yamauchi [1,2], Farhana Ferdousi [3,4,5], Satoshi Fukumitsu [1,2,3] and Hiroko Isoda [1,3,4,5,6,*]

[1] Tsukuba Life Science Innovation Program (T-LSI), University of Tsukuba, 1-1-1 Tennodai, Tsukuba 305-8577, Japan; s2030297@s.tsukuba.ac.jp (Y.Y.); fukumitsu.satoshi.gn@u.tsukuba.ac.jp (S.F.)

[2] Central Research Laboratory Innovation Center, Nippn Corporation, 5-1-3 Midorigaoka, Atsugi 243-0041, Japan

[3] Alliance for Research on the Mediterranean and North Africa (ARENA), University of Tsukuba, 1-1-1 Tennodai, Tsukuba 305-8572, Japan; ferdousi.farhana.fn@u.tsukuba.ac.jp

[4] AIST-University of Tsukuba Open Innovation Laboratory for Food and Medicinal Resource Engineering (FoodMed-OIL), University of Tsukuba, Tsukuba 305-8572, Japan

[5] Faculty of Life and Environmental Sciences, University of Tsukuba, Tsukuba 305-8575, Japan

[6] R&D Center for Tailor-Made QOL, University of Tsukuba, Tsukuba 305-8550, Japan

* Correspondence: isoda.hiroko.ga@u.tsukuba.ac.jp; Tel.: +81-298-53-5775

Abstract: Maslinic acid (MA) is a pentacyclic triterpene abundant in olive peels. MA reportedly increases skeletal muscle mass and strength in older adults; however, the underlying mechanism is unknown. This study aimed to investigate the effects of MA on denervated muscle atrophy and strength and to explore the underlying molecular mechanism. Mice were fed either a control diet or a 0.27% MA diet. One week after intervention, the sciatic nerves of both legs were cut to induce muscle atrophy. Mice were examined 14 days after denervation. MA prevented the denervation-induced reduction in gastrocnemius muscle mass and skeletal muscle strength. Microarray gene expression profiling in gastrocnemius muscle demonstrated several potential mechanisms for muscle maintenance. Gene set enrichment analysis (GSEA) revealed different enriched biological processes, such as myogenesis, PI3/AKT/mTOR signaling, TNFα signaling via NF-κB, and TGF-β signaling in MA-treated mice. In addition, qPCR data showed that MA induced *Igf1* expression and suppressed the expressions of *Atrogin-1*, *Murf1* and *Tgfb*. Altogether, our results suggest the potential of MA as a new therapeutic and preventive dietary ingredient for muscular atrophy and strength.

Keywords: maslinic acid; muscle atrophy; muscle strength; denervation; olive peel

Citation: Yamauchi, Y.; Ferdousi, F.; Fukumitsu, S.; Isoda, H. Maslinic Acid Attenuates Denervation-Induced Loss of Skeletal Muscle Mass and Strength. *Nutrients* **2021**, *13*, 2950. https://doi.org/10.3390/nu13092950

Academic Editor: Yoshinori Katakura

Received: 30 July 2021
Accepted: 23 August 2021
Published: 25 August 2021

Publisher's Note: MDPI stays neutral with regard to jurisdictional claims in published maps and institutional affiliations.

Copyright: © 2021 by the authors. Licensee MDPI, Basel, Switzerland. This article is an open access article distributed under the terms and conditions of the Creative Commons Attribution (CC BY) license (https://creativecommons.org/licenses/by/4.0/).

1. Introduction

Skeletal muscle makes up about 40% of body weight and is crucial for maintaining an ideal quality of life and achieving superior athletic performance [1]. Skeletal muscle constitutes a highly plastic tissue that easily adapts to environmental and physiological changing conditions [2]. Muscle atrophy is caused by inactivity, sarcopenia, and various neuromuscular diseases. Sarcopenia is a condition in which muscle mass and strength decrease due to the gradual decline in skeletal muscle content with aging, which reduces overall muscle quality [3]. It is a major health problem among the elderly and increases the risk of disability, falls, fall-related injuries, hospitalization, dependence, and mortality [4].

The constituent proteins of skeletal myocytes are constantly being degraded and synthesized. Muscle atrophy is the loss of muscle mass due to the imbalance between protein degradation and synthesis [5]. Insulin-like growth factor 1 (IGF-1) induces protein synthesis in skeletal muscle and induces muscle hypertrophy through modulating PI3K/AKT/mTOR signaling. Additionally, IGF-1 prevents skeletal muscle atrophy by inhibiting the phosphorylation of forkhead box O (FOXO) downstream of AKT. Thus, two different AKT signaling pathways, with IGF-1 as a starting point, are responsible for the dynamic balance of muscle protein synthesis and degradation [6,7].

Transforming growth factor beta (TGF-β) is considered the vital factor that modulates the cross-talk between these two AKT pathways. TGF-β overexpression stimulates muscle atrophy by downregulating the phosphorylation of AKT [8]. The cellular mechanisms for muscle atrophy are indeed complex. Activation of the nuclear factor-kappa B (NF-κB) pathway and inflammatory cytokines also mediates muscle atrophy. NF-kB stimulates protein degradation by enhancing the expressions of muscle RING finger 1 (MuRF1) and muscle atrophy F-box protein 32 (Atrogin-1); both are ubiquitin ligases specific to skeletal muscle and act downstream of the FOXO family transcription factor [9].

The Mediterranean diet is known to provide a variety of health benefits; olives and their oils are considered essential ingredients in this diet [10–12]. The residues from olive oil extraction contain a lot of pentacyclic triterpenes, including maslinic acid (MA; 2α,3β-dihydroxyolean-12-en-28-oic acid). MA has potent anti-inflammatory effects that inhibit the action of inflammatory cytokines. It also promotes the formation of synovial membrane tissue and repairs arthritis-induced cartilage damage via the NF-κB canonical signaling pathway [13,14]. A recent clinical study showed that MA supplements improved knee muscle strength and inhibited knee joint inflammation in older women with knee osteoarthritis who participated in a whole-body vibration training program [15]. Furthermore, another clinical trial showed that the combination of resistance training and MA intake effectively increased skeletal muscle mass in the elderly [16]. However, the comprehensive molecular mechanisms underlying the preventive effects of MA on muscle atrophy have not yet been investigated.

Muscle atrophy is caused by several factors, including an imbalance between protein synthesis and breakdown, lack of physical activity, aging, and denervation. While there are several overlapping signatures in the transcriptomic responses to different atrophic stimuli, each of the stimuli, including the denervation-induced atrophy, may also display unique transcriptional signatures [17–19]. The present study aimed to evaluate the effects of MA on denervation-induced loss of muscle mass and strength in a mice model and examine the underlying mechanism of the beneficial effects of MA through whole-genome microarray analysis.

2. Materials and Methods

2.1. Preparation of the MA Fraction in Olive Pomace

The MA fraction in olive pomace was obtained by solvent extraction with 90% (*v/v*) ethanol for 3 h at 85 °C. Aqueous ethanol extracts were evaporated and then dissolved in chloroform. The dissolved extract was retained in a silica column (Chromatorex FL100D; Fuji Silysia Chemical Ltd., Aichi, Japan) with chloroform. Subsequently, the purified fraction (MA purity: 97.9%) was eluted with a combination of chloroform and methanol (49:1, *v/v*) solution. There is no detectable peak of other triterpenes such as ursolic acid and oleanolic acid in the purified fraction. MA was identified using the Shimadzu Nexera UHPLC system (Shimadzu Co., Kyoto, Japan) equipped with an Infinitylab poroshell 120 column (4.6 × 100 mm, 2.7 μm; Agilent Technologies Japan Ltd., Tokyo, Japan) set at 30 °C. The mobile phase consisted of acetonitrile/methanol/water/phosphoric acid (500:400:100:0.5, *v/v/v/v*) with a flow rate of 1 mL/min. MA was detected using a UV detector (Shimadzu) at 210 nm. MA standard reagent was purchased from Funakoshi Co., Ltd. (Tokyo, Japan).

For the mouse experimental group, we prepared the diet containing 0.27% MA. The concentration of MA intake was determined following the previous reports [20].

2.2. Animal Experimental Procedures

All animal experimental procedures were approved by the Ethical Committee of NIPPN Corporation (permission number: 2020-6). Seven-week-old male Slc: ICR mice (*n* = 30) were purchased from Japan SLC Inc. (Shizuoka, Japan) and acclimated under conventional conditions (room temperature at 24 °C ± 1 °C, humidity 50 ± 10%, a 12 h light-dark cycle). The mice were fed the AIN-93G diet (Funabashi Farm Ltd., Chiba, Japan) and were provided the free access to the diet and water. After acclimation for one week, the mice were randomly assigned to two groups: control group that were fed the usual diet ($n = 15$; 0, 7, 14 days, $n = 3, 6, 6$, respectively) and MA group that were fed the usual diet containing 0.27% MA ($n = 15$; 0, 7, 14 days, $n = 3, 6, 6$, respectively). After one week, the sciatic nerves of both legs were severed. Then, a 5 mm section was removed under anesthesia. During the process of disuse muscle atrophy, we continued to feed the mice the control or MA diet until the end of the experiment. Whole grip strength was tested using a grip strength meter (GPM-100; Melquest, Toyama, Japan). Mice were placed on a wire mesh to tightly grip the wire mesh using whole limbs. The tail of each mouse was pulled directly toward the tester with the same force. The test was conducted five times and the mean of the three central points was calculated. On days 0, 7, and 14 after surgery, the mice were anesthetized. The gastrocnemius, plantaris, and soleus muscles were harvested, weighed, quickly frozen in liquid nitrogen, and stored at $-80\,^\circ$C until needed for analyses.

2.3. Total RNA Extraction from Gastrocnemius Muscle

We used Isogen reagent (Nippon Gene Co., Ltd., Tokyo, Japan) to isolate the total RNA of the gastrocnemius muscle. RNA samples were then purified using the Qiagen's RNeasy Mini Kit (Qiagen K.K., Tokyo, Japan). RNA quantity and quality were determined with NanoDrop 2000 spectrophotometer (Thermo Fisher Scientific, Tokyo, Japan).

2.4. DNA Microarray

A whole-genome microarray analysis was conducted using the Affymetrix's Gene-Atlas® System following the user manual (Affymetrix Inc., Santa Clara, CA, USA). Firstly, 100 ng of total RNA samples were prepared from the gastrocnemius muscles of both control and MA-treated mice groups (denervated). RNA quality and quantity were determined using the NanoDrop 2000 spectrophotometer (Thermo SCIENTIFIC, Wilmington, DE, USA). Then, the amplified and biotin-labeled cRNA samples (complementary RNAs) were generated from poly(A) RNAs in the total RNA samples following a reverse transcription priming method using the GeneChip 3′ IVT PLUS Reagent Kit (902415, Affymetrix Inc., Santa Clara, CA, USA). Next, the fragmented and labeled cRNA samples were prepared for hybridization using the GeneAtlas® Hybridization, Wash, and Stain Kit for 3′ IVT Array Strips (901531). We used Affymetrix's Mouse Genome 430 PM array strips (901570) in our study. The samples were placed in the array strips and were hybridized on the GeneAtlas® Hybridization Station for 16 h at 45 °C. Finally, washing, staining, and scanning of the hybridized arrays were performed on the GeneAtlas® Fluidics Station and the GeneAtlas® Imaging Station (Affymetrix).

For raw data processing, we used Affymetrix Expression Console™ Software. The Robust Multi-array Analysis (RMA) algorithm was employed for gene-level normalization and signal summarization (http://www.affymetrix.com, accessed on 28 August 2020). Next, Transcriptome Analysis Console (TAC) software version 4 (Thermo Fisher Scientific Inc.) was used for subsequent differential expression analysis. Differentially expressed genes (DEGs) were identified as the genes with a fold change over 1.2 in linear space and a p-value less than 0.05 in one-way between-subjects Analysis of Variance (ANOVA). To determine the Hallmark gene sets, we used the Molecular Signature Database (MSigDB) of Gene Set Enrichment Analysis (GSEA) web tool (https://software.broadinstitute.org/gsea/index.jsp, accessed on 10 June 2021) [21,22]. An online data mining tool, Database for Annotation, Visualization and Integrated Discovery (DAVID) ver. 6.8 was used for gene ontology (GO) analysis to identify the significantly enriched biological processes by the DEGs [23]. Finally, heat maps were created using an online data visualization software Morpheus (https://software.broadinstitute.org/morpheus, accessed on 21 July 2021). The microarray data, including the CEL and CHP files, have been deposited to the NCBI Gene Expression Omnibus database and are available under the accession number

GSE181031 (https://www.ncbi.nlm.nih.gov/geo/query/acc.cgi?acc=GSE181031, accessed on 29 July 2021).

2.5. Real-Time RT-PCR (qRT-PCR)

Reverse transcription of purified total RNA samples (1.0 µg) was performed using the PrimeScript RT Reagent Kit (RR037A, TaKaRa Bio Inc., Shiga, Japan). Amplification of cDNAs was performed using SYBR Premix EX Taq (RR041A, TaKaRa Bio) on a Thermal Cycler Dice TP950 (TaKaRa Bio). The thermal cycling conditions were 95 °C for 30 s, and then 40 cycles of 95 °C for 5 s, followed by 60 °C for 30 s. The relative amount of each gene transcript was normalized to that of TATA-binding protein (TBP). In real-time PCR analysis, the following primers were used: *Igf1* (primer sequences: forward, TGCTCTTCAGTTCGTGTG; reverse, ACATCTCCAGTCTCCTCAG), *Atrogin-1(Fbxo32)* (primer sequences: forward, ACTTCTCGACTGCCATCCTG; reverse,), *Murf1* (primer sequences: forward, GGGCCATTGACTTTGGGACA; reverse, TGGTGTTCTTCTTTAC-CCTCTGTG), *Tgfb* (primer sequences: forward, GAGACGGAATACAGGGCTTTC; reverse, TCTCTGTGGAGCTGAAGCAAT), *Tbp* (primer sequences: forward, ACCTTAT-GCTCAGGGCTTGG; reverse, GGTGTTCTGAATAGGCTGTGGA).

2.6. Statistical Analysis

The results are presented in terms of the mean ± standard error of the mean (SEM). In vivo experimental data were statistically compared using ANOVA and Welch's *t*-test. SPSS Statistics ver. 25 for Windows (IBM Inc., Tokyo, Japan) was used for statistical analyses. A significant difference was defined as $p < 0.05$.

3. Results

3.1. Effects of MA on Denervation-Induced Skeletal Muscle Atrophy

After one week of ingestion of MA diet, muscle atrophy was induced by denervation. There were no significant differences in the average daily intake of food (Figure 1A) and body weight (BW) (Figure 1B) between the two groups of mice. The gastrocnemius muscle wet weight in the MA group (2.29 ± 0.11 mg/g BW) was significantly higher than that of the control group (1.95 ± 0.09 mg/g BW) on day 14 after denervation (Figure 1C, $p = 0.045$). MA intake also attenuated the denervation-induced reduction in plantaris muscle weight on day 7 after denervation (Figure 1D, 0.43 ± 0.01 mg/g BW and 0.36 ± 0.02 mg/g BW in MA and control groups, respectively, $p = 0.036$). The soleus muscle wet weight tended to be higher in the MA group (0.17 ± 0.01 mg/g BW and 0.17 ± 0.01 mg/g BW on day 7 and 14, respectively) compared to that of the control group (0.16 ± 0.01 mg/g BW and 0.15 ± 0.01 mg/g BW on day 7 and 14, respectively), but the difference was not significant (Figure 1E). There was no significant difference in grip strength between the groups at the start of the feeding period; however, after denervation, grip strength in the MA group was significantly higher than that of the control group on both day 7 (5.29 ± 0.57 mg/g BW and 3.68 ± 0.17 mg/g BW in MA and control groups, respectively, $p = 0.036$) and day 14 (4.54 ± 0.23 mg/g BW and 3.82 ± 0.18 mg/g BW in MA and control groups, respectively, $p = 0.035$) (Figure 1F). Overall, these results indicate that MA ingestion attenuated denervation-induced skeletal muscle loss and increased skeletal muscle strength.

3.2. Effect of MA on Gene Expression in Denervation-Induced Gastrocnemius Muscle

We analyzed the microarray experiments of gene expression profiling to clarify the effects of MA treatment on the gene expression pattern in the gastrocnemius muscle on day 14 post denervation. A total of 45,078 probe sets could be identified. There were 3172 unique DEGs in the MA group (fold change > ± 1.2 and *p*-value < 0.05 vs. control). The volcano plot is showing the magnitude of significance and fold changes of the DEGs (Figure 2A). Up and downregulated DEGs are shown in red and green dots, respectively.

A total of 1491 DEGs were upregulated, whereas 1681 DEGs were downregulated in the MA group when compared with the control group (Figure 2B).

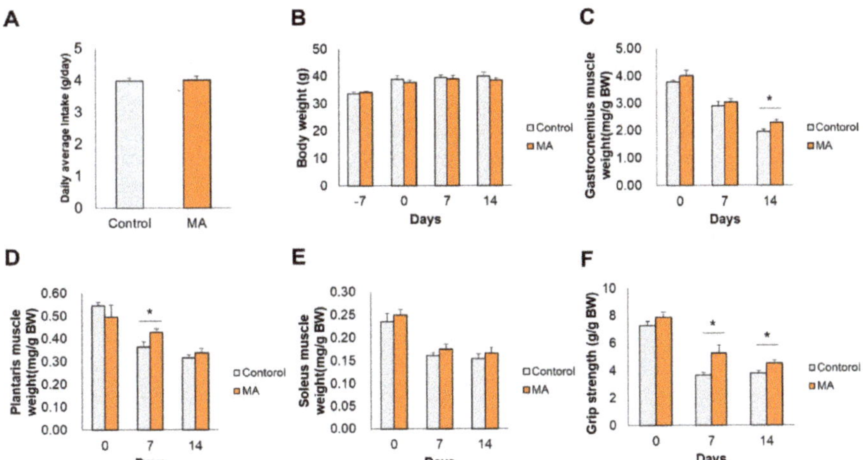

Figure 1. Maslinic acid (MA) prevented skeletal muscle atrophy following denervation. (**A**) Food intake, (**B**) body weight, (**C**) gastrocnemius muscle weight, (**D**) plantaris muscle weight, (**E**) soleus muscle weight, and (**F**) grip strength. Values are shown as mean $\pm$ SEM (n = 3-15), * $p < 0.05$ vs. control.

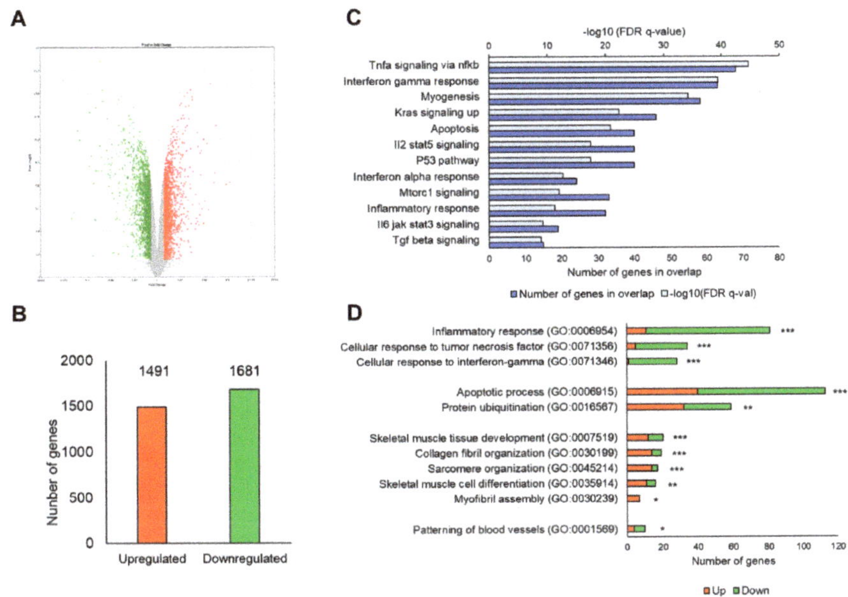

Figure 2. Microarray gene expression profile in gastrocnemius muscle 14 days after denervation. (**A**) Volcano plot showing the DEGs between MA group and control group. The Y-axis corresponds to -log10 ANOVA *p*-value and the X-axis displays the fold change. Up and downregulated DEGs are presented in the red and green dots, respectively. (**B**) Column graph displaying the number of DEGs. (**C**) Bar graph showing the significantly enriched Hallmark gene sets by the DEGs (MsigDB of GSEA). Significance was considered at false discovery rate (FDR) q-value < 0.05. (**D**) Bar graph showing the significantly enriched biological processes by the DEGs (from DAVID), with *p*-value as * $p < 0.05$, ** $p < 0.01$, *** $p < 0.001$.

3.3. Effects of MA on Biological Functions in Denervation-Induced Skeletal Muscle Atrophy Model

Figure 2C shows the significantly enriched Hallmark gene sets in the MA group when compared with the control group. By definition, Hallmark gene sets are 'the gene sets that summarize and represent specific well-defined biological states or processes and display coherent expression' (MSigDB of GSEA). Hallmark gene sets are preferable because they are considered to have relatively less noise and redundancy. We found several significantly enriched Hallmark gene sets in the MA group compared to the control group, such as muscle differentiation-related mTORC1 signaling and KRAS signaling, and myogenesis and muscle degradation-associated TGF-β signaling, inflammatory response, TNFα signaling via NF-κB, IL6/JAK/STAT3 as well as IL2/STAT5 signaling pathways, interferon α response, interferon γ response, P53 pathway, and apoptosis.

3.4. Effects of MA on Biological Processes in Denervation-Induced Skeletal Muscle Atrophy Model

GO analysis shows that several biological processes (from DAVID) related to muscle differentiation, protein degradation, and inflammatory response were regulated in the MA group (Figure 2D). The overlapped genes were highly associated with skeletal muscle tissue development (GO:0007519), collagen fibril organization (GO:0030199), sarcomere organization (GO:0045214), skeletal muscle cell differentiation (GO:0035914), myofibril assembly (GO:0030239), apoptotic process (GO:0006915), protein ubiquitination (GO:0016567), inflammatory response (GO:0006954), cellular response to tumor necrosis factor (GO:0071356), cellular response to interferon-gamma (GO:0071346), and the patterning of blood vessels (GO:0001569) in the MA group compared to the control group.

Additionally, KEGG pathway analysis showed that FoxO signaling, NF-κB signaling and TGF-β signaling pathways were significantly enriched in the MA group compared to the control group (Figure 3A).

3.5. Effect of MA on Atrophy-Related Gene Expression

Next, we examined the effects of MA on individual gene expressions of the significant pathways. Heat maps are showing the row z-scores of genes related to FoxO signaling (Figure 3B), TGF-β signaling (Figure 3C), and inflammatory response (Figure 3D) from microarray analysis. We found that MA downregulated the expressions of F-box protein 32 (*Fbxo32 / Atrogin-1*), forkhead box O1 (*Foxo1*), Kruppel-like factor 1 (*Klf1*), growth arrest and DNA-damage-inducible 45 (*Gadd45*), and phosphatase and tensin homolog (*Pten*), whereas it upregulated the expressions of *Igf1*, superoxide dismutase 2, mitochondrial (*Sod2*), and catalase (*Cat*), all of which are implicated in FoxO signaling. We also found that several inflammation-related genes were highly expressed in denervation-induced control mice, while MA downregulated their expressions, such as thrombospondin 1 (*Thbs1*), transforming growth factor beta 2 (*Tgfb2*), transforming growth factor beta receptor I (*Tgfbr1*), nuclear factor of kappa light polypeptide gene enhancer (*Nfkb2*), chemokine (C-C motif) ligand 5 (*Ccl5*), and chemokine (C-X-C motif) ligand 10 (*Cxcl10*), which are implicated in TGF-β signaling or inflammatory response (Figure 3C,D). We also found that the expression of SMAD-specific E3 ubiquitin protein ligase 2 (*Smurf2*), as a negative regulator of TGF-β signaling, was upregulated (Figure 3C).

Microarray results were further confirmed by qPCR quantitative analysis of the representative gene of signaling pathways. We confirmed that MA induced the expression of *Igf1* on day 14 after denervation ($p = 0.049$) (Figure 3D). On the other hand, MA suppressed the expressions of catabolic genes such as *Atrogin-1* ($p = 0.021$ and $p = 0.010$ on day 7 and day 14 after denervation, respectively) and *Murf1* ($p = 0.017$ on day 7 after denervation). In addition, MA significantly suppressed the *Tgfb* expression on day 14 after denervation ($p = 0.032$), which is upstream of the catabolic genes.

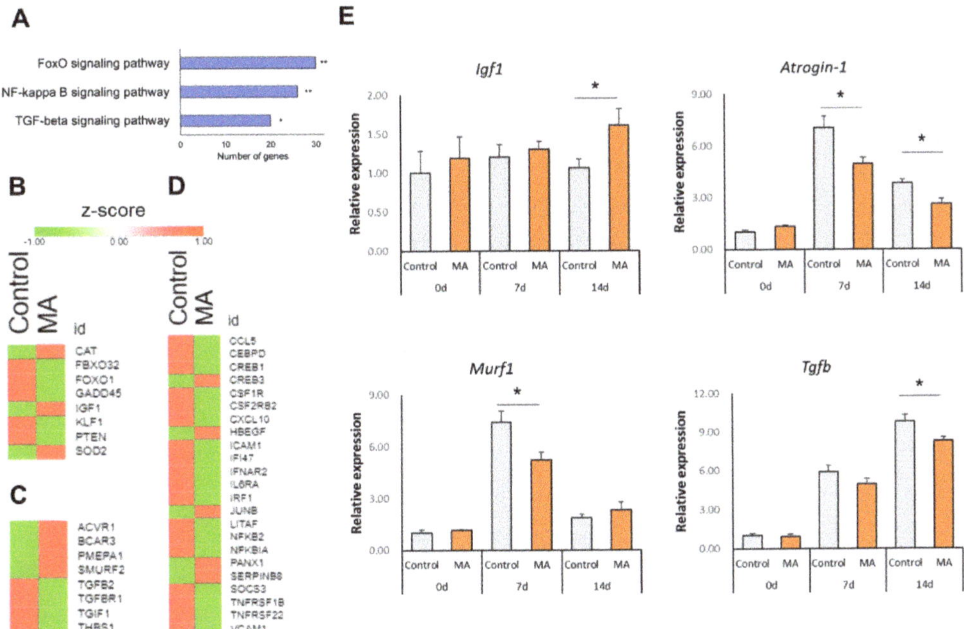

Figure 3. Changes in muscle protein degradation-related factors. (**A**) Bar graph showing significantly enriched selected KEGG pathways. Heat maps showing row z-score of DEGs involved in (**B**) FoxO signaling, (**C**) TGF-β signaling, and inflammatory response (**D**). (**E**) Relative mRNA expression levels of *Igf1*, *Atrogin-1*, *Murf1*, and *Tgfb* by qPCR analysis. Data are shown as mean ± SEM (n = 3–6), * $p < 0.05$, ** $p < 0.01$ vs. control.

4. Discussion

In this study, we have shown that daily ingestion of an MA-supplemented diet could prevent denervation-induced loss of skeletal muscle mass and strength in a mouse model, demonstrating its efficacy against muscle atrophy.

The sciatic-nerve transection model is commonly employed to induce denervation [24]. In the present study, we confirmed that dietary intake of MA effectively prevented the loss of muscle weight and strength on day 14 after sciatic nerve denervation (Figure 1C–E). Investigators previously reported that oral intake of MA in combination with resistance training maintained skeletal muscle mass in the older people [16], consistent with our findings in vivo. It has also been proposed that MA suppresses muscular atrophy through anti-inflammatory effects [13,14]. In accordance with these previous reports, we also found that MA alleviated denervation-induced inflammation response.

In fact, denervation-induced skeletal muscle atrophy is associated with a complex list of physiological and biochemical changes. The gene expression profiling of denervated muscle at different time points revealed four distinct transcriptional stages designated as "oxidative stress stage", "inflammation stage", "atrophy stage" and "atrophic fibrosis stage", respectively [25]. Therefore, to understand MA's underlying molecular mechanisms in preventing denervation-induced muscle atrophy, we conducted a whole-genome microarray analysis of the gastrocnemius muscle on day 14 post denervation. Multiple potential mechanisms of MA for muscle maintenance were identified. GSEA revealed biological processes such as myogenesis, PI3/AKT/mTOR signaling, TNFα signaling via NF-κB, and TGF-β signaling, which were significantly enriched in MA-treated mice compared to non-treated control mice (Figure 2C). Inflammation is considered to be a central factor in the physiological response to muscle atrophy. As originally hypothesized, MA suppressed inflammation caused by denervation, as evidenced by the marked variation in the gene sets related to inflammatory responses, such as Hallmark sets "TNFα signaling via NF-κB", "Interferon Gamma Response", "TGF-β signaling" as well as GOs such as cellular response to TNFα and IFNγ(Figure 2C,D). Reportedly, TGF-β could directly induce muscle fiber atrophy. TGF-β1 overexpression in skeletal muscles results in muscle fiber atrophy and fibrosis and increases the expression of MuRF1 [26]. Furthermore, TGF-β activity in cancer cachexia-induced muscle atrophy confirmed its contribution to muscle weakness [27]. In this study, we found that MA suppressed the expression of *Thbs1*, a major activator of TGF-β signaling (Figure 3B). In addition, the mRNA expression level of *Tgfb* in gastrocnemius muscle was significantly suppressed by MA intake compared to controls (Figure 3D). A previous study reported the downregulation of TGF-β by MA treatment in a colorectal cancer mice model [28]; however, the effect of MA on TGF-β regulation has never been studied in muscle atrophy models before. Thus, our finding suggests potential preventative and therapeutic applications of MA in muscle atrophy through the downregulation of the TGF-β signaling pathway (Figures 2C and 3D).

Protein metabolism plays an important role in the regulation of skeletal muscle mass. Skeletal muscle mass is increased when protein synthesis exceeds protein degradation, and, conversely, skeletal muscle mass is decreased when protein degradation exceeds protein synthesis. It has been reported that MA promotes protein synthesis and muscle mass gain in rainbow trout [29], but the detailed molecular mechanism was not clarified. In this regard, another member of pentacyclic triterpenoid, namely, ursolic acid, has also been reported to promote muscle hypertrophy [20]. From microarray data, we found that MA intake may have promoted myogenesis and regulated PI3/AKT/mTOR signaling, thereby playing a role in muscle maintenance (Figure 2C,D). Additionally, qPCR analysis in the gastrocnemius muscle on day 14 after denervation showed that MA significantly induced the expression of *Igf1*, a key anabolic growth factor known for regulating muscle hypertrophy through stimulating PI3K/Akt signaling [30]. Interestingly, Hennebry et al. showed that IGF1 could stimulate more significant muscle hypertrophy in mice in the absence of myostatin [31], a secreted member of the TGF-β superfamily. Therefore, simultaneous upregulation of *Igf1* and downregulation of *Tgfb* in the denervated muscle by MA might have contributed to muscle maintenance. Moreover, IGF1 suppresses protein degradation and mRNA expressions of *Atrogin-1* and *Murf1* [32,33]. The downregulation of PTEN, a negative regulator of p-Akt, increases Akt signaling, suppresses FoxO1 and ubiquitin E3 ligases, and consequently suppresses muscle degradation [34]. In this study, we found that MA decreased the mRNAs of *Pten*, *Atrogin-1* and *Murf1* in denervated mice and may thereby suppressed muscle atrophy (Figure 3B,E).

It is also worth noting that MA decreased the expression of *Gadd45* in microarray experiments (Figure 3B), as previous microarray studies reported induced Gadd45a mRNA expression in atrophying skeletal muscles of rodents, pigs, and humans [35–39]. *Gadd45a* is induced in response to three distinct atrophic stimuli: fasting, muscle fixation, and muscle denervation [36]. However, interestingly, Gadd45-induced transcriptomic changes showed a strong overlap (over 40%) with the muscle denervation signature and even mimicked the classical ultrastructural changes observed in early muscle denervation [36]. A recent study also confirmed very high expression of *Gadd45* in denervated muscle; nevertheless, in contrast to Ebert et al., the authors proposed that *Gadd45* induction is mediated by a protective negative feedback response to denervation [40]. Therefore, the mechanism by which MA treatment regulates denervation induced Gadd45 expression is worth further exploration.

Interestingly, we also found that the grip strength in the MA group mice was significantly higher than that of the control group on day 7 post denervation, even though there was no significant difference in the gastrocnemius muscle mass on day 7 (Figure 1C,F). It has been reported that the capillary network is also altered in disuse atrophic muscles [41]. Moreover, muscle strength can be improved, without any increase in muscle mass weight, through a combination of exercise and caffeine intake, which exerts a vasodilatory effect [42]. In this regard, MA intake has been reported to promote nitric oxide production and exert a vasodilatory effect in rats [43]. We found that MA intake affected patterning of blood vessels (GO:0001569) from GO analysis. Hence, dietary MA intake might improve blood flow and increased muscle strength regardless of muscle mass weight (Figure 1C,F).

Previous studies have shown similarities in the aging processes of neuromuscular or motor functions between humans and mice [44–46]. Sarcopenia is studied in mice models because of its short life span, low cost, and relative ease of genetic manipulation [47,48]. Muscle atrophy could be of three types: diffuse deconditioning, such as denervation, microgravity, or natural aging, immobilization, and chronic diseases [49–51]. The present study investigates the effect of MA on denervation-induced muscle atrophy; therefore, further studies on different muscle atrophy mice models are required to validate and clarify the effect of MA on preventing muscle atrophy and sarcopenia. Considering more clinical applications, it would also be interesting to investigate whether muscle recovery could be expected by administering MA without a pre-treatment after denervation. In this study, we confirmed the effects of MA intake on muscle mass and strength. Still, we did not perform a histological evaluation of skeletal muscle fibers and protein expression of skeletal muscle. We have previously reported that MA exerted anti-inflammatory effects through inactivation of NF-κB by inhibiting the phosphorylation of IκB-α, as detected by Western blot analysis in RAW 264.7 cells [13]. Additionally, Kunkel et al. have reported that ursolic acid, a type of triterpene, could promote muscle maintenance through enhancing IGF-1-mediated signaling cascades, as detected in the serum-starved C2C12 myotubes by immunoblot analyses [13,20]. Therefore, the effect of MA treatment on the active protein levels of these selected markers is required to be further examined in the in vivo microenvironment.

In conclusion, dietary supplementation of MA prevents disuse muscle atrophy in denervated mice. Our study is the first to report that MA is likely to regulate muscle protein synthesis and degradation by promoting IGF-1 production and suppressing TGF-β, as well as mediating anti-inflammatory effects (Figure 4). MA treatment especially attenuated the denervation-induced expressions of atrophic genes, such as *Atrogin-1* and *Murf1*. These findings support the potential of the anti-atrophy effect of MA on the sciatic nerve denervation-induced muscle atrophy model. Pentacyclic triterpenes, including MA and its derivatives, are attracting attention as dietary supplements [52]. Regular intake of MA derived from sources such as olive fruit may help prevent sarcopenia and extend healthy life expectancy.

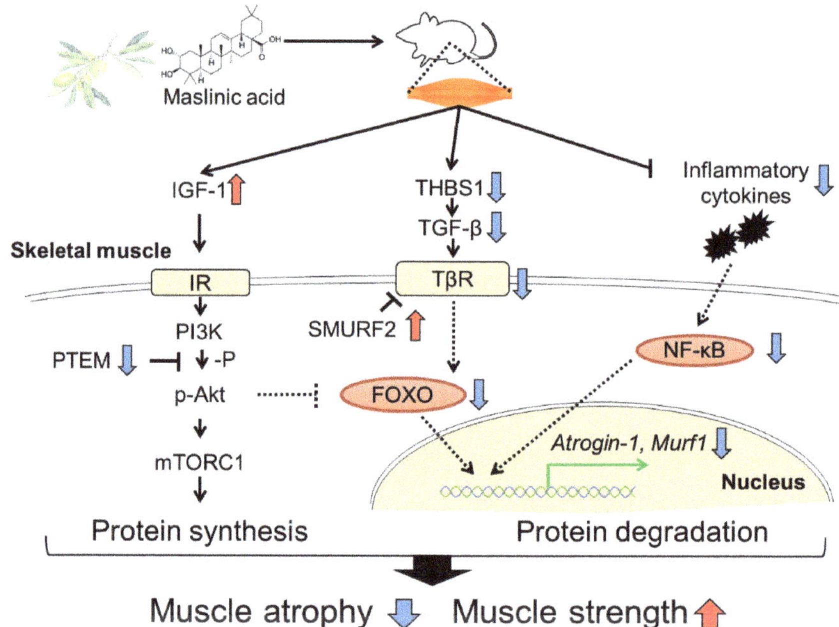

Figure 4. Schematic diagram of the role of preventing muscle atrophy. Dietary MA induces IGF-1, which can regulate the dynamic balance of muscle protein degradation and synthesis. IGF-1 stimulation leads to increased protein synthesis via PI3/AKT/mTOR signaling. In addition, MA acts on anti-inflammation, inhibition of the TGF-β expression, resulting in the regulation of protein degradation, and alleviates muscle atrophy and muscle strength. Abbreviations: IGF-1, insulin-like growth factor 1; IR, insulin-like growth factor receptor; PI3K, phosphatidylinositol 3-kinase; Akt, protein kinase B; PTEM, phosphatase and tensin homolog; mTORC1, mammalian target of rapamycin complex 1; THBS1, thrombospondin 1; LTBP2, latent transforming growth factor-β binding protein 2; TGF-β, transforming growth factor β; TβR, TGF-β receptor type; SMURF2, SMAD-specific E3 ubiquitin protein ligase 2; FoxO, forkhead box O transcription factors; Murf1, muscle ring finger 1; NF-κB, nuclear factor-kappa B.

Author Contributions: Y.Y., F.F., S.F. and H.I.: conceptualization and methodology. Y.Y.: investigation, data curation, formal analysis, visualization, and writing—original draft. F.F., S.F. and H.I.: writing—review and editing. H.I.: supervision, project administration and writing—review and editing. All authors have read and agreed to the published version of the manuscript.

Funding: This research was partially supported by the NIPPN CORPORATION, and Japan Science and Technology Agency (JST), Science and Technology Research Partnership for Sustainable Development (SATREPS, Grant No. JPMJSA1506).

Institutional Review Board Statement: This study was conducted according to the guidelines of Act on Welfare and Management of Animals and Standards relating to the Care and Keeping and Reducing Pain of Laboratory Animals, and approved by the Ethical Committee of NIPPN Corporation (permission number: 2020-6).

Data Availability Statement: The microarray data have been deposited to the NCBI Gene Expression Omnibus database and are available under the accession number GSE181031 (https://www.ncbi. nlm.nih.gov/geo/query/acc.cgi?acc=GSE181031, accessed on July 29, 2021).

Conflicts of Interest: Y.Y. and S.F. are employees of NIPPN CORPORATION. The authors have declared no conflict of interest.

References

1. Tsekoura, M.; Kastrinis, A.; Katsoulaki, M.; Billis, E.; Gliatis, J.; Vlamos, P. Sarcopenia and Its Impact on Quality of Life. *Adv. Exp. Med. Biol.* **2017**, *987*, 213–218. [CrossRef] [PubMed]
2. Stewart, C.; Rittweger, J. Adaptive processes in skeletal muscle: Molecular regulators and genetic influences. *J. Musculoskelet. Neuronal Interact.* **2006**, *6*, 73–86.
3. Evans, W.J. What Is Sarcopenia? *J. Gerontol. Ser. A Boil. Sci. Med. Sci.* **1995**, *50A*, 5–8. [CrossRef]
4. Landi, F.; Liperoti, R.; Fusco, D.; Mastropaolo, S.; Quattrociocchi, D.; Proia, A.; Russo, A.; Bernabei, R.; Onder, G. Prevalence and Risk Factors of Sarcopenia Among Nursing Home Older Residents. *J. Gerontol. Ser. A Boil. Sci. Med. Sci.* **2011**, *67A*, 48–55. [CrossRef] [PubMed]
5. Léger, B.; Cartoni, R.; Praz, M.; Lamon, S.; Dériaz, O.; Crettenand, A.; Gobelet, C.; Rohmer, P.; Konzelmann, M.; Luthi, F.; et al. Akt signalling through GSK-3β, mTOR and Foxo1 is involved in human skeletal muscle hypertrophy and atrophy. *J. Physiol.* **2006**, *576*, 923–933. [CrossRef]
6. Rüegg, M.A.; Glass, D.J. Molecular Mechanisms and Treatment Options for Muscle Wasting Diseases. *Annu. Rev. Pharmacol. Toxicol.* **2011**, *51*, 373–395. [CrossRef] [PubMed]
7. Glass, D.J. Skeletal muscle hypertrophy and atrophy signaling pathways. *Int. J. Biochem. Cell Biol.* **2005**, *37*, 1974–1984. [CrossRef]
8. Lee, S.-J.; McPherron, A.C. Regulation of myostatin activity and muscle growth. *Proc. Natl. Acad. Sci. USA* **2001**, *98*, 9306–9311. [CrossRef]
9. Li, Y.; Schwartz, R.J.; Waddell, I.D.; Holloway, B.R.; Reid, M.B. Skeletal muscle myocytes undergo protein loss and reactive oxygen-mediated NF-κB activation in response to tumor necrosis factorα. *FASEB J.* **1998**, *12*, 871–880. [CrossRef]
10. Willett, W.C.; Sacks, F.; Trichopoulou, A.; Drescher, G.; Ferro-Luzzi, A.; Helsing, E.; Trichopoulos, D. Mediterranean diet pyramid: A cultural model for healthy eating. *Am. J. Clin. Nutr.* **1995**, *61*, 1402S–1406S. [CrossRef]
11. Rosillo, M.A.; Sánchez-Hidalgo, M.; Sanchez-Fidalgo, S.; Aparicio-Soto, M.; Villegas, I.; Alarcón-De-La-Lastra, C. Dietary extra-virgin olive oil prevents inflammatory response and cartilage matrix degradation in murine collagen-induced arthritis. *Eur. J. Nutr.* **2015**, *55*, 315–325. [CrossRef]
12. Keys, A. Mediterranean diet and public health: Personal reflections. *Am. J. Clin. Nutr.* **1995**, *61*, 1321S–1323S. [CrossRef]
13. Fukumitsu, S.; Villareal, M.O.; Fujitsuka, T.; Aida, K.; Isoda, H. Anti-inflammatory and anti-arthritic effects of pentacyclic triterpenoids maslinic acid through NF-κB inactivation. *Mol. Nutr. Food Res.* **2015**, *60*, 399–409. [CrossRef]
14. Shimazu, K.; Fukumitsu, S.; Ishijima, T.; Toyoda, T.; Nakai, Y.; Abe, K.; Aida, K.; Okada, S.; Hino, A. The Anti-Arthritis Effect of Olive-Derived Maslinic Acid in Mice is Due to its Promotion of Tissue Formation and its Anti-Inflammatory Effects. *Mol. Nutr. Food Res.* **2018**, *63*, e1800543. [CrossRef]
15. Yoon, J.; Kanamori, A.; Fujii, K.; Isoda, H.; Okura, T. Evaluation of maslinic acid with whole-body vibration training in elderly women with knee osteoarthritis. *PLoS ONE* **2018**, *13*, e0194572. [CrossRef]
16. Nagai, N.; Yagyu, S.; Hata, A.; Nirengi, S.; Kotani, K.; Moritani, T.; Sakane, N. Maslinic acid derived from olive fruit in combination with resistance training improves muscle mass and mobility functions in the elderly. *J. Clin. Biochem. Nutr.* **2019**, *64*, 224–230. [CrossRef]
17. Sacheck, J.M.; Hyatt, J.K.; Raffaello, A.; Jagoe, R.T.; Roy, R.R.; Edgerton, V.R.; Lecker, S.H.; Goldberg, A.L. Rapid disuse and denervation atrophy involve transcriptional changes similar to those of muscle wasting during systemic diseases. *FASEB J.* **2006**, *21*, 140–155. [CrossRef]
18. Lecker, S.H.; Jagoe, R.T.; Gilbert, A.; Gomes, M.; Baracos, V.; Bailey, J.; Price, S.R.; Mitch, W.E.; Goldberg, A.L. Multiple types of skeletal muscle atrophy involve a common program of changes in gene expression. *FASEB J.* **2003**, *18*, 39–51. [CrossRef]
19. Wijngaarden, M.A.; Van Der Zon, G.C.; van Dijk, K.W.; Pijl, H.; Guigas, B. Effects of prolonged fasting on AMPK signaling, gene expression, and mitochondrial respiratory chain content in skeletal muscle from lean and obese individuals. *Am. J. Physiol. Metab.* **2013**, *304*, E1012–E1021. [CrossRef]
20. Kunkel, S.D.; Suneja, M.; Ebert, S.M.; Bongers, K.S.; Fox, D.K.; Malmberg, S.E.; Alipour, F.; Shields, R.; Adams, C.M. mRNA Expression Signatures of Human Skeletal Muscle Atrophy Identify a Natural Compound that Increases Muscle Mass. *Cell Metab.* **2011**, *13*, 627–638. [CrossRef]
21. Liberzon, A. A Description of the Molecular Signatures Database (MSigDB) Web Site. *Methods Mol. Biol.* **2014**, *1150*, 153–160. [CrossRef]
22. Subramanian, A.; Tamayo, P.; Mootha, V.K.; Mukherjee, S.; Ebert, B.L.; Gillette, M.A.; Paulovich, A.; Pomeroy, S.L.; Golub, T.R.; Lander, E.S.; et al. Gene set enrichment analysis: A knowledge-based approach for interpreting genome-wide expression profiles. *Proc. Natl. Acad. Sci. USA* **2005**, *102*, 15545–15550. [CrossRef]
23. Huang, D.W.; Sherman, B.T.; Lempicki, R. Systematic and integrative analysis of large gene lists using DAVID bioinformatics resources. *Nat. Protoc.* **2008**, *4*, 44–57. [CrossRef]
24. Sarukhanov, V.; Van Andel, R.; Treat, M.D.; Utz, J.; Van Breukelen, F. A refined technique for sciatic denervation in a golden-mantled ground squirrel (Callospermophilus lateralis) model of disuse atrophy. *Lab. Anim.* **2014**, *43*, 203–206. [CrossRef]
25. Shen, Y.; Zhang, R.; Xu, L.; Wan, Q.; Zhu, J.; Gu, J.; Huang, Z.; Ma, W.; Shen, M.; Ding, F.; et al. Microarray Analysis of Gene Expression Provides New Insights Into Denervation-Induced Skeletal Muscle Atrophy. *Front. Physiol.* **2019**, *10*, 1298. [CrossRef]
26. Narola, J.; Pandey, S.; Glick, A.; Chen, Y.-W. Conditional Expression of TGF-β1 in Skeletal Muscles Causes Endomysial Fibrosis and Myofibers Atrophy. *PLoS ONE* **2013**, *8*, e79356. [CrossRef]

27. Waning, D.L.; Mohammad, K.S.; Reiken, S.; Xie, W.; Andersson, D.; John, S.K.; Chiechi, A.; Wright, L.; Umanskaya, A.; Niewolna, M.; et al. Excess TGF-β mediates muscle weakness associated with bone metastases in mice. *Nat. Med.* **2015**, *21*, 1262–1271. [CrossRef]

28. Sánchez-Tena, S.; Reyes-Zurita, F.J.; Diaz-Moralli, S.; Vinardell, M.P.; Reed, M.; Garcia-Garcia, F.; Dopazo, J.; Lupiáñez, J.A.; Günther, U.; Cascante, M. Maslinic Acid-Enriched Diet Decreases Intestinal Tumorigenesis in ApcMin/+ Mice through Transcriptomic and Metabolomic Reprogramming. *PLoS ONE* **2013**, *8*, e59392. [CrossRef]

29. Fernández-Navarro, M.; Peragón, J.; Amores, V.; De La Higuera, M.; Lupiáñez, J.A. Maslinic acid added to the diet increases growth and protein-turnover rates in the white muscle of rainbow trout (Oncorhynchus mykiss). *Comp. Biochem. Physiol. Part. C Toxicol. Pharmacol.* **2008**, *147*, 158–167. [CrossRef]

30. Timmer, L.T.; Hoogaars, W.M.H.; Jaspers, R.T. The Role of IGF-1 Signaling in Skeletal Muscle Atrophy. *Muscle Atrophy* **2018**, *1088*, 109–137. [CrossRef]

31. Hennebry, A.; Oldham, J.; Shavlakadze, T.; Grounds, M.; Sheard, P.; Fiorotto, M.L.; Falconer, S.; Smith, H.K.; Berry, C.; Jeanplong, F.; et al. IGF1 stimulates greater muscle hypertrophy in the absence of myostatin in male mice. *J. Endocrinol.* **2017**, *234*, 187–200. [CrossRef] [PubMed]

32. Frost, R.A.; Huber, D.; Pruznak, A.; Lang, C.H. Regulation of REDD1 by insulin-like growth factor-I in skeletal muscle and myotubes. *J. Cell. Biochem.* **2009**, *108*, 1192–1202. [CrossRef]

33. Sacheck, J.M.; Ohtsuka, A.; McLary, S.C.; Goldberg, A.L. IGF-I stimulates muscle growth by suppressing protein breakdown and expression of atrophy-related ubiquitin ligases, atrogin-1 and MuRF1. *Am. J. Physiol. Metab.* **2004**, *287*, E591–E601. [CrossRef]

34. Xu, J.; Li, R.; Workeneh, B.; Dong, Y.; Wang, X.; Hu, Z. Transcription factor FoxO1, the dominant mediator of muscle wasting in chronic kidney disease, is inhibited by microRNA-486. *Kidney Int.* **2012**, *82*, 401–411. [CrossRef]

35. Stevenson, E.J.; Giresi, P.G.; Koncarevic, A.; Kandarian, S.C. Global analysis of gene expression patterns during disuse atrophy in rat skeletal muscle. *J. Physiol.* **2003**, *551*, 33–48. [CrossRef]

36. Ebert, S.M.; Dyle, M.C.; Kunkel, S.D.; Bullard, S.A.; Bongers, K.S.; Fox, D.K.; Dierdorff, J.M.; Foster, E.D.; Adams, C.M. Stress-induced Skeletal Muscle Gadd45a Expression Reprograms Myonuclei and Causes Muscle Atrophy. *J. Biol. Chem.* **2012**, *287*, 27290–27301. [CrossRef]

37. Banduseela, V.C.; Ochala, J.; Chen, Y.-W.; Göransson, H.; Norman, H.; Radell, P.; Eriksson, L.I.; Hoffman, E.; Larsson, L. Gene expression and muscle fiber function in a porcine ICU model. *Physiol. Genom.* **2009**, *39*, 141–159. [CrossRef] [PubMed]

38. Welle, S.; Brooks, A.I.; Delehanty, J.M.; Needler, N.; Bhatt, K.; Shah, B.; Thornton, C.A. Skeletal muscle gene expression profiles in 20–29 year old and 65–71 year old women. *Exp. Gerontol.* **2004**, *39*, 369–377. [CrossRef]

39. Welle, S.; Brooks, A.I.; Delehanty, J.M.; Needler, N.; Thornton, C.A. Gene expression profile of aging in human muscle. *Physiol. Genom.* **2003**, *14*, 149–159. [CrossRef]

40. Ehmsen, J.T.; Kawaguchi, R.; Kaval, D.; Johnson, A.E.; Nachun, D.; Coppola, G.; Höke, A. GADD45A is a protective modifier of neurogenic skeletal muscle atrophy. *JCI Insight* **2021**, *6*, e149381. [CrossRef]

41. Fujino, H.; Kohzuki, H.; Takeda, I.; Kiyooka, T.; Miyasaka, T.; Mohri, S.; Shimizu, J.; Kajiya, F. Regression of capillary network in atrophied soleus muscle induced by hindlimb unweighting. *J. Appl. Physiol.* **2005**, *98*, 1407–1413. [CrossRef]

42. Goldstein, E.; Jacobs, P.L.; Whitehurst, M.; Penhollow, T.; Antonio, J. Caffeine enhances upper body strength in resistance-trained women. *J. Int. Soc. Sports Nutr.* **2010**, *7*, 18. [CrossRef]

43. Shaik, A.H.; Rasool, S.; Kareem, M.A.; Krushna, G.S.; Akhtar, P.M.; Devi, K.L. Maslinic Acid Protects Against Isoproterenol-Induced Cardiotoxicity in Albino Wistar Rats. *J. Med. Food* **2012**, *15*, 741–746. [CrossRef] [PubMed]

44. Barreto, G.; Huang, T.-T.; Giffard, R. Age-related Defects in Sensorimotor Activity, Spatial Learning, and Memory in C57BL/6 Mice. *J. Neurosurg. Anesthesiol.* **2010**, *22*, 214–219. [CrossRef] [PubMed]

45. Parks, R.J.; Fares, E.; Macdonald, J.K.; Ernst, M.C.; Sinal, C.J.; Rockwood, K.; Howlett, S. A Procedure for Creating a Frailty Index Based on Deficit Accumulation in Aging Mice. *J. Gerontol. Ser. A Boil. Sci. Med. Sci.* **2011**, *67*, 217–227. [CrossRef]

46. Graber, T.; Ferguson-Stegall, L.; Kim, J.-H.; Thompson, L.V. C57BL/6 Neuromuscular Healthspan Scoring System. *J. Gerontol. Ser. A Boil. Sci. Med. Sci.* **2013**, *68*, 1326–1336. [CrossRef]

47. Yuan, R.; Peters, L.L.; Paigen, B. Mice as a mammalian model for research on the genetics of aging. *ILAR J.* **2011**, *52*, 4–15. [CrossRef]

48. Yuan, R.; Tsaih, S.-W.; Petkova, S.B.; De Evsikova, C.M.; Xing, S.; Marion, M.A.; Bogue, M.A.; Mills, K.D.; Peters, L.L.; Bult, C.J.; et al. Aging in inbred strains of mice: Study design and interim report on median lifespans and circulating IGF1 levels. *Aging Cell* **2009**, *8*, 277–287. [CrossRef]

49. Cohen, S.; Nathan, J.A.; Goldberg, A.L. Muscle wasting in disease: Molecular mechanisms and promising therapies. *Nat. Rev. Drug Discov.* **2014**, *14*, 58–74. [CrossRef]

50. Dutt, V.; Gupta, S.; Dabur, R.; Injeti, E.; Mittal, A. Skeletal muscle atrophy: Potential therapeutic agents and their mechanisms of action. *Pharmacol. Res.* **2015**, *99*, 86–100. [CrossRef]

51. Pokorski, M. Preface. Neuroscience and Respiration. *Adv. Exp. Med. Biol.* **2016**, *878*, 5–6.

52. Sheng, H.; Sun, H. Synthesis, biology and clinical significance of pentacyclic triterpenes: A multi-target approach to prevention and treatment of metabolic and vascular diseases. *Nat. Prod. Rep.* **2011**, *28*, 543–593. [CrossRef] [PubMed]

Article

Inhibitory Effect of Tangeretin and Cardamonin on Human Intestinal SGLT1 Activity In Vitro and Blood Glucose Levels in Mice In Vivo

Hideo Satsu [1,*], Ryosuke Shibata [2], Hiroto Suzuki [1], Shimon Kimura [1] and Makoto Shimizu [3]

[1] Department of Biotechnology, Faculty of Engineering, Maebashi Institute of Technology, Gunma 371-0816, Japan; h.suzuki.1306@gmail.com (H.S.); m2166006@maebashi-it.ac.jp (S.K.)

[2] Department of Applied Biological Chemistry, Graduate School of Agricultural and Life Sciences, The University of Tokyo, Tokyo 113-8657, Japan; ryosuke.shibata.1016@gmail.com

[3] Department of Nutritional Science, Tokyo University of Agriculture, Tokyo 156-8502, Japan; ms205346@nodai.ac.jp

* Correspondence: satsu@maebashi-it.ac.jp; Tel.: +81-27-265-7374

Abstract: Rapid postprandial blood glucose elevation can cause lifestyle-related diseases, such as type II diabetes. The absorption of food-derived glucose is primarily mediated by sodium/glucose cotransporter 1 (SGLT1). Moderate SGLT1 inhibition can help attenuate postprandial blood glucose elevation and prevent lifestyle-related diseases. In this study, we established a CHO cell line stably expressing human SGLT1 and examined the effects of phytochemicals on SGLT1 activity. Among the 50 phytochemicals assessed, tangeretin and cardamonin inhibited SGLT1 activity. Tangeretin and cardamonin did not affect the uptake of L-leucine, L-glutamate, and glycyl-sarcosine. Tangeretin, but not cardamonin, inhibited fructose uptake, suggesting that the inhibitory effect of tangeretin was specific to the monosaccharide transporter, whereas that of cardamonin was specific to SGLT1. Kinetic analysis suggested that the suppression of SGLT1 activity by tangeretin was associated with a reduction in V_{max} and an increase in K_m, whereas suppression by cardamonin was associated with a reduction in V_{max} and no change in K_m. Oral glucose tolerance tests in mice showed that tangeretin and cardamonin significantly suppressed the rapid increase in blood glucose levels. In conclusion, tangeretin and cardamonin were shown to inhibit SGLT1 activity in vitro and lower blood glucose level in vivo.

Keywords: SGLT1; transporter; tangeretin; cardamonin; intestinal epithelial cell

Citation: Satsu, H.; Shibata, R.; Suzuki, H.; Kimura, S.; Shimizu, M. Inhibitory Effect of Tangeretin and Cardamonin on Human Intestinal SGLT1 Activity In Vitro and Blood Glucose Levels in Mice In Vivo. *Nutrients* **2021**, *13*, 3382. https://doi.org/10.3390/nu13103382

Academic Editor: Yoshinori Katakura

Received: 13 July 2021
Accepted: 23 September 2021
Published: 26 September 2021

Publisher's Note: MDPI stays neutral with regard to jurisdictional claims in published maps and institutional affiliations.

Copyright: © 2021 by the authors. Licensee MDPI, Basel, Switzerland. This article is an open access article distributed under the terms and conditions of the Creative Commons Attribution (CC BY) license (https://creativecommons.org/licenses/by/4.0/).

1. Introduction

Diabetes is a lifestyle-related disease with increasing global incidence. According to the reports of the International Diabetes Federation, in 2019, the number of patients with diabetes was approximately 463 million worldwide, and by 2045, it is predicted to increase to approximately 700 million [1]. Diabetes is classified as insulin-dependent type I and insulin-non-dependent type II, and approximately 95% of patients with diabetes are diagnosed with type II diabetes. In addition to genetic factors, environmental factors such as excessive nutrient intake and lack of physical activity are important contributors to the onset of type II diabetes. Chronic hyperglycemia, insulin resistance, and obesity are typical characteristics of patients with type II diabetes, which further causes complications such as retinopathy, nephropathy, and neuropathy [2,3].

The excessive intake of carbohydrates, particularly saccharides, is a leading cause of the increasing incidence of type II diabetes and obesity. α-amylase digests saccharides consumed with meals to oligosaccharides, which are further digested to monosaccharides by α-glucosidases, such as sucrase-isomaltase or maltase, in the intestinal tract. In intestinal epithelial cells, glucose is primarily absorbed by sodium-dependent glucose transporter

1 (SGLT1), expressed at the apical membrane of intestinal epithelial cells [4,5]. The incorporated glucose is transported to the blood via glucose transporter 2 (GLUT2), which is expressed at the basolateral membrane of intestinal epithelial cells. Recently, it has been reported that GLUT2 present in cells is translocated to the apical membrane and contributes to glucose uptake under high glucose concentrations [6–8].

To prevent diabetes, it is advisable to inhibit α-amylase or α-glucosidase activity, or to suppress SGLT1 activity in moderation, by using food components to reduce rapid spikes in blood glucose levels after meals and prevent chronic hyperglycemia. Several food substances, such as wheat albumin [9,10] and guava tea polyphenols [11], have been reported to inhibit α-amylase or α-glucosidase activity, which helps suppress rapid spikes in blood glucose levels after meals.

Intestinal epithelial cells play an important role in the absorption of nutrients and functional components from food. Furthermore, since intestinal epithelial cells are exposed to food components at the highest concentration and frequency, their functions are considered to be controlled and regulated by food components. We previously reported that the functions of intestinal epithelial cells, such as transporter activity, cytokine secretion, and gene expression of detoxification enzymes, are regulated or modulated by various food components [12,13]. Among intestinal transporters, the taurine transporter was shown to be suppressed by sesame extracts, with lysophosphatidylcholine being one of the inhibitory compounds [14,15]. The intestinal fructose transporter (GLUT5) is also suppressed by flavonoids [16]. Furthermore, MDR1 activity was inhibited by 2-monopalmitin [17]. These studies were mostly performed using in vitro colon adenocarcinoma-derived intestinal epithelial-like Caco-2 cells. However, it is difficult to assess SGLT1 activity using these cell lines, as Caco-2 cells have often been reported to exhibit extremely low or no SGLT1 activity. Steffansen et al. [18,19] also reported that SGLT1-mediated transport is highly dependent on the cell bank's origin. Therefore, a simple in vitro evaluation system to assess SGLT1 activity is necessary to exhaustively identify food components that can inhibit SGLT1 activity.

In the present study, we constructed a stable human SGLT1-expresing CHO cell line and used this evaluation system to search for and analyze food components, especially phytochemicals, which inhibit SGLT1 activity. To further evaluate the suppressive effect of phytochemicals on SGLT1 activity in vivo, the effect of phytochemicals on glucose absorption in mice was examined using oral glucose tolerance tests (OGTTs).

2. Materials and Methods

2.1. Materials

The following materials were purchased from the listed sources: CHO cell line and Caco-2 cells from the American Type Culture Collection (Rockville, MD, USA); Dulbecco's modified Eagle's medium (DMEM) containing 4 mM L-glutamine from Wako Chemicals (Osaka, Japan); penicillin–streptomycin (10,000 U/mL and 10 mg/mL in 0.9% sodium chloride, respectively) from Gibco (Gaithersburg, MD, USA); fetal calf serum (FCS) from Life Technologies (Grand Island, NY, USA); non-essential amino acids (NEAA) from Cosmobio (Tokyo, Japan); [^{3}H]-glucose (specific radioactivity: 21.2 Ci/mmol), [^{3}H]-fructose (5.0 Ci/mmol), [^{3}H]-L-leucine (142 Ci/mmol) from GE Healthcare (Fairfield, CT, USA), [^{3}H]-L-glutamic acid (46 Ci/mmol), and [^{3}H]-glycyl-sarcosine ([^{3}H]-Gly-Sar) (0.2 Ci/mmol) from Moravek (Brea, CA, USA); and a type-I collagen solution from Nitta Gelatin (Osaka, Japan). The phytochemicals used for this study are shown in Table A1. All other chemicals used were of reagent grade.

2.2. Transfection of Human SGLT1 Expression Vectors for Development of a Stable Cell Line

Chinese hamster ovary-K1 (CHO-K1) cells were cultured in Ham's F-12 medium supplemented with 10% FCS and penicillin–streptomycin. A human SGLT1 expression vector (pcDNA3.1-SGLT1) was constructed as follows: the open reading frame of human SGLT1 was amplified from a plasmid encoding human SGLT1 (Flexi clone, Promega) using the primer 5′-GGGAAGCTTATGGACAGTAGCACCTGGAG-3′, which introduced a *Hind*III site, and the primer 5′-GGGGAATTCGGCAAAATATGCATGGCAAAAG-3′, which introduced an *Eco*RI site. The *Hind*III/*Eco*RI-digested PCR fragment was ligated into the digested pcDNA3.1A (Invitrogen, Carlsbad, CA, USA), thereby creating pcDNA-hSGLT1, and sequenced.

The plasmid vector was transfected into CHO-K1 cells using Lipofectamine 2000 (Invitrogen) according to the manufacturer's instructions. hSGLT1-expressing cells (transfected cells) were selected with G418, and limiting dilution was performed in 96-well plates. Several single clones were selected from the 96-well plates. The clone that exhibited the strongest sodium-dependent [^{3}H]-glucose uptake activity was selected and the hSGLT1 stable cell line was established.

2.3. Assay of Glucose Uptake in a Stable Cell Line Expressing hSGLT1

For the glucose uptake assay, the cells were seeded in a 24-well plate at 1.0×10^5 cells per well in the Ham's F-12 medium supplemented with 10% FCS, 100 U/mL penicillin, 100 µg/mL streptomycin, and 2 mg/mL G418. After seeding for two days, the glucose uptake experiment was performed. The cells were washed twice with 700 µL of phosphate-buffered saline (PBS) and incubated once in 300 µL of uptake buffer (140 mM NaCl, 0.34 mM Na$_2$HPO$_4$, 0.44 mM KH$_2$PO$_4$, 5.33 mM KCl, 1.26 mM CaCl$_2$, 0.49 mM MgCl$_2$, 0.41 mM MgSO$_4$, and 4.16 mM NaHCO$_3$; pH adjusted to 7.4 with KOH) for 10 min. The cells were then incubated with 50 nM [^{3}H]-glucose in 300 µL of the uptake buffer, without or with each phytochemical at 37 °C for 10 min. After the incubation, the buffer was removed, and the cells in each well were carefully washed three times with 700 µL of ice-cold PBS containing 0.05% sodium azide. Subsequently, 250 µL of 0.1% Triton X-100 was added to each well, following which the lysed cells were collected in 3 mL of a scintillation cocktail, and the tritium content of the cells from each well was measured using an LSC 5100 liquid scintillation analyzer (Aloka, Tokyo, Japan).

2.4. Caco-2 Cell Culture

Caco-2 cells were cultured in plastic dishes of 78.5 cm^2 containing a culture medium composed of DMEM, 10% FCS, 1% NEAA, 2% glutamine, 100 U/mL penicillin, and 100 µg/mL streptomycin. The cells were incubated at 37 °C in a humidified atmosphere of 5% CO$_2$ in air, and the culture medium was replaced at a split ratio of 4:8 by trypsinization with a solution containing 0.1% trypsin and 0.02% EDTA in PBS. For uptake experiments, Caco-2 cells were seeded in a 24-well plate at 1.0×10^5 cells per well. After 14 days of culture, the [^{3}H]-fructose or [^{3}H]-Gly-Sar uptake by Caco-2 cells was measured as previously described [16].

2.5. OGTTs in Mice

The animal care procedures and methods adopted were approved by the Animal Care and Use Committee of The University of Tokyo (permission number: P13-840).

ICR mice (6–8 weeks old, male, purchased from CLEA Japan, Japan) were fasted overnight for 18 hours. The tangeretin or cardamonin was suspended in an 0.3% *w*/*v* solution of carboxymethyl cellulose sodium salt, respectively. The mice were orally administered the vehicle or phytochemical (tangeretin or cardamonin) at 250 or 400 mg/kg body weight (BW). The mice were orally administered a 20% glucose solution (1 g/kg BW). Blood drawn from the orbital venous sinus under isoflurane anesthesia at 30, 60, and 120 min after the glucose's administration was collected in heparinized tubes. The plasma glucose concentration was measured using the glucose CII-test WAKO according to the manufacturer's instructions (WAKO, Osaka, Japan). The area under the curve (AUC) was calculated using the trapezoidal rule.

2.6. Statistical Analysis

Data are expressed as the mean ± standard error of the mean from at least three independent experiments performed in triplicate. Statistical comparisons were performed using Student's *t*-test, Dunnett's test, or Tukey's test. Differences were considered statistically significant at $p < 0.05$.

3. Results

3.1. Construction of the Stable hSGLT1-Expressed Cell Line

Ninety-four G418-resitant clones in 96-well plates were passed to 48-well plates and then to 24-well plates. After confluence was achieved in the 24-well plates, the glucose uptake activity of each cell was measured. Among 91 clones, 10 clones that exhibited high glucose uptake activity were selected, and glucose uptake in the selected 10 clones was measured in the absence or presence of sodium ions. Among the 10 clones, the clone with the highest sodium-dependent glucose uptake (4G8) was selected as the hSGLT1 clone. We confirmed that glucose uptake in the 4G8 clone was significantly suppressed by phlorizin, a typical SGLT inhibitor, suggesting that glucose uptake in the 4G8 clone occurred via SGLT1. Therefore, the 4G8 clone was used in the subsequent experiments.

3.2. Characterization of Sodium-Dependent Glucose Uptake Activity in Stable hSGLT1-Expressing Cells

To determine the optimal conditions for assessing SGLT1 activity, the glucose uptake activity was measured in the absence or presence of sodium ions at different incubation times. Figure 1A shows that glucose uptake in the absence or presence of sodium ions increased in a time-dependent manner. Sodium-dependent glucose uptake also increased in a time-dependent manner (Figure 1B). Based on this result, we fixed the incubation time at 30 min to assess sodium-dependent glucose uptake (SGLT1) activity, as SGLT1 activity was greater than sodium-independent glucose uptake activity.

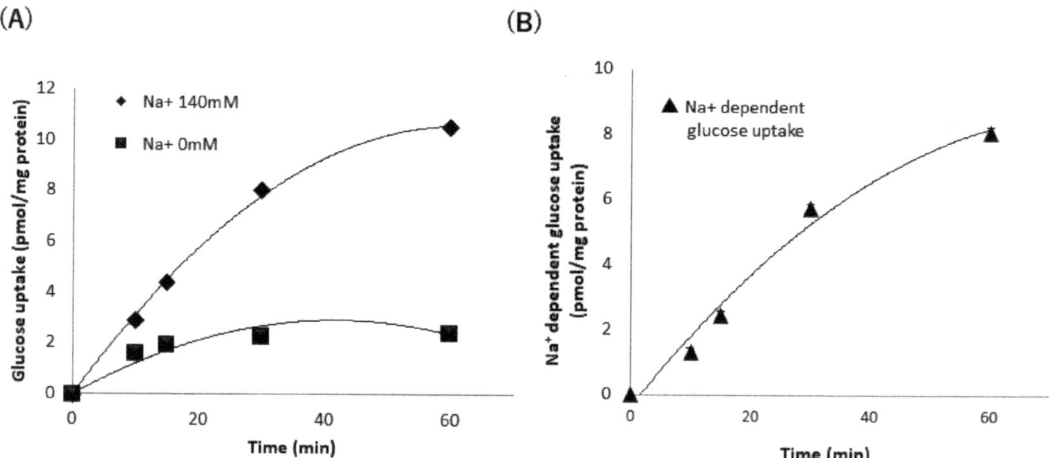

Figure 1. Glucose uptake in stable hSGLT1-expressing Chinese Hamster Ovary (CHO) cells in the absence or presence of sodium ions (**A**). (**B**) Sodium-dependent glucose uptake in stable hSGLT1-expressing CHO cells. Glucose uptake in the absence or presence of sodium ions was measured (**A**). Sodium-dependent glucose uptake was calculated by subtracting the glucose uptake in the presence of sodium ions and in the absence of sodium ions. Each value represents the mean ± standard error of mean ($n = 3$).

3.3. Effect of Phytochemicals on Glucose Uptake in hSGLT1-Expressing CHO Cell Line

Using the developed hSGLT1-expressing CHO cell line, we attempted to identify phytochemicals that inhibit glucose uptake activity. Among about 50 types of phytochemicals, we found that cardamonin, tangeretin, and nobiletin significantly inhibited glucose uptake in hSGLT1-expressing CHO cells (Figure 2). Among these phytochemicals, we focused on tangeretin and cardamonin, as tangeretin and nobiletin have similar structures, and tangeretin exhibited greater inhibitory activity than nobiletin. Next, we examined the effect of these phytochemicals on glucose uptake when administered at various concentrations (0–50 μM). Figure 3 clearly shows that glucose uptake was inhibited in a dose-dependent manner by tangeretin (Figure 3A) and cardamonin (Figure 3B).

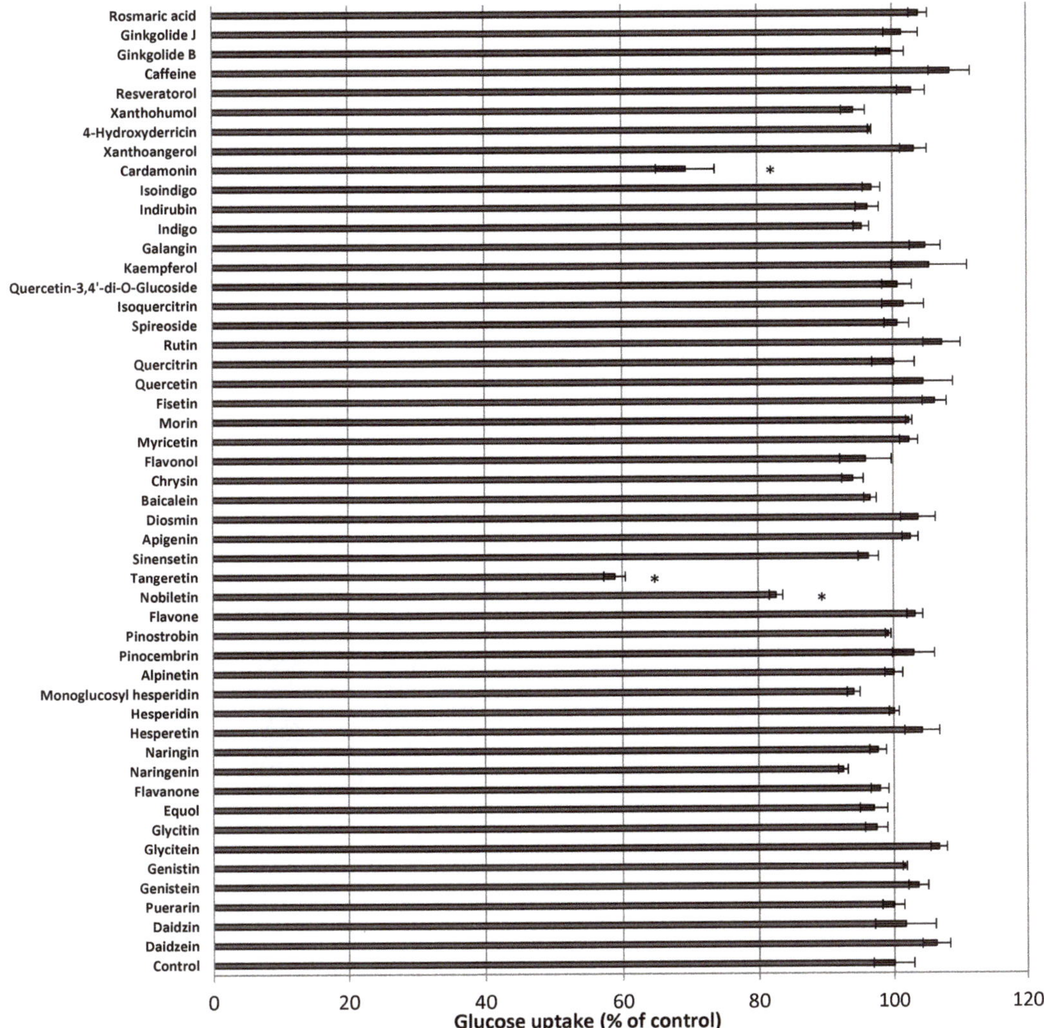

Figure 2. Effect of the phytochemicals on glucose uptake in stable hSGLT1-expressing CHO cells. Glucose uptake was measured in the absence or presence of the phytochemicals (1 μM). The values are expressed in terms of mean ± standard error of mean ($n = 3$); * $p < 0.05$ vs. the control value (Student's t-test).

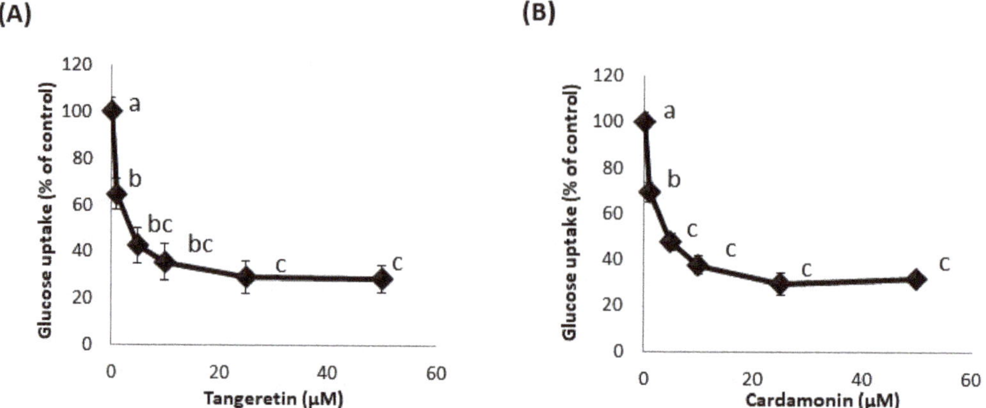

Figure 3. Concentration-dependent inhibition of glucose uptake by tangeretin (**A**) and cardamonin (**B**) in stably hSGLT1-expressed CHO cells. Glucose uptake was measured in the absence or presence of 0–50 μM tangeretin (**A**) and cardamonin (**B**). The values are the mean ± SE (n = 3), and the values indicated by different characters are significantly different from each other (Tukey's test; $p < 0.05$).

3.4. Effect of Tangeretin and Cardamonin on Nutrient Uptake in hSGLT1/CHO Cells and Human Intestinal-like Caco-2 Cells

We examined the effect of tangeretin and cardamonin on the uptake of [^{3}H]-glucose (50 nM), [^{3}H]-L-leucine (6.25 nM), and [^{3}H]-L-glutamic acid (22.2 nM). At 10 μM, tangeretin and cardamonin did not affect the uptake of L-leucine or L-glutamic acid, but suppressed glucose uptake (Figure 4A,B). These results suggested that the phytochemicals exerted no effect on amino acid transporters involved in L-leucine or L-glutamic acid uptake.

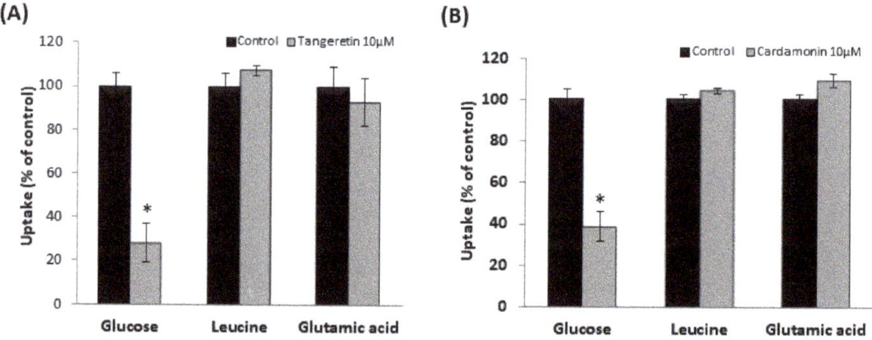

Figure 4. Effect of tangeretin (**A**) and cardamonin (**B**) on leucine and glutamic acid uptake in stable hSGLT1-expressing CHO cells. The uptake of glucose, L-leucine, and L-glutamic acid uptake was measured in the absence or presence of 10 μM tangeretin (**A**) or cardamonin (**B**). The values are expressed in terms of mean ± standard error of mean (n = 3); * $p < 0.05$ vs. the control value (Student's *t*-test).

Caco-2 cells are often used as an in vitro model for human intestinal epithelial cells and express fructose transporter (GLUT5) and di- and tri-peptide transporter (PepT1). Therefore, we investigated the effects of treatment with the phytochemicals on the uptake of fructose and Gly-Sar, a typical substrate of PepT1. As shown in Figure 5A, tangeretin significantly inhibited fructose uptake, but not Gly-Sar uptake. Conversely, cardamonin did not inhibit fructose or Gly-Sar uptake (Figure 5B).

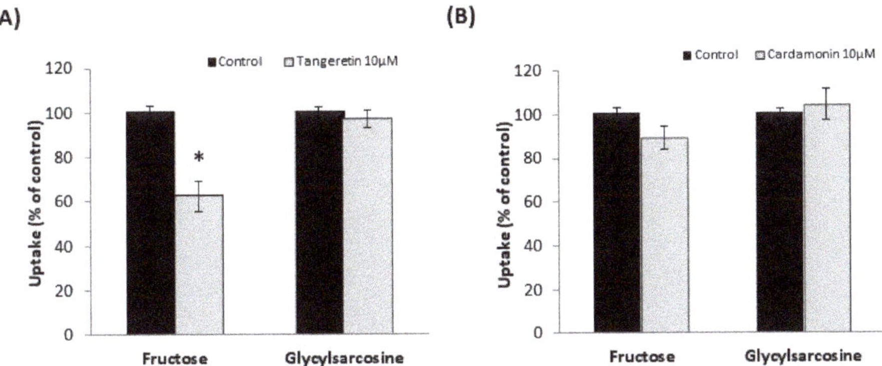

Figure 5. Effect of tangeretin (**A**) and cardamonin (**B**) on fructose and glycyl-sarcosine uptake in Caco-2 cells. Fructose and glycyl-sarcosine uptake were measured in the absence or presence of 10 µM tangeretin (**A**) or cardamonin (**B**). The values are expressed in terms of mean ± standard error of mean (*n* = 3); * *p* < 0.05 vs. the control value (Student's *t*-test).

3.5. Effect of Tangeretin-Related Compounds on Glucose Uptake in hSGLT1/CHO Cells

We investigated the inhibitory effect of tangeretin-related compounds on glucose uptake in hSGLT1-expressing cells. As shown in Figure 6A, tangeretin and nobiletin significantly inhibited the glucose uptake, whereas sinensetin did not affect the glucose uptake in hSGLT1/CHO cells. Tangeretin and nobiletin dose-dependently inhibited sodium-dependent glucose uptake (Figure 6B), suggesting that these phytochemicals inhibit SGLT1 activity.

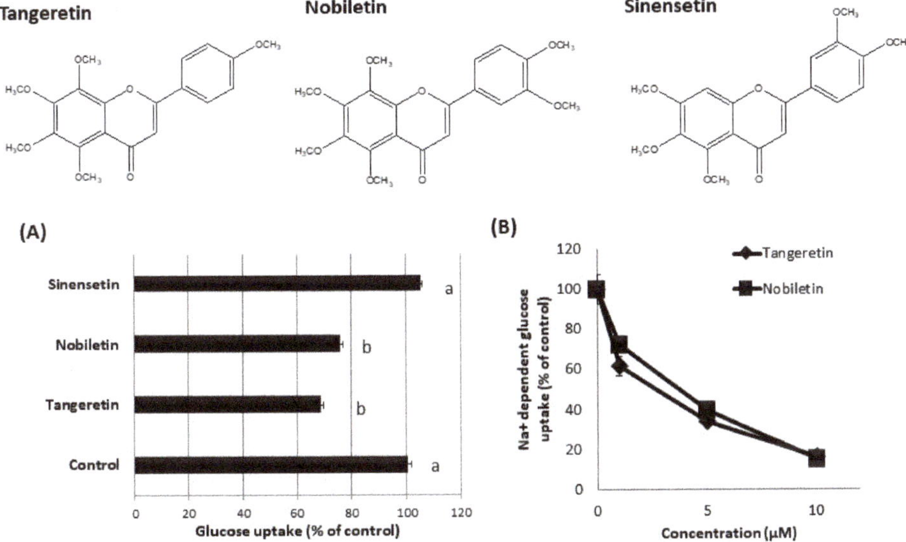

Figure 6. Effect of tangeretin-related compounds on glucose uptake in stable hSGLT1-expressing CHO cells. Glucose uptake was measured in the absence or presence of 1 µM tangeretin, nobiletin, and sinensetin (**A**). The values are expressed in terms of mean ± standard error of mean (SE) (*n* = 3), and the values indicated by the different characters are significantly different from each other (Tukey's test; *p* < 0.05). Sodium-dependent glucose uptake was assessed by subtracting glucose uptake in the presence of sodium ions and in the absence of sodium ions, further, in the absence or presence of tangeretin-related compounds (**B**). Each value represents mean ± SE (*n* = 3).

3.6. Effect of Cardamonin-Related Compounds on Glucose Uptake in hSGLT1/CHO Cells

We also investigated the inhibitory effect of cardamonin-related compounds on the glucose uptake in hSGLT1-expressed cells. Figure 7A shows that flavokawain B, naringenin chalcone, and cardamonin markedly inhibited glucose uptake. Naringenin and alpinetin also significantly inhibited glucose uptake, but to a lesser extent. To confirm whether cardamonin inhibits SGLT1 activity, we evaluated the effect of cardamonin and related compounds on sodium-dependent glucose uptake. Cardamonin and flavokawain B inhibited sodium-dependent glucose uptake in a dose-dependent manner (Figure 7B), suggesting that these phytochemicals inhibit SGLT1 activity.

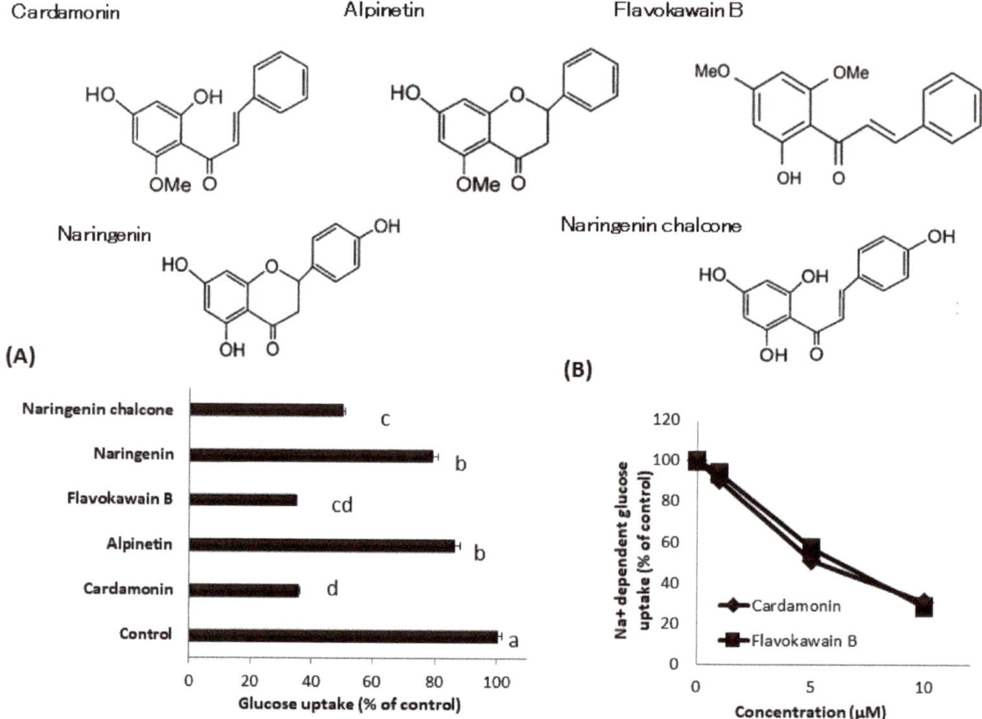

Figure 7. Effect of cardamonin-related compounds on glucose uptake in stable hSGLT1-expressing CHO cells. Glucose uptake was measured in the absence or presence of 10 µM cardamonin and its related compound, respectively (**A**). The values represent the mean ± standard error of mean (SE) ($n = 3$), and the values indicated by different characters are significantly different from each other (Tukey's test; $p < 0.05$). Sodium-dependent glucose uptake was assessed by subtracting glucose uptake in the presence of sodium ions and in the absence of sodium ions, and further, in the absence or presence of cardamonin-related compounds (**B**). Each value represents the mean ± SE ($n = 3$).

3.7. Kinetic Analysis of Sodium-Dependent Glucose Uptake by hSGLT1/CHO Cells in the Absence or Presence of Tangeretin and Cardamonin

A kinetic analysis of SGLT1 activity was performed, and Lineweaver–Burk plots were constructed to calculate the maximal velocity (V_{max}) and K_m values for glucose uptake in cells in the absence (Figure 8A) or presence of 10 µM tangeretin (Figure 8B) and cardamonin (Figure 8C). The V_{max} values in the presence of tangeretin and cardamonin decreased compared to that of control cells. However, the K_m values were not significantly different between the control and cardamonin-treated groups, although the K_m value increased upon tangeretin treatment. These results suggest that the inhibition of SGLT1 activity was primarily attributed to the decrease in the V_{max} value in the case of treatment with cardamonin, and to the decrease in the V_{max} value and increase in the K_m value in the case of treatment of tangeretin.

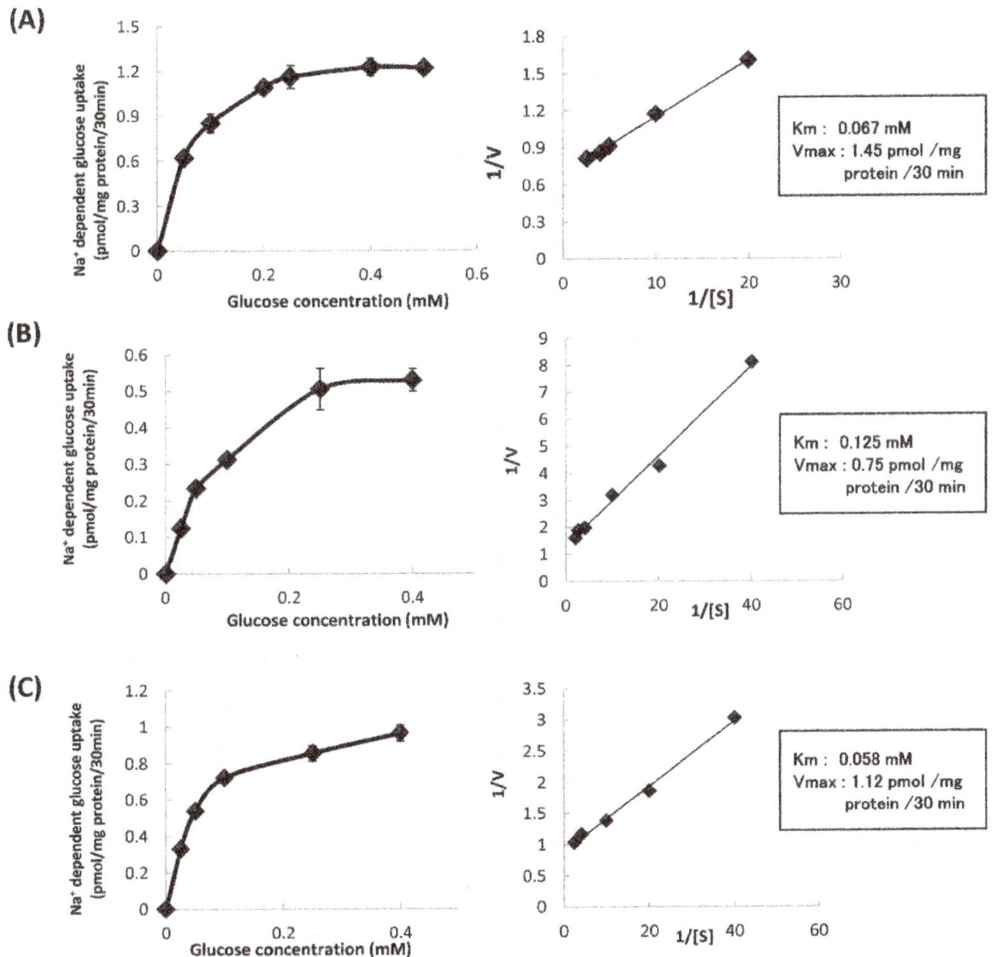

Figure 8. Kinetic analysis of sodium-dependent glucose uptake in the absence (**A**) or presence of tangeretin (**B**) and cardamonin (**C**). Sodium-dependent glucose uptake was measured after the administration of 0–0.4 mM glucose without (**A**) or along with 10 μM tangeretin (**B**) or cardamonin (**C**). The values are expressed in terms of mean ± standard error of mean (*n* = 3). Lineweaver–Burk plots were constructed to calculate the Vmax and Km values for sodium-dependent glucose uptake in the absence or presence of 10 μM tangeretin or cardamonin.

3.8. Effect of Tangeretin and Cardamonin on the Increase in Blood Glucose Levels In Vivo Following Oral Glucose Administration

To confirm the inhibitory effect of tangeretin and cardamonin on SGLT1 activity in vivo, we performed OGTTs in ICR mice. The oral administration of tangeretin significantly suppressed the rapid spike in the blood glucose levels (Figure 9A). The AUC value also decreased upon the co-administration of tangeretin (Figure 9B). Cardamonin suppressed the rapid spike in the blood glucose levels and AUC values (Figure 10A,B). These findings suggested that tangeretin and cardamonin suppress blood glucose spikes in vivo.

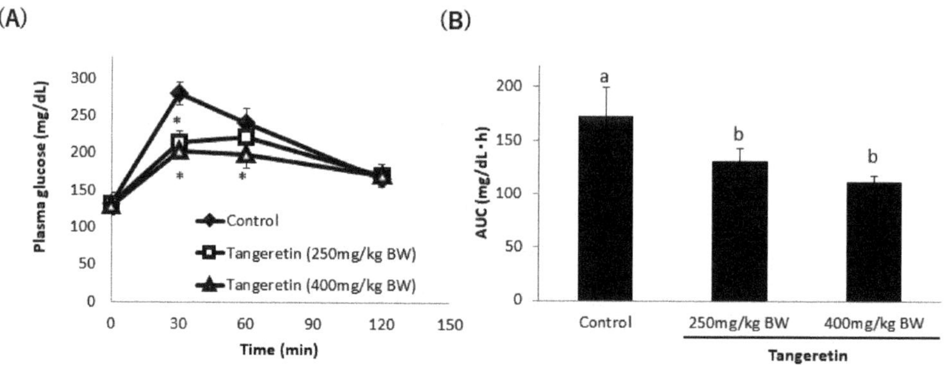

Figure 9. Effect of tangeretin on plasma glucose levels after oral glucose administration in ICR mice. ICR mice fasted overnight were orally administered 1 g glucose/kg body weight (BW) (20% glucose solution) along with or without 250 mg or 400 mg tangeretin/kg BW, and blood was drawn at 0, 30, 60, and 120 min. The plasma glucose concentration was measured (**A**), and the AUC under each experimental condition was calculated based on the values obtained (**B**). The values are expressed in terms of mean ± standard deviation of mean ($n = 6$); (**A**) * $p < 0.05$ vs. the control value (Dunnett's test). (**B**) The values indicated by the different characters are significantly different from each other (Tukey's test; $p < 0.05$).

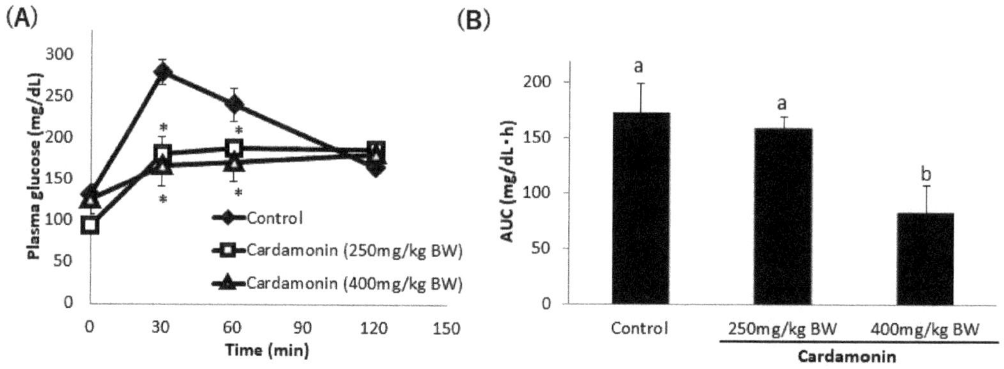

Figure 10. Effect of cardamonin on plasma glucose levels after oral glucose administration in ICR mice. ICR mice fasted overnight were orally administered 1 g glucose/kg body weight (BW) (20% glucose solution) along with or without 250 mg or 400 mg cardamonin/kg BW, and blood was drawn at 0, 30, 60, and 120 min. The plasma glucose concentration was measured (**A**), and the AUC under each experimental condition was calculated based on the values obtained (**B**). The values are expressed in terms of mean ± standard deviation of mean ($n = 6$); (**A**) * $p < 0.05$ vs. the control value (Dunnett's test). (**B**) The values indicated by the different characters are significantly different from each other (Tukey's test; $p < 0.05$).

4. Discussion

In the present study, we established a stable hSGLT1-expressing CHO cell line and attempted to identify phytochemicals that could inhibit SGLT1 activity. We found that two phytochemicals, tangeretin and cardamonin, significantly inhibited SGLT1 activity via different mechanisms. OGTTs were performed using the mice, which further confirmed that tangeretin and cardamonin significantly suppressed rapid blood glucose spikes, suggesting the in vivo SGLT1 inhibitory activity of the compounds.

In this study, we found that tangeretin and nobiletin markedly inhibited SGLT1 activity. Tangeretin and nobiletin are both flavonoids present in citrus fruits such as *Citrus unshiu*, *C. depressa*, *C. tangerina*, and *C. hassaku*. As the hydroxyl groups of these compounds are methoxylated, these are also referred to as methoxyflavonoids. Methoxyflavonoids contain a methoxy group in the flavone backbone, and therefore, exhibit high hydrophobicity. In recent years, various functional properties of methoxyflavonoids present in citrus fruits have been reported, including anti-inflammatory, anti-tumor, and anti-obesity effects [20–22]. *C. unshiu* contains various bioactive components, such as auraptene, β-cryptoxanthin, and limonine. Many methoxyflavonoids, such as tangeretin, nobiletin, hesperetin, and sinensetin, are also present in the form of glycosides and contribute to the functionality of *C. unshiu*. However, at least based on our findings, the relationship between SGLT1 activity and tangeretin or nobiletin remains unreported, and the SGLT1 inhibitory activities of tangeretin and nobiletin may be considered novel.

Interestingly, even though tangeretin and nobiletin exhibited significant inhibitory activity against SGLT1, sinensetin did not exhibit an inhibitory effect, as shown in Figure 6. However, tangeretin, nobiletin, and sinensetin significantly inhibited fructose uptake via GLUT5 in Caco-2 cells [16]. Tangeretin and sinensetin both possess a flavone structure with five methoxy groups, and nobiletin possesses a flavone structure with six methoxy groups. The structures of the compounds are considerably similar, but only sinensetin failed to exhibit inhibitory activity. Therefore, these results clearly indicate that the inhibitory effects of the three methoxyflavonoids against SGLT1 and GLUT5 are different. Based on these findings, we assumed that the methoxy group at position 8 of the A ring, which is common only to tangeretin and nobiletin, is essential for SGLT1 inhibition, but not for GLUT5 inhibition. Further, the analysis of various tangeretin-related chemical compounds is necessary for the thorough evaluation of the structure–activity relationship between methoxyflavonoids and SGLT1.

We also confirmed that cardamonin inhibits SGLT1 activity. Cardamonin is a type of methoxychalcone that is abundantly present in the seeds of *Alpinia katsumadai* (ginger plant) and the rhizome of *Boesenbergia pandurata* (Chinese bamboo shoot). In recent years, several functional polyphenols have been identified in plants native to Southeast Asia, and among these, cardamonin has been reported to exhibit several bioactive properties, including antitumor, anti-mutagenic, anti-inflammatory, and antioxidant properties [23–25]. With respect to its action on glucose transporters, cardamonin was reported to increase glucose uptake in L6 myotubular cells by promoting GLUT4 translocation to the plasma membrane [26]. However, to the best of our knowledge, there have been no reports on the effect of cardamonin on intestinal glucose absorption. The SGLT1 inhibitory activity of cardamonin is considerably novel. The effect of structural analogues of cardamonin on SGLT1 activity was examined, and flavokawain B, which has the same methoxychalcone structure as cardamonin, was also found to exhibit significant inhibitory activity against SGLT1. In addition, naringenin chalcone also exhibited strong inhibitory activity, whereas alpinetin and naringenin, which possess flavanone structures, exhibited limited inhibitory activity (Figure 7). These findings suggested that the methoxychalcone structure is important for the inhibition of SGLT1 activity. We intend to examine the effects of other compounds with a methoxychalcone structure on SGLT1 activity.

Kinetic analysis was performed to analyze the characteristics of the SGLT1 inhibitory activity of the phytochemicals (Figure 8). Cardamonin did not significantly alter the K_m value, but decreased the V_{max} values to approximately three-fourths of the original value. Conversely, tangeretin increased the K_m value by approximately two-fold and decreased the V_{max} value by approximately half (Figure 8). The results suggested that cardamonin exerted a non-competitive inhibitory effect, since cardamonin significantly reduced the V_{max} without significantly altering the K_m value. In the case of non-competitive inhibition, the inhibitor acts at a site different from the active site to inhibit transporter activity. Glucose transporters are considered to transport substrates as glucose-gated channels. Cardamonin was found to act at a site different from the substrate recognition site of SGLT1. Conversely, tangeretin considerably increased the Km value and decreased the Vmax value, suggesting that it exhibits both competitive and non-competitive inhibitory activity. Thorough analysis based on the crystal structure of human SGLT1 may help understand the inhibitory mechanism at the molecular level using docking simulation.

Reportedly, tiliroside, present in rosehip extract, inhibited the increase in blood glucose level caused by simultaneous administration of glucose in ICR mice [27,28]. Additionally, the long-term administration of nobiletin was reported to improve postprandial blood glucose levels in diabetic mouse models [29–31]. Furthermore, although the glucose transporter in intestinal epithelial cells was not targeted, the oral administration of *Ashitaba* extract, which is known to promote glucose uptake in skeletal muscles, in ICR mice was reported to significantly suppress postprandial blood glucose levels [32]. In the present study, the OGTTs performed in ICR mice showed that 250 mg of tangeretin significantly suppressed the increase in blood glucose 30 min after administration, 400 mg of tangeretin significantly suppressed the increase in blood glucose 30 and 60 min after administration, and both 250 mg and 400 mg of tangeretin significantly decreased the AUC. Meanwhile, both 250 mg and 400 mg of cardamonin significantly suppressed the increase in blood glucose levels at 30 and 60 min after administration, whereas the blood glucose level at 120 min after administration was almost equal to that in the control group. Furthermore, the AUC decreased significantly upon the administration of 400 mg of cardamonin, suggesting that cardamonin delayed the blood glucose spike.

Further, to confirm the in vivo inhibition of SGLT1 by these phytochemicals, intraperitoneal glucose tolerance tests should be conducted. When administered intraperitoneally rather than orally, glucose enters the bloodstream of the body via the portal vein without being absorbed from the intestinal tract. Therefore, it can be determined whether the inhibitory effects of phytochemicals on elevated blood glucose levels is attributed to the inhibition of intestinal glucose absorption. For example, a significant decrease in the blood glucose level and AUC was noted after the oral administration of tiliroside, whereas there was no decrease in the blood glucose level and AUC after intraperitoneal administration of tiliroside, confirming that tiliroside suppressed the increase in blood glucose levels by inhibiting intestinal epithelial glucose transporters [27].

5. Conclusions

In the present study, we developed a stable hSGLT1-expressing CHO cell line and used it to identify phytochemicals that inhibit SGLT1 activity using a simple in vitro method. We found tangeretin and cardamonin to be SGLT1 inhibitory compounds and confirmed their novelty. Further, these two compounds effectively prevented rapid blood glucose spikes in mice in vivo. Our findings indicate the potential of these compounds as functional food components with inhibitory effects on blood glucose elevation. Further, the developed cell line can be used to identify other functional food components that can inhibit SGLT1 activity.

Author Contributions: H.S. (Hideo Satsu) and R.S. conceived and designed the experiments; H.S. (Hideo Satsu), R.S. and H.S. (Hiroto Suzuki) performed the experiments; H.S. (Hideo Satsu), R.S. and S.K. analyzed the data; H.S. (Hideo Satsu), R.S. and M.S. discussed the data; H.S. (Hideo Satsu) and S.K. wrote the paper. All authors have read and agreed to the published version of the manuscript.

Funding: This work was funded by a Grant-in-Aid for Scientific Research (C) from the Ministry of Education, Culture, Sports, Science and Technology (MEXT) (24580176).

Institutional Review Board Statement: All animals received humane care, and the study protocols were approved by the Committee for the Care of Laboratory Animals in the Graduate School of Agricultural and Life Sciences at the University of Tokyo (permission number: P13-840).

Informed Consent Statement: Not applicable.

Data Availability Statement: The data that support the findings of this study are available from the corresponding author, H.S., upon reasonable request.

Acknowledgments: We acknowledge Norio Yamamoto for providing us with experimental techniques and much useful discussion.

Conflicts of Interest: The authors have declared no conflict of interest.

Appendix A

Table A1. Phytochemicals used in this experiment.

Compounds	*Company*
Myricetin	Sigma-Aldrich
Morin	Sigma-Aldrich
Puerarin	Sigma-Aldrich
Rutin	Wako
Fisetin	Sigma-Aldrich
Kaempherol	Extrasynthese
Quercitrin	Tokyo Chemical Industry
Quercetin	Tokyo Chemical Industry
Genistin	Fujicco
Diosmin	Sigma-Aldrich
Ginkgolide J	Tama Biochemical
Ginkgolide B	Tama Biochemical
Baicalein	Wako
Apigenin	Sigma-Aldrich
Flavonol	Sigma-Aldrich
Flavanone	Wako
Flavone	Sigma-Aldrich
Hesperetin	Sigma-Aldrich
Galangin	Sigma-Aldrich
Genistein	Sigma-Aldrich
Tangeretein	Wako
Daidzein	Sigma-Aldrich
Naringin	Sigma-Aldrich
Naringenin	Sigma-Aldrich
Flavone	Sigma-Aldrich
Daidzin	Wako
Equol	LC laboratories
Nobiletin	Wako
Glycitin	Wako
Glycitein	Wako
Caffeine	Sigma-Aldrich
Sinensetin	Wako
Rosmaric acid	Wako
Resveratrol	Wako
Xanthohumol	Tokyo Chemical Industry
4-Hydroxyderricin	Medchemexpress
Xanthoangerol	Medchemexpress
Cardamonin	Medchemexpress
Isoindigo	Cayman
Indirubin	Tokyo Chemical Industry
Indigo	Wako
Quercetin-3,4'-di-O-Glucoside	Extrasynthese
Isoquercitrin	Cayman
Spireoside	Sigma-Aldrich
Chrysin	Tokyo Chemical Industry
Pinostrobin	Sigma-Aldrich
Pinocembrin	Wako
Alpinetin	Selleck
Monoglucosyl hesperidin	Wako
Hesperidin	Tokyo Chemical Industry

References

1. IDF. Diabetes Atlas. Available online: https://www.diabetesatlas.org/en/sections/worldwide-toll-of-diabetes.html (accessed on 12 July 2021).
2. Gerich, J. Pathogenesis and management of postprandial hyperglycemia: Role of incretin-based therapies. *Int. J. Gen. Med.* **2013**, *6*, 877–895. [CrossRef]
3. Peterson, S.B.; Hart, G.W. New insights: A role for O-GlcNAcylation in diabetic complications. *Crit. Rev. Biochem. Mol. Biol.* **2016**, *51*, 150–161. [CrossRef]
4. Wright, E.M.; Loo, D.D.; Hirayama, B.A. Biology of human sodium glucose transporters. *Physiol. Rev.* **2011**, *91*, 733–794. [CrossRef] [PubMed]
5. Sano, R.; Shinozaki, Y.; Ohta, T. Sodium-glucose cotransporters: Functional properties and pharmaceutical potential. *J. Diabetes Investig.* **2020**, *11*, 770–782. [CrossRef] [PubMed]
6. Gouyon, F.; Caillaud, L.; Carriere, V.; Klein, C.; Dalet, V.; Citadelle, D.; Kellett, G.L.; Thorens, B.; Leturque, A.; Brot-Laroche, E. Simple-sugar meals target GLUT2 at enterocyte apical membranes to improve sugar absorption: A study in GLUT2-null mice. *J Physiol.* **2003**, *552*, 823–832. [CrossRef] [PubMed]
7. Kellett, G.L.; Brot-Laroche, E. Apical GLUT2: A major pathway of intestinal sugar absorption. *Diabetes* **2005**, *54*, 3056–3062. [CrossRef] [PubMed]
8. Gromova, L.V.; Fetissov, S.O.; Gruzdkov, A.A. Mechanisms of glucose absorption in the small intestine in health and metabolic diseases and their role in appetite regulation. *Nutrients* **2021**, *13*, 2474. [CrossRef] [PubMed]
9. Piasecka-Kwiatkowska, D.; Warchalewski, J.R.; Zielińska-Dawidziak, M.; Michalak, M. Digestive enzyme inhibitors from grains as potential components of nutraceuticals. *J. Nutr. Sci. Vitaminol.* **2012**, *58*, 217–220. [CrossRef]
10. Saito, S.; Oishi, S.; Shudo, A.; Sugiura, Y.; Yasunaga, K. Glucose response during the night is suppressed by wheat albumin in healthy participants: A randomized controlled trial. *Nutrients* **2019**, *11*, 187. [CrossRef] [PubMed]
11. Shabbir, H.; Kausar, T.; Noreen, S.; Rehman, H.U.; Hussain, A.; Huang, Q.; Gani, A.; Su, S.; Nawaz, A. In vivo screening and antidiabetic potential of polyphenol extracts from guava pulp, seeds and leaves. *Animals* **2020**, *10*, 1714. [CrossRef]
12. Satsu, H. Molecular and cellular studies on the absorption, function, and safety of food components in intestinal epithelial cells. *Biosci. Biotechnol. Biochem.* **2017**, *81*, 419–425. [CrossRef]
13. Satsu, H. Regulation of detoxification enzymes by food components in intestinal epithelial cells. *Food Sci. Technol. Res.* **2019**, *25*, 149–156. [CrossRef]
14. Ishizuka, K.; Kanayama, A.; Satsu, H.; Miyamoto, Y.; Furihata, K.; Shimizu, M. Identification of a taurine transport inhibitory substance in sesame seeds. *Biosci. Biotechnol. Biochem.* **2000**, *64*, 1166–1172. [CrossRef] [PubMed]
15. Ishizuka, K.; Miyamoto, Y.; Satsu, H.; Sato, R.; Shimizu, M. Characteristics of lysophosphatidylcholine in its inhibition of taurine uptake by human intestinal Caco-2 cells. *Biosci. Biotechnol. Biochem.* **2002**, *66*, 730–736. [CrossRef] [PubMed]
16. Satsu, H.; Awara, S.; Unno, T.; Shimizu, M. Suppressive effect of nobiletin and epicatechin gallate on fructose uptake in human intestinal epithelial Caco-2 cells. *Biosci. Biotechnol. Biochem.* **2018**, *82*, 636–646. [CrossRef] [PubMed]
17. Konishi, T.; Satsu, H.; Hatsugai, Y.; Aizawa, K.; Inakuma, T.; Nagata, S.; Sakuda, S.; Nagasawa, H.; Shimizu, M. Inhibitory effect of a bitter melon extract on the P-glycoprotein activity in intestinal Caco-2 cells. *Br. J. Pharmacol.* **2004**, *143*, 379–387. [CrossRef]
18. Steffansen, B.; Pedersen, M.D.L.; Laghmoch, A.M.; Nielsen, C.U. SGLT1-mediated transport in Caco-2 cells is highly dependent on cell bank origin. *J. Pharm. Sci.* **2017**, *106*, 2664–2670. [CrossRef]
19. Kulkarni, C.P.; Thevelein, J.M.; Luyten, W. Characterization of SGLT1-mediated glucose transport in Caco-2 cell monolayers, and absence of its regulation by sugar or epinephrine. *Eur. J. Pharmacol.* **2021**, *897*, 173925. [CrossRef]
20. Assini, J.M.; Mulvihill, E.E.; Huff, M.W. Citrus flavonoids and lipid metabolism. *Curr. Opin. Lipidol.* **2013**, *24*, 34–40. [CrossRef]
21. Lee, Y.S.; Cha, B.Y.; Choi, S.S.; Choi, B.K.; Yonezawa, T.; Teruya, T.; Nagai, K.; Woo, J.T. Nobiletin improves obesity and insulin resistance in high-fat diet-induced obese mice. *J. Nutr. Biochem.* **2013**, *24*, 156–162. [CrossRef]
22. Kang, S.I.; Shin, H.S.; Ko, H.C.; Kim, S.J. Effects of sinensetin on lipid metabolism in mature 3T3-L1 adipocytes. *Phytother Res.* **2013**, *27*, 131–134. [CrossRef]
23. Gonçalves, L.M.; Valente, I.M.; Rodrigues, J.A. An overview on cardamonin. *J. Med. Food* **2014**, *17*, 633–640. [CrossRef]
24. Qin, Y.; Sun, C.Y.; Lu, F.R.; Shu, X.R.; Yang, D.; Chen, L.; She, X.M.; Gregg, N.M.; Guo, T.; Hu, Y. Cardamonin exerts potent activity against multiple myeloma through blockade of NF-κB pathway in vitro. *Leuk. Res.* **2012**, *36*, 514–520. [CrossRef] [PubMed]
25. Sung, B.; Prasad, S.; Yadav, V.R.; Aggarwal, B.B. Cancer cell signaling pathways targeted by spice-derived nutraceuticals. *Nutr. Cancer* **2012**, *64*, 173–197. [CrossRef] [PubMed]
26. Yamamoto, N.; Kawabata, K.; Sawada, K.; Ueda, M.; Fukuda, I.; Kawasaki, K.; Murakami, A.; Ashida, H. Cardamonin stimulates glucose uptake through translocation of glucose transporter-4 in L6 myotubes. *Phytother. Res.* **2011**, *25*, 1218–1224. [CrossRef]
27. Goto, T.; Horita, M.; Nagai, H.; Nagatomo, A.; Nishida, N.; Matsuura, Y.; Nagaoka, S. Tiliroside, a glycosidic flavonoid, inhibits carbohydrate digestion and glucose absorption in the gastrointestinal tract. *Mol. Nutr. Food Res.* **2012**, *56*, 435–445. [CrossRef]
28. Yuca, H.; Özbek, H.; Demirezer, L.Ö.; Kasil, H.G.; Güvenalp, Z. Trans-tiliroside: A potent α-glucosidase inhibitor from the leaves of *Elaeagnus angustifolia* L. *Phytochmistry* **2021**, *188*, 112795. [CrossRef]
29. Lee, Y.S.; Cha, B.Y.; Saito, K.; Yamakawa, H.; Choi, S.S.; Yamaguchi, K.; Yonezawa, T.; Teruya, T.; Nagai, K.; Woo, J.T. Nobiletin improves hyperglycemia and insulin resistance in obese diabetic ob/ob mice. *Biochem. Pharmacol.* **2010**, *79*, 1674–1683. [CrossRef] [PubMed]

30. Kim, Y.J.; Choi, M.S.; Woo, J.T.; Jeong, M.J.; Kim, S.R.; Jung, U.J. Long-term dietary supplementation with low-dose nobiletin ameliorates hepatic steatosis, insulin resistance, and inflammation without altering fat mass in diet-induced obesity. *Mol. Nutr. Food Res.* **2017**, *61*, 1600889. [CrossRef] [PubMed]

31. Gandhi, G.R.; Vasconcelos, A.B.S.; Wu, D.T.; Li, H.B.; Antony, P.J.; Li, H.; Geng, F.; Gurgel, R.Q.; Narain, N.; Gan, R.Y. Citrus flavonoids as promising phytochemicals targeting diabetes and related complications: A systematic review of in vitro and in vivo studies. *Nutrients* **2020**, *12*, 2907. [CrossRef]

32. Kawabata, K.; Sawada, K.; Ikeda, K.; Fukuda, I.; Kawasaki, K.; Yamamoto, N.; Ashida, H. Prenylated chalcones 4-hydroxyderricin and xanthoangelol stimulate glucose uptake in skeletal muscle cells by inducing GLUT4 translocation. *Mol. Nutr. Food Res.* **2011**, *55*, 467–475. [CrossRef] [PubMed]

Article

Hepatocyte-Specific *Phgdh*-Deficient Mice Culminate in Mild Obesity, Insulin Resistance, and Enhanced Vulnerability to Protein Starvation

Momoko Hamano [1,2,*], Kayoko Esaki [3], Kazuki Moriyasu [4], Tokio Yasuda [4], Sinya Mohri [4], Kosuke Tashiro [4,5,6], Yoshio Hirabayashi [7,8] and Shigeki Furuya [2,4,6,*]

1   Department of Bioscience and Bioinformatics, Faculty of Computer Science and Systems Engineering, Kyushu Institute of Technology, Fukuoka 820-8502, Japan
2   Laboratory of Functional Genomics and Metabolism, Faculty of Agriculture, Kyushu University, Fukuoka 819-0395, Japan
3   Laboratory for Neural Cell Dynamics, RIKEN Center for Brain Science, Wako 351-0198, Japan; kayoko.esaki@riken.jp
4   Department of Bioscience and Biotechnology, Graduate School of Bioresource and Bioenvironmental Sciences, Kyushu University, Fukuoka 819-0395, Japan; momotarosan28@gmail.com (K.M.); yasuda.tokio.221@s.kyushu-u.ac.jp (T.Y.); mintonazure@gmail.com (S.M.); ktashiro@grt.kyushu-u.ac.jp (K.T.)
5   Laboratory of Molecular Gene Technology, Faculty of Agriculture, Kyushu University, Fukuoka 819-0395, Japan
6   Innovative Bio-Architecture Center, Faculty of Agriculture, Kyushu University, Fukuoka 819-0395, Japan
7   Cellular Informatics Laboratory, RIKEN, Wako 351-0198, Japan; hirabaya@riken.jp
8   Institute for Environmental and Gender-Specific Medicine, Juntendo University Graduate School of Medicine, Chiba 279-0021, Japan
*   Correspondence: momoko@bio.kyutech.ac.jp (M.H.); furuya.shigeki.805@m.kyushu-u.ac.jp (S.F.)

Citation: Hamano, M.; Esaki, K.; Moriyasu, K.; Yasuda, T.; Mohri, S.; Tashiro, K.; Hirabayashi, Y.; Furuya, S. Hepatocyte-Specific *Phgdh*-Deficient Mice Culminate in Mild Obesity, Insulin Resistance, and Enhanced Vulnerability to Protein Starvation. *Nutrients* **2021**, *13*, 3468. https://doi.org/10.3390/nu13103468

Academic Editor: Yoshinori Katakura

Received: 7 September 2021
Accepted: 27 September 2021
Published: 29 September 2021

Publisher's Note: MDPI stays neutral with regard to jurisdictional claims in published maps and institutional affiliations.

Copyright: © 2021 by the authors. Licensee MDPI, Basel, Switzerland. This article is an open access article distributed under the terms and conditions of the Creative Commons Attribution (CC BY) license (https:// creativecommons.org/licenses/by/ 4.0/).

Abstract: L-Serine (Ser) is synthesized de novo from 3-phosphoglycerate via the phosphorylated pathway committed by phosphoglycerate dehydrogenase (*Phgdh*). A previous study reported that feeding a protein-free diet increased the enzymatic activity of Phgdh in the liver and enhanced Ser synthesis in the rat liver. However, the nutritional and physiological functions of Ser synthesis in the liver remain unclear. To clarify the physiological significance of de novo Ser synthesis in the liver, we generated liver hepatocyte-specific *Phgdh* KO (LKO) mice using an albumin-Cre driver. The LKO mice exhibited a significant gain in body weight compared to Floxed controls at 23 weeks of age and impaired systemic glucose metabolism, which was accompanied by diminished insulin/IGF signaling. Although LKO mice had no apparent defects in steatosis, the molecular signatures of inflammation and stress responses were evident in the liver of LKO mice. Moreover, LKO mice were more vulnerable to protein starvation than the Floxed mice. These observations demonstrate that *Phgdh*-dependent de novo Ser synthesis in liver hepatocytes contributes to the maintenance of systemic glucose tolerance, suppression of inflammatory response, and resistance to protein starvation.

Keywords: *Phgdh*; liver; L-serine deficiency; insulin signaling; glucose tolerance

1. Introduction

L-Serine (Ser), a nutritionally dispensable amino acid, serves as an indispensable metabolite essential for mammalian fetal development [1,2]. Ser is synthesized de novo from 3-phosphoglycerate, which is catalyzed by the phosphorylated pathway composed of 3-phosphoglycerate dehydrogenase (*Phgdh*), phosphoserine aminotransferase 1 (*Psat1*), and phosphoserine phosphatase (*Psph*). Ser is utilized for the synthesis of important metabolic components, such as glycine, nucleotides, glutathione, tetrahydrofolate derivatives, and membrane lipids. We have previously demonstrated the physiological significance of de novo Ser synthesis at the cellular level. Extracellular Ser limitation leads to cell growth arrest and cell death, which is associated with the enhanced phosphorylation of p38MAPK and SAPK/JNK in *Phgdh*-deficient embryonic fibroblasts (KO-MEFs) under conditions of reduced intracellular Ser availability [3]. Simultaneously, p38MAPK is activated in part by 1-deoxy-sphinganine (doxSA), an atypical sphingolipid, which is synthesized by serine palmitoyltransferase [3,4]. Intracellular Ser deficiency induces the rapid activation of the integrated stress response (ISR) pathway, altered expression profiles of Atf4-dependent in KO-MEFs, and enhanced vulnerability to oxidative stress with the simultaneous upregulation of inflammatory gene expression [5]. Moreover, we demonstrated that extracellular Ser limitation led to a transient transcriptional activation of Atf4-target genes, including cation transport regulator-like protein 1 (*Chac1*) in a mouse hepatocarcinoma cell line expressing Phgdh [6]. These observations indicate that reduced availability of Ser by *Phgdh* disruption elicits a wide variety of stress and injury responses in non-malignant cells, at least in an in vitro cell culture setting.

In parallel with these cell culture studies, our in vivo studies demonstrated that systemic *Phgdh* deletion resulted in severe intrauterine growth retardation and embryonic lethality, which recaptures the major symptoms of human Neu-Laxova syndrome patients. Unlike conventional KO mice, brain-specific *Phgdh* knockout (KO) mice were able to escape from embryonic lethality but exhibited marked reductions in both L- and D-Ser levels in the brain, which were accompanied by a diminished N-methyl-D-aspartate (NMDA) receptor function [7]. These observations reinforce the importance of de novo Ser synthesis in physiological processes, including embryonic development and mature brain functions. However, Phgdh is expressed at higher levels in certain tissues, including the heart, kidney, muscle, and liver [8]. Among these tissues, the expression and activity of the phosphorylated pathway in the liver are regulated by systemic levels of protein/amino acid nutrition and hormones [7,8]. Recent studies have implicated that Phgdh-dependent Ser synthesis supports general lipid homeostasis [9] and appears to prevent non-alcoholic fatty liver disease [10–12]. Nonetheless, it remains unclear how de novo Ser synthesis via the phosphorylated pathway contributes to the physiological function of the liver at steady state.

To clarify the physiological significance of de novo Ser synthesis in the liver, we generated liver hepatocyte-specific *Phgdh* KO (LKO) mice using an albumin-Cre driver. In this study, we demonstrated that hepatocyte-specific *Phgdh*-deficient mice led to mild obesity, deteriorated glucose tolerance, and increased mortality when fed a protein-free diet.

2. Materials and Methods

2.1. Hepatocyte-Specific Phgdh Knockout Mice

Mice with homozygous conditional *Phgdh* alleles (*Phgdh Phgdh*$^{flox/flox}$), hereafter called Floxed, were obtained as previously described [7]. The presence of a conditional allele (*Phgdh*^{flox}) was identified by PCR using tail DNA with a primer pair directed against the third intron (forward primer, 5′-CATGAGGAA CTGAACTGAAGGATTGA-3′; reverse primer, 5′-CAAGGAGGCTCACACATCCCAGAAC-3′), which generated a 310 bp amplicon, and the fourth and fifth intron (forward primer, 5′-CATGAGGAACTGAACTGAAGGATTGA-3′; reverse primer, 5′-CTTCAGCTCTCATGGCAGACGAGCA-3′), which generated a 350 bp amplicon. Mice conditionally lacking *Phgdh* in hepatocytes (*Albumin* (Alb)$^{+/Cre}$;*Phgdh*$^{flox/flox}$), hereafter called LKO, were obtained by interbreeding female Floxed mice with male Floxed mice carrying the Alb-Cre transgene, which were generated by crossing Floxed mice with Alb-Cre transgenic mice. To detect the Cre transgene used in this study, PCR was carried out with the primers 5′-AATTTGCCTGCATTACCGGTCGATGCAACG-3′ and 5′-CCATTTCCGGTTATTCAACTTGCACCATGC-3′, which generated a 190 bp amplicon of part of the Cre-coding region. Unlike conventional *Phgdh* knockout mice (*Phgdh*$^{-/-}$), LKO pups were born at the expected Mendelian ratio when female Floxed mice were crossed with male LKO mice (data not shown). Littermates with the Floxed genotype (*Phgdh*$^{flox/flox}$) were used as controls.

LKO and Floxed mice were maintained in a 12-h light/dark cycle with unlimited access to water and laboratory chow containing 20% casein. The animal experimental protocols for this study were approved by the Animal Ethics Committees of the RIKEN Center for Brain Science (H25-2-241) and Kyushu University (A27-103).

2.2. Glucose Tolerance Test

Glucose solution was prepared as 2.5 g/10 mL in saline and sterilized by filtration. A glucose tolerance test was performed by intraperitoneally injecting glucose (2 g/kg body weight) into mice after overnight fasting. Tail-vein blood samples were collected at 0 (prior to administration) and 30, 60, 90, and 120 min after glucose administration. Blood glucose levels were measured using an ACCU-CHEK system (Roche Diagnostics, Tokyo, Japan).

2.3. Feeding of Protein-Free Diet

Mice were maintained in a 12-h light/dark cycle with unlimited access to food and water. Male Floxed and LKO mice (10 weeks old, $n = 5$ each) were fed a protein-free diet (TestDiet, #5765; casein-vitamin free 0%, sucrose 36.15%) and normal diet (TestDiet, #5755; Casein-vitamin free 21%, Sucrose 15%) for 3 weeks, and their body weights were measured daily.

2.4. RNA Isolation and Microarray Analysis

Total RNA was extracted from the livers of LKO and littermate Floxed mice (male, 30 weeks old, $n = 6$, each genotype) using the RiboPure^{TM} Kit (Thermo Fisher Scientific, Tokyo, Japan) according to the manufacturer's instructions. After the extraction of total RNA, the concentration of RNA was measured using a NanoDrop LITE spectrophotometer (Thermo Fisher Scientific). Total RNA (5 μg) was treated with DNase (TURBO DNA-free^{TM}, Thermo Fisher Scientific). After DNase treatment, the quality of treated RNA was assessed using an RNA Nano Chip and a 2100 Bioanalyzer (Agilent Technologies, Santa Clara, CA, USA). Then, 200 ng of total RNA was reverse-transcribed into double-stranded cDNA, then transcribed and labeled with Cyanine 3-CTP using T7 RNA Polymerase. Next, 1.65 mg of the purified cyanine 3-labeled cRNA was hybridized to mouse GE 4 × 44 K v2 Microarrays (Agilent Technologies, Tokyo, Japan) according to the manufacturer's protocol. The signal intensity was measured using a G2565CA Microarray Scanner System (Agilent Technologies). The processed intensities were normalized across the samples and loaded using quantile normalization. All microarray data were submitted to the Gene Expression Omnibus (accession number GSE179912). The raw signal intensities of all samples were $\log_2$-transformed and normalized by quantile algorithm with 'preprocessCore' library package [13] on Bioconductor software. We selected the probes, excluding the control probes, where the detection p-values of all samples were less than 0.01 and used them to identify differentially expressed genes. We then applied the Linear Models for Microarray Analysis (limma) package [14] of Bioconductor software. The criteria were limma $p < 0.05$, between LKO and Flox liver samples.

2.5. KEGG Pathway Enrichment Analysis for Differentially Expressed Genes

The Database for Annotation, Visualization and Integrated Discovery (DAVID) (https://david.ncifcrf.gov/ (accessed on 14 March 2020)) [15] was used for KEGG pathway enrichment analysis of differentially expressed genes (DEGs) in LKO mice whose expression was upregulated or downregulated compared to Floxed mice. The top GO terms and KEGG pathway in the annotation clusters that ranked in the functional annotation clustering function with statistical significance ($p < 0.05$) were extracted. The enrichment p-values of all extracted GO terms and KEGG pathways for each module were calculated using DAVID.

2.6. Gene Ontology Enrichment Analysis for Differentially Expressed Genes

Gene set enrichment analysis (GSEA) was used to determine whether a priori defined sets of genes showed significantly enriched GO terms. GSEA was also used to identify the

GO terms associated with significantly enriched upregulated or downregulated genes in the liver of LKO mice.

2.7. Ingenuity Pathways Analysis

Biologically relevant networks were created using the Ingenuity Pathways Analysis (IPA) program (http://www.Ingenuity.com (accessed on 15 April 2015)) [16] as previously described. Based on algorithmically generated connectivity between gene–gene, gene–protein, and protein–protein interactions, the program develops functional molecular networks that overlay genes in the dataset. This program calculated p-values for each network by comparing the number of focus genes that were mapped in a given network relative to the total number of occurrences of those genes in all networks. The score for each network is shown as the negative log of the p-value, which indicates the likelihood of finding a set of genes in the network by random chance.

2.8. Quantitative Analysis of mRNA Expression

Total RNA was extracted from the liver at 30 weeks using the Ribo Pure kit (Thermo Fisher Scientific, Waltham, MA, USA), as described above. Following isolation, 1 μg of DNase-treated RNA was used to generate cDNA by reverse transcription using the High-Capacity cDNA Reverse Transcription kit (Applied Biosystems, Life Technologies Japan Ltd., Tokyo, Japan). Quantitative real-time PCR was performed using a Model Mx3000P Real Time PCR system (Agilent Technologies Japan Ltd, Tokyo, Japan) containing Thunderbird SYBR qPCR Mix (TOYOBO, Osaka, Japan) and the reference dye ROX according to the manufacturer's recommendations. Primer sequences used were *Phgdh* (forward, 5′-TGGAGGAGATCTGGCCTCTC-3′, and reverse, 5′-GCCTCCTCGAGCACAGTTCA-3′), 6-phosphofructo-2-kinase/fructose-2,6-biphosphatase 3 (*Pfkfb3*) (forward, 5′-CTACGAGCA-GTGGAAGGCACTC-3′, and reverse, 5′-AATTCCATGATCACAGGCTCCA-3′), pyruvate dehydrogenase lipoamide kinase isozyme 4 (*Pdk4*) (forward, 5′-ACCGCATTTCTACTCGG-ATGC-3′, and reverse, 5′-CGCAGAGCATCTTTGCACACT-3′), early growth response protein 1 (*Egr1*) (forward, 5′-CCGAGCGAACAACCCTATGA-3′, and reverse, 5′-GTCATGCTC-ACGAGGCCACT-3′), and fatty acid synthase (*Fasn*) (forward, 5′-TTCCAAGACGAAAATG-ATGC-3′, and reverse, 5′-AATTGTGGGATCAGGAGAGC-3′). All reactions were performed in triplicate. Data analysis was carried out using the cycle threshold values of target gene expression normalized to actin as the internal control.

2.9. Western Blot Analysis

Samples of liver at 30 weeks were homogenized in a buffer containing 1.25 mM Tris-HCl (pH 7.6), 150 mM NaCl, 1% NP40, 1% sodium deoxycholate, 0.1% SDS, a protease inhibitor cocktail (Nacalai Tesque), and a phosphatase inhibitor cocktail (Nacalai Tesque). Homogenates were centrifuged at $20,000 \times g$ for 10 min to obtain total protein extracts, and concentrations were determined using a Protein Assay Bicinchoninate kit (Nacalai Tesque). Protein samples were fractionated by 7.5% SDS-polyacrylamide gel electrophoresis and transferred onto a polyvinylidene fluoride membrane (Bio-Rad, Hercules, CA, USA). Blotted proteins were probed with the following primary antibodies: anti-Phgdh (rabbit, 0.3 g/mL, provided by Dr. M. Watanabe at Hokkaido University), which is cross-reactive with the mouse homolog, anti-NF-κB (also known as Nuclear factor kappa-light-chain-enhancer of activated B cells) (rabbit monoclonal, 1:1000; Cell Signaling Technology, Danvers, MA, USA), anti-phospho NF-κB Ser 536 (rabbit monoclonal, 1:1000; Cell Signaling Technology, Danvers, MA, USA), anti-Akt (also known as protein kinase B) (rabbit polyclonal, 1:1000; Cell Signaling Technology, Danvers, MA, USA), anti-phospho Akt Thr 308 (rabbit polyclonal, 1:1000; Cell Signaling Technology, Danvers, MA, USA), anti-GSK3β (also known as glycogen synthase kinase 3β) (rabbit monoclonal, 1:1000; Cell Signaling Technology, Danvers, MA, USA), anti-phospho GSK3β Ser 9 (rabbit polyclonal, 1:1000; Cell Signaling Technology, Danvers, MA, USA), anti-IRS1 (also known as insulin receptor substrate 1) (rabbit monoclonal, 1:1000; Cell Signaling Technology, Danvers, MA,

USA), anti-phospho IRS1 Ser 612 (rabbit monoclonal, 1:1000; Cell Signaling Technology, Danvers, MA, USA), anti-phospho IRS1 Ser 636/639 (mouse 632/635) (rabbit polyclonal, 1:1000; Cell Signaling Technology, Danvers, MA, USA), anti-Erk1/2 (also known as extracellular signal-regulated kinase 1/2 and p44/42 MAPK) (rabbit monoclonal, 1:1000; Cell Signaling Technology, Danvers, MA, USA), anti-phospho Erk1/2 Thr 202/Thr 204 (rabbit monoclonal, 1:1000; Cell Signaling Technology, Danvers, MA, USA), anti-SAPK/JNK (also known as stress-activated protein kinase (SAPK)/jun amino terminal kinase (JNK)) (rabbit monoclonal, 1:250; Cell Signaling Technology, Danvers, MA, USA), anti-phospho SAPK/JNK (rabbit monoclonal, 1:250; Cell Signaling Technology, Danvers, MA, USA), anti-Egr1 (also known as early growth response protein 1) (rabbit monoclonal, 1:200; Cell Signaling Technology, Danvers, MA, USA), anti- β-Actin (mouse monoclonal, 1:500; FUJIFILM Wako Pure Chemical Corporation, Japan), and anti-Gapdh (mouse monoclonal, Chemicon, 1:50,000; Merck Millipore, Billerica, MA, USA). The bound antibodies were visualized with the Pierce SuperSignal West Pico Chemiluminescence Detection System (SuperSignal; Thermo Fisher Scientific) after incubation with the appropriate secondary antibodies conjugated with horseradish peroxidase (Cell Signaling Technology Japan K.K., Tokyo, Japan). The chemiluminescent signal was detected by exposure to X-ray films (FUJIFILM, Tokyo, Japan), and signal intensities were quantified using the CS Image Analyzer 3 software (ATTO Corp., Tokyo, Japan).

2.10. Histological Evaluation

Mice (male, 28–32 weeks old: $n = 5$ in each genotype) were anesthetized with isoflurane and perfused with 4% paraformaldehyde in 0.1 M sodium phosphate buffer (pH 7.2) after removing blood with 0.1 M sodium phosphate buffer (pH 7.2). The livers were post-fixed overnight in the same fixative. Hematoxylin and eosin staining was performed by Soshiki Kagaku, Lad, Inc. (Yokohama, Japan).

2.11. Serum Biochemical Test

Blood was collected after decapitation under anesthesia with isoflurane. Serum was prepared by centrifuging blood at $3000 \times g$ for 15 min after standing still for 2 h. Sample analysis of serum (male mice, 28–32 weeks old) was performed by the Health Sciences Research Institute East Japan Co., Ltd. (Saitama, Japan). The serum concentrations of lipoprotein (Lipopro), total cholesterol (Total-Cho), LDL cholesterol (LDL-C), HDL cholesterol (HDL-C), aspartate transaminase (AST), alanine transaminase (ALT), and non-esterified fatty acid (NEFA) were measured.

2.12. Amino Acid Analysis

The liver and kidney at 30 weeks were homogenized in 5 volumes of ultrapure water and centrifuged at 15,000 rpm for 30 min at 4 °C. The supernatant was mixed with a 1/10 volume of 5% perchloric acid ($HClO_4$). The mixture was incubated on ice for 25 min and then centrifuged for 30 min at 15,000 rpm at 4 °C. The supernatant was neutralized by adding 1/10 volume of 8 N KOH. The sera were diluted 3-fold in sterilized water. Proteins in the supernatant and diluted sera were removed by adding a 1/10 volume of 60% perchloric acid, and the solution was adjusted to pH 7–8 with 8 M KOH. The amino acid composition of the supernatant was determined using an Acquity UPLC H-class system (Waters, Milford, MA, USA).

2.13. Statistical Analysis

To detect DEGs in the liver of LKO mice compared to Floxed mice, the fold change of gene expression was calculated compared to Floxed mice, using the thresholds of $|\log_2\text{Fold change}| \geq 1$. To visualize the distributions of gene expression levels, a volcano plot was generated in R. Unpaired two-tailed Student's t-test was applied using adjusted $p < 0.05$. Differences between two groups were examined using Student's t-test were considered statistically significant at $p < 0.05$. All statistical analyses were performed using KaleidaGraph 4.0 (Synergy Software, Tokyo, Japan).

3. Results

To inactivate *Phgdh* in a liver-hepatocyte-specific manner, the *Phgdh* allele was disrupted by crossing the female Floxed mice (*Phgdh*$^{flox/flox}$) with male mice expressing Cre under the control of the *albumin* promoter (referred to as Alb-Cre mice). We obtained mice in subsequent generations with the Alb$^{+/Cre}$;*Phgdh*$^{flox/flox}$ genotype, referred to as LKO. The Floxed allele of *Phgdh* without exon deletion (*Phgdh*^{flox}) was detected in the livers of both Floxed and LKO mice, while the null allele lacking the fourth and fifth exons of *Phgdh* (*Phgdh*$^-$) was detected only in the liver (Figure 1A), but not in the kidney of LKO mice (data not shown). The efficacy of Cre-mediated deletion of *Phgdh* was assessed by qRT-PCR and Western blotting. The mRNA and protein levels of *Phgdh* in the liver of LKO mice were lower than those in Floxed mice (Figure 1B,C). The expression of *Phgdh* mRNA and protein was significantly reduced but not completely abolished in the liver of LKO mice because the Cre-mediated recombination by the Alb-Cre transgene occurred only in hepatocytes, and the *Phgdh* expression was maintained in other cell types in the liver of LKO mice.

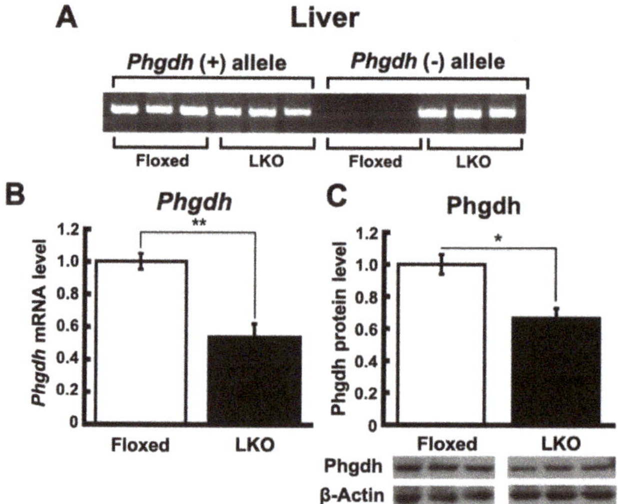

Figure 1. Targeted inactivation of *Phgdh* in liver. (**A**) Genotyping PCR from *Phgdh*(+) allele and *Phgdh*(−) allele in the liver of Floxed and LKO mice at 30 weeks ($n = 6$ each). (**B**) mRNA level of Phgdh in the liver of Floxed ($n = 3$) and LKO ($n = 4$) mice at 30 weeks. (**C**) Protein level of Phgdh in the liver of Floxed ($n = 4$) and LKO ($n = 6$) mice at 30 weeks. Comparable staining of β-actin was used to verify equivalent protein loading. Student's *t*-test, * $p < 0.05$, ** $p < 0.005$.

To evaluate the in vivo phenotypes caused by *Phgdh* deletion in hepatocytes, we compared the body and organ weights of LKO and Floxed mice. Interestingly, LKO mice exhibited a subtle but significant weight gain compared to the Floxed mice after 23 weeks of age (Figure 2A). Body weight, organ weight of liver, epididymal white adipose tissue (eWAT), parametrial white adipose tissue (pWAT), and mesenteric white adipose tissue (mWAT) were significantly higher than those of Floxed mice at 30 weeks of age (Figures 2B and S1A,B). However, steatosis was not observed in the liver of LKO mice (Figure 2C,D). To clarify the alteration of liver function in LKO mice, we performed a biochemical analysis of liver markers in the serum. There were no significant changes in the concentrations of cholesterol, non-esterified fatty acid (NEFA), aspartate aminotransferase (AST), and alanine aminotransferase (ALT) in LKO mice (Figure 3A). To evaluate the effect on glucose metabolism, we performed a glucose tolerance test. The blood glucose levels in the LKO mice were significantly higher than those in the Floxed mice after intraperitoneal glucose injection (Figure 3B). Thus, hepatocyte-specific *Phgdh* deletion impaired glucose clearance and led to mild obesity without affecting the levels of liver biochemical markers in the serum.

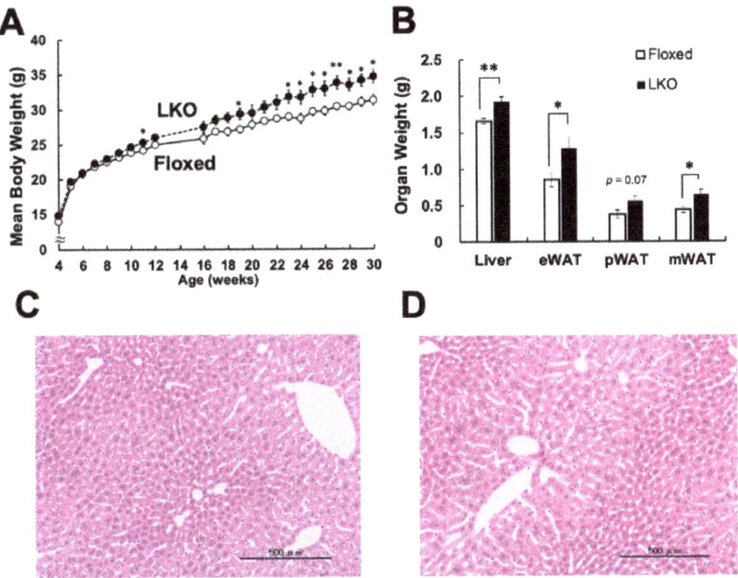

Figure 2. Hepatocyte-specific deletion of Phgdh causes the weight gain. (**A**) Growth curve of Floxed (white node) and LKO (black node) mice from 4 to 30 weeks ($n = 6$ each). (**B**) The body weight and the organ weight of the liver, epididymal white adipose tissue (eWAT), parametrial white adipose tissue (pWAT), and mesenteric white adipose tissue (mWAT) were measured in Floxed and LKO mice at 30 weeks ($n = 6$ each). (**C,D**) Histological evaluation of the liver in Floxed (**C**) and LKO (**D**) mice at 28 weeks ($n = 5$ each). Representative images are shown. Scale bar, 500 μm. Student's *t*-test, * $p < 0.05$, ** $p < 0.005$.

We then examined the alterations in the amino acid metabolism in LKO mice. To our surprise, the Ser concentration in the liver of LKO mice was not altered compared to that of Floxed mice (Table 1A), while Ser concentration in LKO kidneys was significantly increased compared to Floxed mice (Table 1B). In addition, the concentrations of L-aspartic acid (L-Asp), L-histidine (L-His), L-arginine (L-Arg), L-alanine (L-Ala), L-tyrosine (L-Tyr), L-methionine (L-Met), L-phenylalanine (L-Phe), L-isoleucine, γ-aminobutyric acid, and L-leucine were increased in the kidneys of LKO mice (Table 1B). This suggests that amino acid metabolism in LKO mice was altered in the kidney and Ser availability in the liver of LKO mice was maintained via the supply of non-hepatic cells in the liver and/or kidney.

To assess whether hepatocyte-specific *Phgdh* deletion in the liver modulates gene expression, we performed a microarray analysis of liver mRNA. We identified 2770 genes that were significantly differentially expressed in the liver of LKO mice compared to that of Floxed mice (see Materials and Methods). Among them, 1191 genes were downregulated (<0.90–0.17-fold) and 1579 genes were upregulated (>1.1–4.9-fold) (Figure S2). A KEGG pathway analysis by DAVID demonstrated that the PPAR signaling pathway, which regulates lipid metabolism in the liver, was enriched in the upregulated DEGs in the liver of LKO mice (Table 2A), while the phosphoinositide 3-kinase (PI3K)-Akt signaling pathway, which is involved in intracellular insulin signaling, was enriched in the downregulated DEGs of the liver in LKO mice (Table 2B). An ingenuity pathway analysis (IPA) was also performed to identify the affected gene networks and canonical signaling pathways. By analyzing DEGs in the liver of LKO mice, IPA identified the IL-10 signaling network (Supplementary Figure S3) and generated lists of significantly affected "canonical pathway" (Table 3A), "disease and disorder" (Table 3B), and "hepatotoxicity" (Table 3C). These include IL-12 signaling pathways in the top canonical pathway, multiple hepatotoxicity, and hepatic system diseases. These results suggest that *Phgdh* deletion in hepatocytes induces the dysregulation of multiple canonical pathways in the liver. Since it has been well documented that inflammation deteriorates systemic insulin sensitivity and obesity [17], we tested whether an inflammation-related response occurred in the liver of LKO mice. First, we examined the phosphorylation of nuclear factor-kappa B (NF-κB), which serves as an essential transcription factor for inflammatory responses. The phosphorylated NF-κB at Ser-536 showed an increasing trend in the liver of LKO mice compared to Floxed mice (Figure 4). These observations reveal the occurrence of inflammation-related molecular changes in the liver of LKO mice.

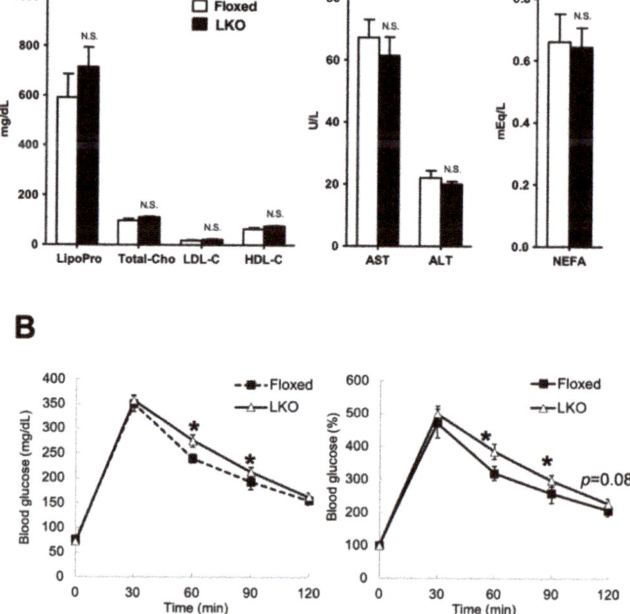

Figure 3. Measurement of serum biochemical markers. (**A**) The values of the serum biochemical test of lipoprotein (Lipopro), total cholesterol (Total-Cho), LDL cholesterol (LDL-C), HDL cholesterol (HDL-C), aspartate transaminase (AST), alanine transaminase (ALT), and non-esterified fatty acid (NEFA) (*n* = 6 each). (**B**) Blood glucose level in Floxed (black square node) and LKO (white triangle node) mice after glucose administration (Floxed: *n* = 4, LKO: *n* = 6). Left graph shows the concentration of blood glucose, and right graph shows the rate of increase of blood glucose. Student's *t*-test, * *p* < 0.05. N.S.: not significant.

Table 1. Evaluation of amino acid metabolism in the liver and kidney of LKO mice.

(A)				
Amino Acid Concentration in Liver				
Concentration (nmol/g Weight Tissue)				
Amino Acid	**Floxed Group**	**LKO Group**	**Ratio (%: LKO/Floxed)**	*p*-**Value**
L-Asp	79.17 ± 11.29	105.21 ± 14.55	132.9	N.S.
L-Glu	713.44 ± 97.8	731.2 ± 107.57	102.5	N.S.
L-Ser	323.55 ± 39.72	343.51 ± 37.4	106.2	N.S.
L-Gln	2946.45 ± 308.2	3167.27 ± 174.77	107.5	N.S.
L-His	364.72 ± 6.91	414.93 ± 28.46	113.8	N.S.
L-Thr	234.27 ± 17.43	224.61 ± 15.36	95.9	N.S.
Gly	1715.62 ± 49.12	1717.33 ± 116	100.1	N.S.
L-Arg	26.97 ± 14.24	29.42 ± 18.42	109.1	N.S.
Tau	9200.13 ± 823.4	7966.99 ± 1591.07	86.6	N.S.
GABA	130.97 ± 2.42	135.75 ± 3.4	103.7	N.S.
L-Ala	2618.32 ± 210.7	2746.58 ± 140.02	104.9	N.S.
L-Tyr	248.25 ± 21.96	238.91 ± 13.9	96.2	N.S.
L-Val	399.78 ± 100.8	455.24 ± 169.22	113.9	N.S.
L-Met	68.41 ± 8.43	64.92 ± 6.41	94.9	N.S.
L-Phe	249.84 ± 16	237.87 ± 12.7	95.2	N.S.
L-Ile	99.47 ± 8.42	104.51 ± 8.33	105.1	N.S.
L-Leu	496.8 ± 36.46	456.04 ± 21.95	91.8	N.S.

(B)				
Amino Acid Concentration in Kidney				
Concentration (nmol/g Weight Tissue)				
Amino Acid	**Floxed Group**	**LKO Group**	**Ratio (%: LKO/Floxed)**	*p*-**Value**
L-Asp	1464.3 ± 162.39	1439.86 ± 89.12	98.3	0.02
L-Glu	4171.94 ± 321.28	3847.12 ± 214.57	92.2	N.S.
L-Ser	502.84 ± 41.5	702.08 ± 41.04	139.6	0.007
L-Gln	847.54 ± 39.02	937.62 ± 40.59	110.6	N.S.
L-His	94.1 ± 10.05	127.82 ± 9.09	135.8	0.009
L-Thr	300.35 ± 25.49	354.49 ± 19.05	118.0	N.S.
Gly	3641.13 ± 182.57	3830.17 ± 273.83	105.2	N.S.
L-Arg	178.67 ± 22.76	250.48 ± 17.65	140.2	0.03
Tau	5757.92 ± 238.44	5924.89 ± 366.5	102.9	N.S.
GABA	81.84 ± 1.09	95.92 ± 6.36	117.2	0.054
L-Ala	983.93 ± 66.02	1126.23 ± 53.74	114.5	0.07
L-Tyr	350.66 ± 27.68	444.34 ± 27.39	126.7	0.04
L-Val	330.69 ± 91.3	406.58 ± 80.47	122.9	N.S.
L-Met	75.69 ± 8.7	106.19 ± 8.98	140.3	0.03
L-Phe	187.23 ± 12.96	248.64 ± 15.72	132.8	0.01
L-Ile	104.43 ± 5.86	135.03 ± 9.53	129.3	0.02
L-Leu	430.37 ± 25.3	562.25 ± 31.6	130.6	0.009

(**A**,**B**) Amino acid concentrations in Floxed and LKO mice in the liver (**A**) and kidney (**B**) (*n* = 6 each). The ratio shows each amino acid concentration level in LKO mice compared to that in Floxed mice. Statistical analysis was performed using the Student's *t*-test. N.S. indicates not statistically significant. L-Asp: L-Asparatic acid, L-Glu: Glutamic acid, L-Ser: Serine, L-Gln, Glutamine, L-His: Histidine, L-Thr: Threonine, Gly: Glycine, L-Arg: L-Arginine, Tau: Taurine, GABA: γ(gamma)-aminobutyric acid, L-Ala: L-alanine, L-Tyr: L-tyrosine, L-Val: L-valine, L-Met: L-methionine, L-Phe: L-phenylalanine, L-Ile: L-isoleucine, L-Leu: l-leucine.

Table 2. Detection of altered signaling pathway in the liver of LKO mice.

(A)	
Enriched KEGG Pathway in Up-Regulated Genes	
Term	***p*-Value**
mmu04740:O1factory transduction	2.90×10^{-18}
mmu03320:PPAR signaling pathway	0.02511031
mmu04360:Axon guidance	0.04367046

(B)	
Enriched KEGG Pathway in Down-Regulated Genes	
Term	***p*-Value**
mmu05211:Renal cell carcinoma	0.00163639
mmu03015:mRNA surveillance pathway	0.00207197
mmu05220:Chronic myeloid leukemia	0.01099769
mmu04510:Focal adhesion	0.01549065
mmu03018:RNA degradation	0.01850316
mmu04630:Jak-STAT signaling pathway	0.01888063
mmu05221:Acute myeloid leukemia	0.02040881
mmu04151:PI3K-Akt signaling pathway	0.03143164
mmu04152:AMPK signaling pathway	0.03272039
mmu05212:Pancreatic cancer	0.03310616
mmu05200:Pathways in cancer	0.03348629
mmuO4713:Circadian entrainment	0.03641698
mmu04015:Rapl signaling pathway	0.04736727

KEGG pathway enrichment analysis of differentially expressed genes (DEGs) in LKO mice whose expression was upregulated (**A**) or downregulated (**B**) compared to Floxed mice.

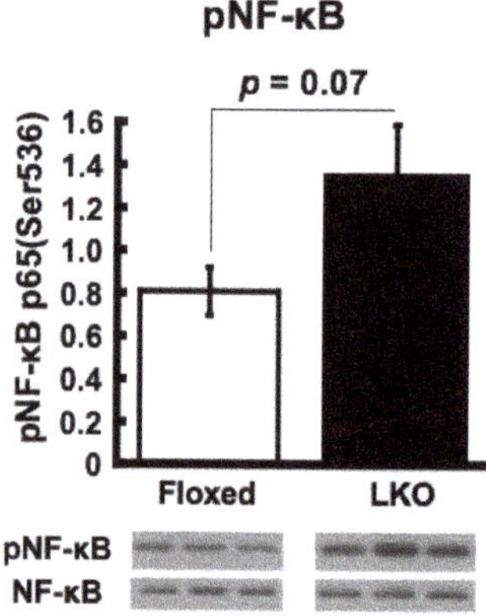

Figure 4. *Phgdh* deletion in hepatocytes induces inflammation and stress response in the liver. Protein levels of phosphorylated NK-κB in the liver of Floxed and LKO mice at 30 weeks (n= 6 each). Comparable staining of NK-κB was used to verify equivalent protein loading.

A gene set enrichment analysis (GSEA) was used to identify functionally related groups of gene sets, annotated use systems, the Gene Ontology (GO) biological processes, and KEGG pathways in the liver of LKO mice. GSEA identified positively and negatively correlated gene sets in GO biological processes (GOBP) (Table 4A,B) and KEGG pathways (Table 4B,D). Several gene sets with higher normalized enrichment scores in positively correlated gene sets in GOBP "Negative regulation of nucleocytoplasmic transport," "Fatty acid beta oxidation," and "Electron transport chain" were considered as liver-relevant positively correlated gene sets (Table 4A and Figure 5A,B). In contrast, "Regulation of cytoplasmic translation," "RNA phosphodiester bond hydrolysis exonucleolytic," and "processes of RNA metabolism "were detected with higher normalized enrichment scores in negatively correlated gene sets of GOBP in the liver of LKO mice (Table 4B and Figure 5C,D). In addition, "Glutathione metabolism" and "Fatty acid metabolism" were considered as liver-relevant positively correlated KEGG pathways with higher normalized enrichment scores in the liver of LKO mice (Table 4C and Figure 5E,F). "RNA degradation" and "Basal transcription factors" were detected as negatively correlated KEGG pathways with higher normalized enrichment scores (Table 4D and Figure 5G,H). These results suggest that multiple biological processes, such as RNA metabolism, mitochondrial function, and energy metabolism, were altered by the hepatocyte-specific deletion of *Phgdh* in the liver.

Table 3. Detection of altered pathway, toxicity, and disease/disorder by IPA in the liver of LKO mice.

(A)		
Top Canonical Pathway		
Name	***p*-Value**	**Overlap**
IL-12 Signaling and Production in Macrophages	5.96×10^{-4}	9.8% (11/112)
Ephrin Receptor Signaling	6.14×10^{-4}	8.3% (14/168)
UVA-Induced MAPK Signaling	9.92×10^{-4}	10.7% (9/84)
FLT3 Signaling in Hematopoietic Progenitor Cells	1.23×10^{-3}	11.4% (8/70)
Calcium Signaling	1.27×10^{-3}	8.1% (13/161)
(B)		
Diseases and Disorders		
Name	***p*-Value**	**# Molecules**
Cancer	$0.0304–3.00 \times 10^{-6}$	328
Hematological Disease	$0.0304–3.00 \times 10^{-6}$	52
Immunological Disease	$0.0304–3.00 \times 10^{-6}$	43
Organismal Injury and Abnormalities	$0.0304–3.00 \times 10^{-6}$	338
Tumor Morphology	$0.0304–3.00 \times 10^{-6}$	11
(C)		
Hepatotoxicity		
Name	***p*-Value**	**# Molecules**
Liver Regeneration	0.440–0.0228	3
Liver Edema	0.0304	1
Liver Fibrosis	0.306–0.0304	7
Liver Necrosis/Cell Death	0.247–0.0304	11
Hepatocellular Carcinoma	1.00–0.0352	12

IPA detected the top canonical pathway (**A**), disease and disorder (**B**), and hepatotoxicity (**C**) following the DEGs in the livers of LKO mice.

Since the current KEGG pathway analysis in DAVID points to an alteration in the intracellular insulin signaling cascade in the liver of LKO mice (Table 2B), we examined protein phosphorylation of components in the cascade. Western blot analysis demonstrated a trend toward decreasing phosphorylation of Akt at Thr-308 (Figure 6A), and a significant reduction in the phosphorylation of GSK3β at Ser-9 (Figure 6B) in the liver of LKO mice. Glycogen synthase is negatively regulated by inhibitory phosphorylation by GSK3β, which is also negatively regulated by Akt phosphorylation at serine-9 [18,19]. Decreased phosphorylation levels of Akt and GSK3β coincide with impaired glucose tolerance in the liver of LKO mice. We then examined the phosphorylation status of insulin receptor substrate 1 (IRS-1) because serine/threonine phosphorylation of IRS-1 is closely related to insulin resistance [20–22]. Among them, it is well documented that the phosphorylation of IRS-1 at Ser-612 and Ser-632/635 residues is negatively correlated with insulin signaling [23,24]. The phosphorylation levels of Ser-612 and Ser-632/635 were significantly higher in the livers of LKO mice than in Floxed mice (Figure 6C,D). Taken together, these observations indicate insulin resistance in the liver of LKO mice.

Table 4. Top 20 GO terms and KEGG pathways of differentially expressed genes in the liver of LKO mice. Top 20 GO terms of upregulated (**A**) and downregulated (**B**) genes in the liver of LKO mice compared with Floxed mice. Top 20 KEGG pathways of upregulated genes (**C**) and downregulated genes (**D**) in liver of LKO mice compared with Floxed mice.

(A)		
Name	**NES**	**NOM *p*-Value**
GOBP_NEGATIVE_REGULATION_OF_NUCLEOCYTOPLASMIC_TRANSPORT	1.7842134	0.00613497
GOBP_BRANCHED_CHAIN_AMINO_ACID_METABOLIC_PROCESS	1.7758015	0.00203666
GOBP_FATTY_ACID_BETA_OXIDATION	1.7479441	0.01434426
GOBP_REGULATION_OF_CAMP_DEPENDENT_PROTEIN_KINASE_ ACTIVITY	1.732202	0
GOBP_ELECTRON_TRANSPORT_CHAIN	1.7117621	0.02362205
GOBP_CELLULAR_METABOLIC_COMPOUND_SALVAGE	1.7083353	0
GOBP_ATP_SYNTHESIS_COUPLED_ELECTRON_TRANSPORT	1.707343	0.04918033
GOBP_NOTOCHORD_DEVELOPMENT	1.6475885	0.00393701
GOBP_COCHLEA_DEVELOPMENT	1.646138	0
GOBP_SECRETION_BY_TISSUE	1.6355267	0.00587084
GOBP_PYRIMIDINE_NUCLEOSIDE_TRIPHOSPHATE_METABOLIC_PROCESS	1.6341659	0.01185771
GOBP_PYRIMIDINE_RIBONUCLEOSIDE_TRIPHOSPHATE_METABOLIC_PROCESS	1.6233511	0.02564103
GOBP.METANEPHRIC_N EPHRON_MORPHOGEN ESIS	1.6204721	0.01996008
GOBP_REGULATION_OF_CARDIAC_CONDUCTION	1.6196082	0.01859504
GOBP_RESPIRATORY_ELECTRON_TRANSPORT_CHAIN	1.6190714	0.076
GOBP_SPERM_EGG_RECOGNITION	1.6184356	0.00592885
GOBP_DNA_UNWINDING_INVOLVED_IN_DNA_REPLICATION	1.618192	0.01030928
GOBP_METANEPHROS_MORPHOGENESIS	1.6089716	0.01629328
GOBP_MONOVALENT_INORGANIC_ANION_HOMEOSTASIS	1.6088727	0.00804829
GOBP_PYRIMIDINE_NUCLEOSIDE_TRIPHOSPHATE_BIOSYNTHETIC_PROCESS	1.6057541	0.01207244

(B)		
Name	**NES**	**NOM *p*-Value**
GOBP_REGULATION_OF_CYTOPLASMIC_TRANSLATION	−1.877478	0
GOBP_RNA_PHOSPHODIESTER_BOND_HYDROLYSIS_EXONUCLEOLYTIC	−1.8400294	0.002
GOBP_MATURATION_OF_5_8S_RRNA_FROM_TRICISTRONIC_RRNA_TRANSCRIPT_SSU_RRNA_5_8S_RRNA_LSU_RRNA	−1.8186126	0.00412371
GOBP_NEGATIVE_REGULATION_OF_PROTEIN_TYROSINE_KINASE_ACTIVITY	−1.8014549	0
GOBP_RNA_PHOSPHODIESTER_BOND_HYDROLYSIS	−1.7509472	0
GOBP_CLEAVAGE_INVOLVED_IN_RRNA_PROCESSING	−1.7341155	0.01649485
GOBP_POSITIVE_REGULATION_OF_VIRAL_TRANSCRIPTION	−1.7214938	0.0260521
GOBP_REGULATION_OF_MACROPHAGE_CHEMOTAXIS	−1.7151726	0
GOBP_PEPTIDYL_LYSINE_ ACETYLATION	−1.7136337	0
GOBP_MRNA_CLEAVAGE	−1.6955862	0.01
GOBP_POSITIVE_REGULATION_OF_HISTONE_DEACETYLATION	−1.684807	0.00984252
GOBP_NUCLEAR_TRANSCRIBED_MRNA_CATABOLIC_PROCESS_EXONUCLEOLYTIC	−1.680216	0.006
GOBP_VIRAL_GENE_EXPRESSION	−1.6742575	0.02615694
GOBP_MATURATION_OF_5_8S_RRNA	−1.6727061	0.02340426
GOBP_TRANSCRIPTION_PREINITIATION_COMPLEX_ASSEMBLY	−1.6675799	0.01757813
GOBP_PROTEIN_LIPID_COMPLEX_ASSEMBLY	−1.6524748	0.01335878
GOBP_NUCLEAR_ENVELOPE_REASSEMBLY	−1.6334432	0.03012048
GOBP_PROTEIN_ACETYLATION	−1.6297097	0
GOBP_PEPTIDYL_ASPARAGINE_MODIFICATION	−1.6176745	0.02985075
GOBP_TRANSEPITHELIAL_TRANSPORT	−1.6128986	0.00395257

Table 4. *Cont.*

(C)		
Name	**NES**	**NOM *p*-Value**
KEGG_ALZHEIMERS_DISEASE	1.7874615	0
KEGG_CARDIAC_MUSCLE_CONTRACTION	1.6276087	0.00796813
KEGG_VALINE_LEUCINE_AND_ISOLEUCINE_DEGRADATION	1.6064283	0.02484472
KEGG_GLUTATHIONE_METABOLISM	1.5816755	0.01020408
KEGG_PARKINSONS_DISEASE	1.5710168	0.10224949
KEGG_HUNTINGTONS_DISEASE	1.4995617	0.03952569
KEGG_FATTY_ACID_METABOLISM	1.4914919	0.04208417
KEGG_GLYCEROLIPID_METABOLISM	1.4762139	0.05633803
KEGG_PEROXISOME	1.4749482	0.125
KEGG_OXIDATIVE_PHOSPHORYLATION	1.4733046	0.14229248
KEGG_ARACHIDONIC_ACID_METABOLISM	1.4536077	0.00626305
KEGG_PROPANOATE_METABOLISM	1.446644	0.11332008
KEGG_OLFACTORY_TRANSDUCTION	1.4194456	0.02385686
KEGG_PPAR_SIGNALING_PATHWAY	1.4145677	0.06412826
KEGG_TRYPTOPHAN_METABOLISM	1.409311	0.12352941
KEGG_REGULATION_OF_AUTOPHAGY	1.4061221	0.07272727
KEGG_GLYCOSAMINOGLYCAN_BIOSYNTHESIS_HEPARAN_SULFATE	1.383326	0.05371901
KEGG_DNA_REPLICATION	1.3792615	0.17693837
KEGG_CALCIUM_SIGNALING_PATHWAY	1.362877	0.03193613
KEGG_GNRH_SIGNALING_PATHWAY	1.3259736	0.07628866

(D)		
Name	**NES**	**NOM *p*-Value**
KEGG_DORSO_VENTRAL_AXIS_FORMATION	−1.6805534	0.00199601
KEGG_RNA_DEGRADATION	−1.5593725	0.02674897
KEGG_NON_SMALL_CELL_LUNG_CANCER	−1.5283813	0.03092784
KEGG_N_GLYCAN_BIOSYNTHESIS	−1.5252374	0.06681035
KEGG_BASAL_TRANSCRIPTION_FACTORS	−1.4940801	0.02443992
KEGG_GLYCOSYLPHOSPHATIDYLINOSITOL_GPI_ANCHOR_BIOSYNTHESIS	−1.4563912	0.05285412
KEGG_PANCREATIC_CANCER	−1.4190394	0.05020081
KEGG_RENAL_CELL_CARCINOMA	−1.3920702	0.0831643
KEGG_ADHERENSJUNCTION	−1.3889191	0.09543569
KEGG_ENDOMETRIAL_CANCER	−1.3684485	0.05702648
KEGG_PROSTATE_CANCER	−1.3127599	0.10655738
KEGG_SMALL_CELL_LUNG_CANCER	−1.2742031	0.12576064
KEGG_TGF_BETA_SIGNALING_PATHWAY	−1.2639772	0.16232465
KEGG_SPLICEOSOME	−1.2469473	0.18526316
KEGG_PROTEIN_EXPORT	−1.2308676	0.32635984
KEGG_CHRONIC_MYELOID_LEUKEMIA	−1.22399	0.1523046
KEGG_MTOR_SIGNALING_PATHWAY	−1.2147567	0.17979798
KEGG_STEROID_HORMONE_BIOSYNTHESIS	−1.2091544	0.13541667
KEGG_RNA_POLYMERASE	−1.1975825	0.27021277
KEGG_PORPHYRIN_AND_CHLOROPHYLL_METABOLISM	−1.196312	0.21991701

To examine the dysregulation of kinases directing IRS-1, we measured the phosphorylation of Erk1/2 and stress-activated protein kinase (SAPK)/Jun amino-terminal kinase (JNK), which phosphorylates the Ser-632/635 residues of IRS1 [20]. The phosphorylation of Erk1/2 and SAPK/JNK was significantly increased in the liver of LKO mice compared to Floxed mice (Figure 7A,B). Since the phosphorylation of Erk1/2 is regulated by early growth response (Egr)-1 in type 2 diabetic mice [25], we compared the Egr-1 expression in the liver of LKO and Floxed mice. The Egr-1 mRNA and protein levels showed a significant increase (Figure 7C) and a trend toward increasing levels, respectively, in the liver of LKO mice (Figure 7D). These results suggest that the activation of Egr-1 the Erk1/2 axis contributes to the negative regulation of insulin signaling at IRS-1 in the liver of LKO mice.

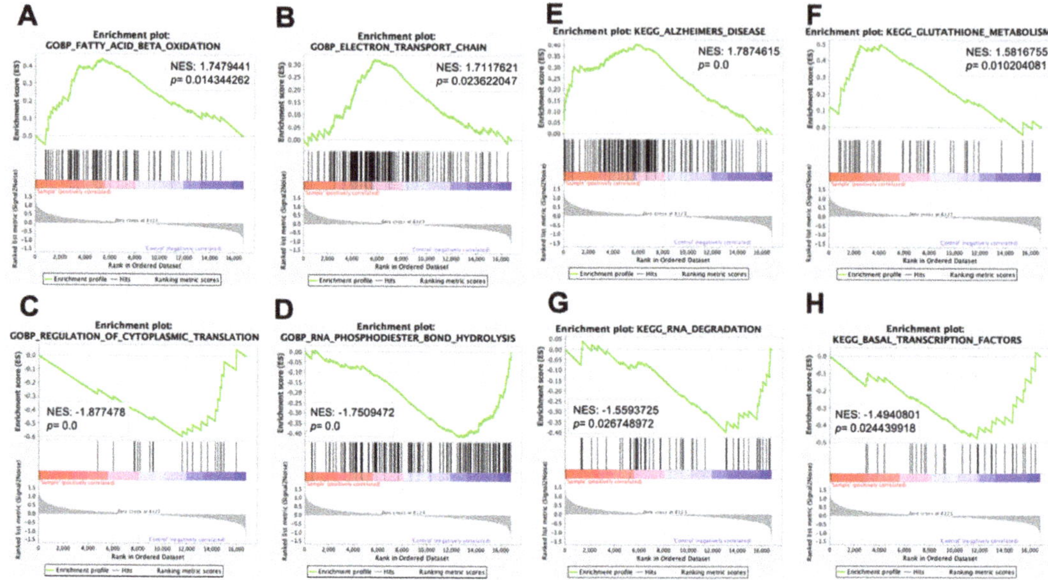

Figure 5. Identification of positively and negatively correlated gene sets in the liver of LKO mice in GO terms and KEGG pathways by gene set enrichment analysis (GSEA). The enrichment plots of GSEA showed positively correlated gene sets in fatty acid beta oxidation (**A**), electron transport chain (**B**), negatively correlated gene sets in regulation of cytoplasmic translation (**C**), and RNA phosphodiester bond hydrolysis (**D**) of GO terms. The enrichment plots of GSEA showed positively correlated gene sets in Alzheimer's disease (**E**) and glutathione metabolism (**F**) and negatively correlated gene sets in RNA degradation (**G**) and basal transcription factors (**H**) of KEGG pathways. Nominal enrichment scores (NESs) and *p*-values are indicated in each enrichment plot.

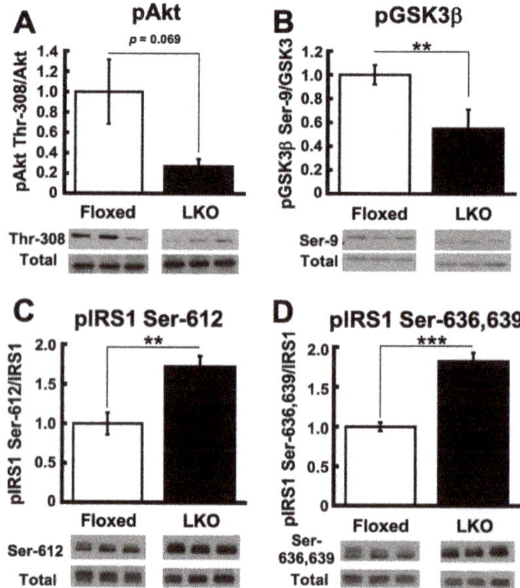

Figure 6. *Phgdh* deletion in hepatocytes impairs insulin signaling in the liver. (**A**) Protein level of phosphorylated Akt in the liver of Floxed and LKO mice at 30 weeks. Comparable staining of Akt was

used to verify equivalent protein loading. (**B**) Protein level of phosphorylated GSK3β in the liver of Floxed and LKO mice at 30 weeks. Comparable staining of GSK3β was used to verify equivalent protein loading. (**C,D**) Protein level of phosphorylated IRS-1 on Ser612 (C) and Ser636/639 (**D**) residue in the liver of Floxed and LKO mice at 30 weeks. Comparable staining of IRS-1 was used to verify equivalent protein loading. $n = 6$ each. Student's t-test, ** $p < 0.005$, *** $p < 0.0005$.

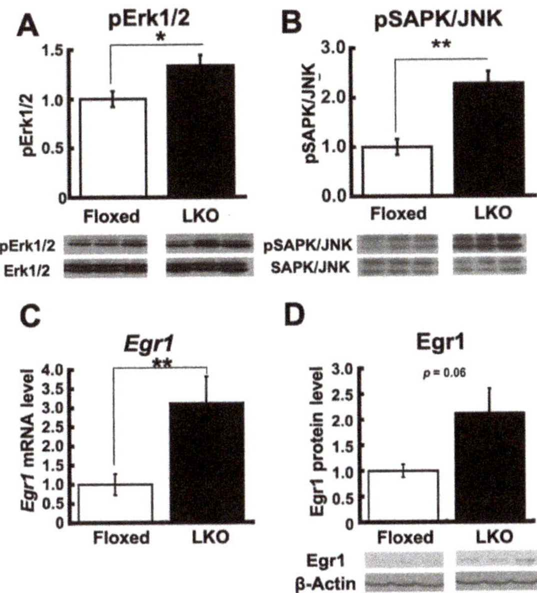

Figure 7. *Phgdh* deletion in hepatocytes alters signaling pathway upstream of IRS1 in liver. (**A**) Protein level of phosphorylated Erk1/2 in the liver of Floxed and LKO mice at 30 weeks. Comparable staining of Erk1/2 was used to verify equivalent protein loading. (**B**) Protein level of phosphorylated SAPK/JNK in the liver of Floxed and LKO mice at 30 weeks. Comparable staining of SAPK/JNK was used to verify equivalent protein loading. (**C,D**) mRNA (**C**) and protein (**D**) level of Egr-1 in the liver of Floxed and LKO mice at 30 weeks. $N = 6$ each. Student's t-test, * $p < 0.05$, ** $p < 0.005$.

To gain an insight into the dysregulation of the enzymes regulating glucose metabolism under diminished insulin signaling in the liver of LKO mice, we examined the mRNA expression of 6-phosphofructo-2-kinase/fructose-2,6-biphosphatase 3 (*Pfkfb3*) and pyruvate dehydrogenase kinase 4 (*Pdk4*) as insulin-regulated regulators of glycolysis and gluconeogenesis, respectively [26,27]. *Pfkfb3* encodes 6-phosphofructo-2-kinase/fructose-2,6-bisphosphatase synthesizing fructose 2,6-bisphosphate, the glycolytic activator that promotes the conversion of fructose 1,6-bisphosphate to fructose 6-phosphate, a rate-limiting reaction in the glycolytic pathway [28]. Pdk4 inhibits the conversion of pyruvate to acetyl-CoA by inhibiting the phosphorylation of pyruvate dehydrogenase, which leads to suppression of glucose oxidation in the glycolytic pathway [29]. The mRNA levels of *Pfkfb3* in the liver were significantly reduced in LKO mice compared to Floxed mice (Figure 8A), while the mRNA levels of *Pdk4* in the liver were significantly increased in LKO mice compared to Floxed mice (Figure 8B). These gene expression profiles suggest diminished glucose oxidation and coincide with diminished insulin signaling in the livers of LKO mice. In addition to these results, we evaluated fatty acid synthesis to measure Fasn mRNA levels in the liver and adipose tissue. The mRNA levels of *Fasn* were not increased in the liver (Figure 8C) but increased in the eWAT (Figure 8D). These results suggested that steatosis did not progress, but fat accumulation in the adipose tissue was enhanced.

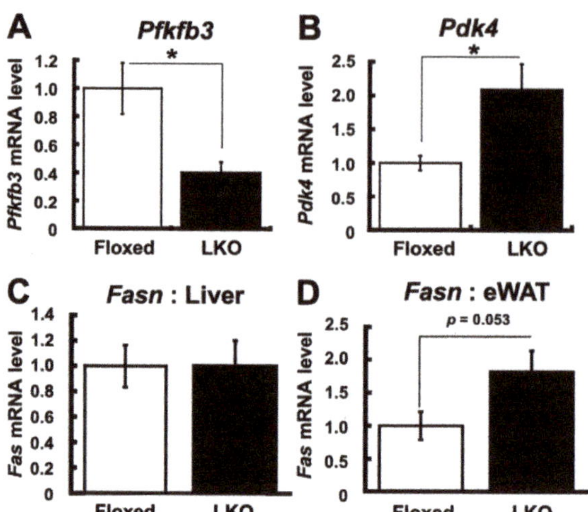

Figure 8. *Phgdh* deletion in hepatocytes induces the reduction of glycolysis in the liver. (**A,B**) mRNA levels of *Pfkfb3* (**A**) and *Pdk4* (**B**) in the liver of Floxed and LKO mice at 30 weeks. (**C,D**) mRNA levels of *Fasn* in liver(**C**) and eWAT (**D**) in the liver of Floxed and LKO mice at 30 weeks. $n = 6$ each. Student's *t*-test, * $p < 0.05$.

We then evaluated the alteration in tolerance to dietary protein starvation in LKO mice. LKO mice were fed a protein-free diet for 21 days, and their body weight (Figure 9A) and survival rate (Figure 9B) were found to change as a result. LKO mice showed a significantly greater weight loss than Floxed mice throughout the feeding of the protein-free diet (Figure 9A). The survival rate after feeding protein-free diet for 21 days was markedly lower in LKO mice than in Floxed mice (Figure 9B). LKO mice exhibited diarrhea and swelling of the small intestine 21 days after feeding the protein-free diet (data not shown). These observations suggest that the loss of *Phgdh* in hepatocytes leads to increased vulnerability to short-term protein malnutrition.

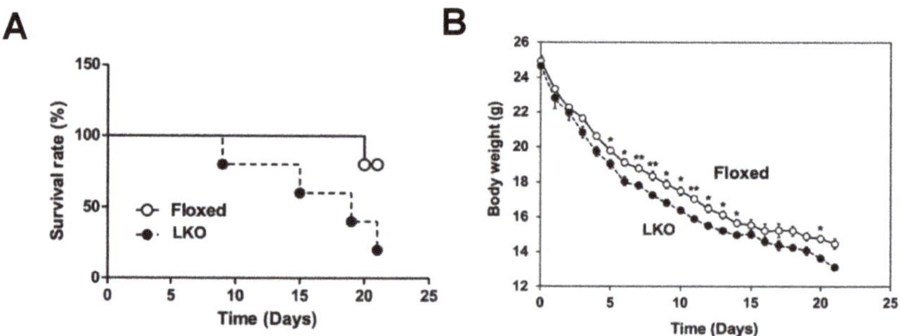

Figure 9. Effect of body weight and survival rate by feeding LKO mice a protein-free diet. (**A**) The transition of body weight in Floxed mice (white node) and LKO mice (black node) after feeding protein-free diet. (**B**) The transition of survival rate in Floxed and LKO mice during feeding protein-free diet. $n = 5$ each. Student's *t*-test, * $p < 0.05$, ** $p < 0.005$.

4. Discussion

Recent studies in mice and humans suggest that the altered expression of enzymes composed of the phosphorylated pathway of de novo Ser synthesis in the liver is associated with fatty liver disease. Bioinformatic approaches have indicated the downregulation of

PHGDH in the liver of patients with non-alcoholic fatty liver disease [10] and alcoholic hepatitis patients [30]. Diet-induced fatty liver model mice also exhibited a reduced expression of Phgdh in the liver [12]. Interestingly, the fatty acid treatment at high concentrations (500–700 μM) caused a downregulation of Phgdh in isolated hepatocytes [12]. Although these observations raise the possibility that the downregulation of hepatic Phgdh may be implicated in the onset and/or progression of fatty liver disease, the pathophysiological consequences of genetic *Phgdh* disruption in liver hepatocytes have not been previously explored experimentally. The present study demonstrates for the first time that hepatocyte-specific *Phgdh* deletion resulted in body weight gain and increased adipose tissue weights in mice at 23–30 weeks of age but did not cause the ectopic accumulation of fatty acids in the liver and/or increases in blood fatty acids and other lipids. These phenotypes implicate a regulatory role for Phgdh expressed in hepatocytes in systemic glucose metabolism, whereas the onset and development of fatty acid deposition in the liver are not directly promoted by downregulation of *Phgdh* in hepatocytes. The present observations strongly suggest that reduced Phgdh expression in hepatocytes alone is not sufficient to induce ectopic fat accumulation in the liver. Presumably, since the liver is composed of several different types of cell populations in addition to hepatocytes, the downregulation of Phgdh in other liver cell types may also be necessary to induce fat accumulation. To evaluate the pathobiological role of Phgdh in fatty liver, it will be necessary to delete or downregulate *Phgdh* in all cell types of the liver and then to identify the cell types involved in using liver cell type-specific mutants.

Clinical and experimental studies have shown that hepatic insulin resistance is strongly associated with non-alcoholic fatty liver disease (NAFLD) [31,32]. It was apparent that the insulin signal cascade was downregulated in the liver of LKO mice (Figure 10), as indicated by the reduced phosphorylation levels of Akt and GSK3β with enhanced phosphorylation of Ser residues in IRS1 (Figure 5) and Pfkfb3 downregulation (Figure 8A). However, ectopic lipid accumulation did not occur in the liver of LKO mice. Accumulating evidence has established a causal role for inflammation in the development of insulin resistance in insulin-responsive tissues, including the liver [33]. A bioinformatics analysis of gene expression profiles by IPA showed that an increased production and signaling of the pro-inflammatory cytokine interleukin (IL)-12 in macrophages was the top canonical pathway of upregulated genes in the liver of LKO mice (Table 3A). The IL-12 family of cytokines is known to be a potential inflammatory mediator linking obesity and insulin resistance. IL-12 family molecules and their receptors have been reported to be upregulated in insulin-responsive tissues, including the liver, under obese conditions in experiments using the Ob/Ob model [34]. The increased phosphorylation of NF-κB was also indicative of inflammation-like changes in the liver of LKO mice (Figure 4). Regarding the link between changes in Ser availability and NF-κB activation, Wang et al. reported that the NF-κB phosphorylation levels were increased in the brains of mice fed a Ser/Gly restricted diet (SGRD) and further increased when D-glucose was administered in combination with SGRD to induce inflammation through the intestinal tract [35]. In addition, macrophages play a multifaceted role in the regulation of inflammatory responses, and it was reported that extracellular Ser restriction leading to a marked reduction of Ser within the cells increases the expression of pro-inflammatory cytokines and suppresses the expression of IL-10, which exerts anti-inflammatory effects [36]; however, other studies observed a seemingly contradictory phenomenon in which the expression of the pro-inflammatory cytokine IL-1β depends on Ser metabolism in macrophages [37,38]. Erg1, an immediate early gene encoding zinc-finger transcription factor, responds quickly to a variety of stimuli, such as injury, growth factors, cytokines, and physical insults, and participates in tissue injury and repair in the liver via its downstream target genes [39]. Indeed, Egr-1 was reported to induce liver inflammation in a cholestatic injury model [40]. Hence, the induction of Egr1 mRNA and protein together with the inflammatory gene signature and enhanced NF-κB phosphorylation supports the notion that inflammation-like changes were elicited in the liver of LKO mice.

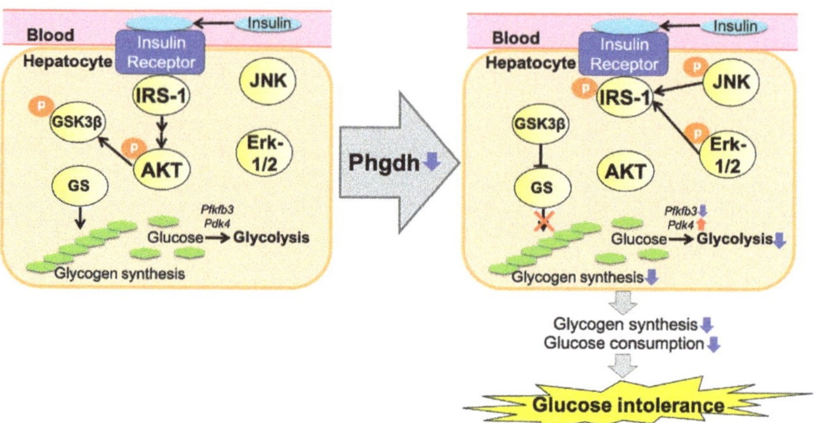

Figure 10. A summary of the molecular mechanisms caused by *Phgdh* deletion in hepatocytes, as inferred by this study. *Phgdh* deletion in hepatocytes impairs insulin signaling via IRS-1 phosphorylation. Akt inactivation by impaired insulin signaling induced the suppression of gluconeogenesis. Enhanced glycolysis resulted in mild obesity and glucose intolerance.

Although the cause of inducing inflammatory-like changes remains unknown, the GSEA analysis of the genes upregulated in the liver of LKO mice indicated significant changes in the gene sets of the glutathione metabolism and the mitochondrial electron transport chain system. Previously, we reported that extracellular Ser restriction decreased total glutathione content and led to the generation of H_2O_2 and induction of inflammation-related gene expression in mouse embryonic fibroblasts deficient in *Phgdh* [5]. Recent studies also demonstrated that *Phgdh* deletion elicited an inflammation-like response through the production of cytokines via an increase in reactive oxygen species in chondrocytes [41], while orally administered Ser was able to protect LPS-induced intestinal inflammation via p53-mediated glutathione synthesis [42]. Hence, it is presumed that a decrease in the glutathione content and/or a dysregulation of the mitochondrial electron transport system lead to oxidative stress in the liver of LKO mice, which may trigger inflammation-like responses in the liver of LKO mice. In particular, Kupffer cells are likely to be candidate cells mediating these responses because they have macrophage-like functions in the liver. Further studies are needed to investigate whether the genetic inactivation of Phgdh elicits oxidative stress and/or inflammatory responses in the different cell types of the liver.

It should be noted that the results of the amino acid analysis performed in this study were unexpected, and no significant differences in Ser content in the liver or free Ser concentration in the serum were observed in LKO mice compared to those in Floxed control mice. On the other hand, the Ser content in the kidney was increased by 1.4-fold compared to that in Floxed mice (Table 1B), and the Phgdh protein levels were also increased significantly by 1.15-fold (data not shown). In addition, Ser levels in the soleus muscle was increasing (Mohri, Hamano, Furuya, data not shown). These changes may reflect the upregulation of de novo Ser synthesis in the kidney and muscle via the phosphorylated pathway to supply the liver. Similarly, cell types other than hepatocytes may also augment de novo Ser synthesis in the liver of LKO mice. A similar phenomenon has been observed in organs other than the liver. Mutations in PHGDH have been identified as the cause of a rare neurodegenerative degenerative retinal disease, macular telangiectasia type 2 (MacTel), which results in photoreceptor degeneration in humans [43]. Shen et al. recently reported that the Müller cell-specific deletion of the *Phgdh* gene in mice caused photoreceptor degeneration but did not cause a reduction in the retinal Ser levels, rather, it markedly increased amino acid levels [44]. This may be due to the enhanced expression of Phgdh in microglia and macrophages in the retina, which were recruited and activated by photoreceptor degeneration, and most likely compensated for Ser in the retina. Likewise, the hepatocyte-specific inactivation of Phgdh was also likely to elicit a compensatory supply of Ser within the liver, kidney, muscle, and other organs. There is a possibility that this compensatory Ser synthesis arises from cells mediating inflammatory responses in the liver of LKO mice in the same way as microglia and macrophages in the retina with Müller cell-specific deletion of Phgdh.

This study demonstrated that LKO mice exhibited increased mortality compared to Floxed mice on a protein-free diet (Figure 8B), which was accompanied by diarrhea (data not shown). These consequences of LKO mice fed the protein-free diet strongly suggest that de novo Ser synthesis via Phgdh in hepatocytes plays a crucial role for adapting to a protein starved nutritional condition. Furthermore, the present observations raise the possibility that Phgdh in hepatocytes preserves intestinal integrity under protein-starved nutritional conditions via an unknown mechanism. It was demonstrated that the enzymatic activity and mRNA of Phgdh increased quickly in rats fed a protein-free diet [45,46], although the functional significance of Phgdh induction by a protein-free diet has remained unclear for a long time. The intake of a protein-free diet was shown to impair the morphological architecture and function of the intestine [47], while a recent study observed that oral supplementation of Ser maintained morphological and functional integrity of the intestine and prevented diarrhea incidence in piglets [42]. Given these previous observations, together with the present findings, the intake of a protein-free diet appears to be rapidly sensed by hepatocytes. Immediately thereafter, intestinal function is maintained via an unknown liver–small intestine correlation mechanism mediated by Ser. More detailed experimental studies are needed to understand the molecular basis of this phenomenon.

In conclusion, this study reveals that de novo serine synthesis initiated by Phgdh in liver hepatocytes contributes to maintaining the insulin/IGF signaling pathway in the liver, systemic glucose tolerance, and resilience to protein deprivation. Since an impairment in systemic glucose tolerance serves as a high risk for developing type 2 diabetes mellitus in humans, it is anticipated that the elucidation of the pathobiological relationship between the decrease in serine synthesis via the phosphorylation pathway, the diminishment of the insulin/IGF signaling pathway, and the impaired glucose tolerance in the human liver will provide new mechanisms related to impaired glucose tolerance and useful opportunities for its diagnosis and prevention.

Supplementary Materials: The following are available online at https://www.mdpi.com/article/10.3390/nu13103468/s1, Table S1: Evaluation of amino acid metabolism in serum of LKO mice, Figure S1: Organs and the accumulation of adipose tissue in abdomen in LKO mice, Figure S2: Volcano plot of DEGs in the liver of LKO mice, Figure S3: IPA identify the phenotypically relevant IL-10 gene network.

Author Contributions: Conceptualization and writing, M.H. and S.F.; generation of hepatocyte-specific *Phgdh* knockout mice, K.E., Y.H. and S.F.; investigation, M.H., K.E., K.M., T.Y. and S.M.; Supervised and contributed to microarray analysis, K.T. All authors have read and agreed to the published version of the manuscript.

Funding: This research was funded in part by the Japan Society for the Promotion of Science KAKENHI Grants (no. 14J05809 to K.E., no. 15K12346 to M.H., no. 26660116 and no. 18K19748 to S.F.) and a grant from the Tojuro Iijima Foundation for Food and Technology (to S.F.).

Institutional Review Board Statement: This article does not contain any studies with human participants performed by any of the authors. All animal experiments were approved by the ethics committee for animals. Experiments at Kyushu University (A27-103), RIKEN Brain Science Institute (2012-043(1) and H25-2-241(1)) adhered strictly to the animal experiment guidelines.

Informed Consent Statement: Not applicable.

Data Availability Statement: Microarray data have been deposited in NCBI's Gene Expression Omnibus and are accessible through GEO series accession number GSE179912.

Conflicts of Interest: The authors declare no conflict of interest.

References

1. Yoshida, K.; Furuya, S.; Osuka, S.; Mitoma, J.; Shinoda, Y.; Watanabe, M.; Azuma, N.; Tanaka, H.; Hashikawa, T.; Itohara, S.; et al. Targeted Disruption of the Mouse 3-Phosphoglycerate Dehydrogenase Gene Causes Severe Neurodevelopmental Defects and Results in Embryonic Lethality. *J. Biol. Chem.* **2004**, *279*, 3573–3577. [CrossRef]
2. Shaheen, R.; Rahbeeni, Z.; Alhashem, A.; Faqeih, E.; Zhao, Q.; Xiong, Y.; Almoisheer, A.; Al-Qattan, S.M.; Almadani, H.A.; Al-Onazi, N.; et al. Neu-laxova syndrome, an inborn error of serine metabolism, is caused by mutations in PHGDH. *Am. J. Hum. Genet.* **2014**, *94*, 898–904. [CrossRef]
3. Sayano, T.; Kawano, Y.; Kusada, W.; Arimoto, Y.; Esaki, K.; Hamano, M.; Udono, M.; Katakura, Y.; Ogawa, T.; Kato, H.; et al. Adaptive response to l-serine deficiency is mediated by p38 MAPK activation via 1-deoxysphinganine in normal fibroblasts. *FEBS Open Bio* **2016**, *6*, 303–316. [CrossRef]
4. Esaki, K.; Sayano, T.; Sonoda, C.; Akagi, T.; Suzuki, T.; Ogawa, T.; Okamoto, M.; Yoshikawa, T.; Hirabayashi, Y.; Furuya, S. L-serine deficiency elicits intracellular accumulation of cytotoxic deoxysphingolipids and lipid body formation. *J. Biol. Chem.* **2015**, *290*, 14595–14609. [CrossRef]
5. Hamano, M.; Haraguchi, Y.; Sayano, T.; Zyao, C.; Arimoto, Y.; Kawano, Y.; Moriyasu, K.; Udono, M.; Katakura, Y.; Ogawa, T.; et al. Enhanced vulnerability to oxidative stress and induction of inflammatory gene expression in 3-phosphoglycerate dehydrogenase-deficient fibroblasts. *FEBS Open Bio* **2018**, *8*, 914–922. [CrossRef]
6. Hamano, M.; Tomonaga, S.; Osaki, Y.; Oda, H.; Kato, H.; Furuya, S. Transcriptional activation of chac1 and other atf4-target genes induced by extracellular l-serine depletion is negated with glycine consumption in hepa1-6 hepatocarcinoma cells. *Nutrients* **2020**, *12*, 3018. [CrossRef]
7. Yang, J.H.; Wada, A.; Yoshida, K.; Miyoshi, Y.; Sayano, T.; Esaki, K.; Kinoshita, M.O.; Tomonaga, S.; Azuma, N.; Watanabe, M.; et al. Brain-specific Phgdh deletion reveals a pivotal role for l-serine biosynthesis in controlling the level of D-serine, an N-methyl-D-aspartate receptor co-agonist, in adult brain. *J. Biol. Chem.* **2010**, *285*, 41380–41390. [CrossRef]
8. Klomp, L.W.J.; De Koning, T.J.; Malingre, H.E.M.; Van Beurden, E.A.C.M.; Brink, M.; Opdam, F.L.; Duran, M.; Jaeken, J.; Pineda, M.; Van Maldergem, L.; et al. Molecular Characterization of 3-Phosphoglycerate Dehydrogenase Deficiency—A Neurometabolic Disorder Associated with Reduced L-Serine Biosynthesis. *Am. J. Hum. Genet.* **2000**, *67*, 1389–1399. [CrossRef]
9. Kang, Y.P.; Falzone, A.; Liu, M.; Saller, J.J.; Karreth, F.A.; DeNicola, G.M. PHGDH supports liver ceramide synthesis and sustains lipid homeostasis. *BioRxiv* **2019**, *15*, 1–13. [CrossRef]
10. Mardinoglu, A.; Agren, R.; Kampf, C.; Asplund, A.; Uhlen, M.; Nielsen, J. Genome-scale metabolic modelling of hepatocytes reveals serine deficiency in patients with non-alcoholic fatty liver disease. *Nat. Commun.* **2014**, *5*, 1–11. [CrossRef]
11. Mardinoglu, A.; Bjornson, E.; Zhang, C.; Klevstig, M.; Söderlund, S.; Ståhlman, M.; Adiels, M.; Hakkarainen, A.; Lundbom, N.; Kilicarslan, M.; et al. Personal model-assisted identification of NAD + and glutathione metabolism as intervention target in NAFLD. *Mol. Syst. Biol.* **2017**, *13*, 916. [CrossRef]
12. Sim, W.C.; Lee, W.; Sim, H.; Lee, K.Y.; Jung, S.H.; Choi, Y.J.; Kim, H.Y.; Kang, K.W.; Lee, J.Y.; Choi, Y.J.; et al. Downregulation of PHGDH expression and hepatic serine level contribute to the development of fatty liver disease. *Metabolism* **2020**, *102*, 154000. [CrossRef] [PubMed]
13. Bolstad, B.M. PreprocessCore: A Collection of Pre-Processing Functions. R Package Version 1.0. 2013. Available online: https://scholar.google.co.uk/citations?view_op=view_citation&hl=en&user=Ntck_10AAAAJ&citation_for_view=Ntck_10AAAAJ:4JMBOYKVnBMC (accessed on 26 September 2021).
14. Smyth, G.K. Limma: Linear models for microarray data. In *Bioinformatics and Computational Biology Solution Using R and Bioconductor*; Springer: New York, NY, USA, 2006; pp. 397–420.
15. Dennis, G.; Sherman, B.T.; Hosack, D.A.; Yang, J.; Gao, W.; Lane, H.C.; Lempicki, R.A. DAVID: Database for Annotation, Visualization, and Integrated Discovery. *Genome Biol.* **2003**, *4*, 1–11. [CrossRef]
16. Krämer, A.; Green, J.; Pollard, J.; Tugendreich, S. Causal analysis approaches in ingenuity pathway analysis. *Bioinformatics* **2014**, *30*, 523–530. [CrossRef] [PubMed]
17. Zeyda, M.; Stulnig, T.M. Obesity, inflammation, and insulin resistance—A mini-review. *Gerontology* **2009**, *55*, 379–386. [CrossRef] [PubMed]
18. Jope, R.S.; Johnson, G.V.W. The glamour and gloom of glycogen synthase kinase-3. *Trends Biochem. Sci.* **2004**, *29*, 95–102. [CrossRef] [PubMed]
19. Martin, M.; Rehani, K.; Jope, R.S.; Michalek, S.M. Toll-like receptor-Mediated cytokine production is differentially regulated by glycogen synthase kinase 3. *Nat. Immunol.* **2005**, *6*, 777–784. [CrossRef]
20. Gual, P.; Le Marchand-Brustel, Y.; Tanti, J.F. Positive and negative regulation of insulin signaling through IRS-1 phosphorylation. *Biochimie* **2005**, *87*, 99–109. [CrossRef] [PubMed]
21. De Fea, K.; Roth, R.A. Protein kinase C modulation of insulin receptor substrate-1 tyrosine phosphorylation requires serine 612. *Biochemistry* **1997**, *36*, 12939–12947. [CrossRef]
22. Ozes, O.N.; Akca, H.; Mayo, L.D.; Gustin, J.A.; Maehama, T.; Dixon, J.E.; Donner, D.B. A phosphatidylinositol 3-kinase/Akt/mTOR pathway mediates and PTEN antagonizes tumor necrosis factor inhibition of insulin signaling through insulin receptor substrate-1. *Proc. Natl. Acad. Sci. USA* **2001**, *98*, 4640–4645. [CrossRef]
23. Boura-Halfon, S.; Zick, Y. Phosphorylation of IRS proteins, insulin action, and insulin resistance. *Am. J. Physiol.-Endocrinol. Metab.* **2009**, *296*, 581–591. [CrossRef] [PubMed]

24. Copps, K.D.; White, M.F. Regulation of insulin sensitivity by serine/threonine phosphorylation of insulin receptor substrate proteins IRS1 and IRS2. *Diabetologia* **2012**, *55*, 2565–2582. [CrossRef] [PubMed]

25. Shen, N.; Yu, X.; Pan, F.Y.; Gao, X.; Xue, B.; Li, C.J. An early response transcription factor, Egr-1, enhances insulin resistance in type 2 diabetes with chronic hyperinsulinism. *J. Biol. Chem.* **2011**, *286*, 14508–14515. [CrossRef]

26. Backman, M.; Flenkenthaler, F.; Blutke, A.; Dahlhoff, M.; Ländström, E.; Renner, S.; Philippou-Massier, J.; Krebs, S.; Rathkolb, B.; Prehn, C.; et al. Multi-omics insights into functional alterations of the liver in insulin-deficient diabetes mellitus. *Mol. Metab.* **2019**, *26*, 30–44. [CrossRef] [PubMed]

27. Kim, Y.I.; Lee, F.N.; Choi, W.S.; Lee, S.; Youn, J.H. Insulin regulation of skeletal muscle PDK4 mRNA expression is impaired in acute insulin-resistant states. *Diabetes* **2006**, *55*, 2311–2317. [CrossRef]

28. Trefely, S.; Khoo, P.S.; Krycer, J.R.; Chaudhuri, R.; Fazakerley, D.J.; Parker, B.L.; Sultani, G.; Lee, J.; Stephan, J.P.; Torres, E.; et al. Kinome screen identifies PFKFB3 and glucose metabolism as important regulators of the insulin/insulin-like growth factor (IGF)-1 signaling pathway. *J. Biol. Chem.* **2015**, *290*, 25834–25846. [CrossRef]

29. Tao, R.; Xiong, X.; Harris, R.A.; White, M.F.; Dong, X.C. Genetic Inactivation of Pyruvate Dehydrogenase Kinases Improves Hepatic Insulin Resistance Induced Diabetes. *PLoS ONE* **2013**, *8*, e71997. [CrossRef]

30. Yao, J.; Cheng, Y.; Zhang, D.; Fan, J.; Zhao, Z.; Li, Y.; Jiang, Y.; Guo, Y. Identification of key genes, MicroRNAs and potentially regulated pathways in alcoholic hepatitis by integrative analysis. *Gene* **2019**, *720*, 144035. [CrossRef]

31. Samuel, V.T.; Shulman, G.I. The pathogenesis of insulin resistance: Integrating signaling pathways and substrate flux. *J. Clin. Investig.* **2016**, *126*, 12–22. [CrossRef]

32. Utzschneider, K.M.; Kahn, S.E. Review: The role of insulin resistance in nonalcoholic fatty liver disease. *J. Clin. Endocrinol. Metab.* **2006**, *91*, 4753–4761. [CrossRef]

33. Shoelson, S.E.; Lee, J.; Goldfine, A.B. Inflammation and insulin resistance. *J. Clin. Investig.* **2006**, *116*, 1793–1801. [CrossRef] [PubMed]

34. Nam, H.; Ferguson, B.S.; Stephens, J.M.; Morrison, R.F. Impact of obesity on IL-12 family gene expression in insulin responsive tissues. *Biochim. Biophys. Acta-Mol. Basis Dis.* **2013**, *1832*, 11–19. [CrossRef] [PubMed]

35. Wang, F.; Zhou, H.; Deng, L.; Wang, L.; Chen, J.; Zhou, X.; Jiang, H. Serine Deficiency Exacerbates Inflammation and Oxidative Stress via Microbiota-Gut-Brain Axis in D-Galactose-Induced Aging Mice. *Mediat. Inflamm.* **2020**, *2020*, 5821428. [CrossRef]

36. Kurita, K.; Ohta, H.; Shirakawa, I.; Tanaka, M.; Kitaura, Y.; Iwasaki, Y.; Matsuzaka, T.; Shimano, H.; Aoe, S.; Arima, H.; et al. Macrophages rely on extracellular serine to suppress aberrant cytokine production. *Sci. Rep.* **2021**, *11*, 1–14. [CrossRef] [PubMed]

37. Rodriguez, A.E.; Ducker, G.S.; Billingham, L.K.; Martinez, C.A.; Mainolfi, N.; Suri, V.; Friedman, A.; Manfredi, M.G.; Weinberg, S.E.; Rabinowitz, J.D.; et al. Serine Metabolism Supports Macrophage IL-1β Production. *Cell Metab.* **2019**, *29*, 1003–1011.e4. [CrossRef] [PubMed]

38. Chen, S.; Xia, Y.; He, F.; Fu, J.; Xin, Z.; Deng, B.; He, L.; Zhou, X.; Ren, W. Serine Supports IL-1β Production in Macrophages Through mTOR Signaling. *Front. Immunol.* **2020**, *11*, 1–15. [CrossRef] [PubMed]

39. Magee, N.; Zhang, Y. Role of early growth response 1 in liver metabolism and liver cancer. *Hepatoma Res.* **2017**, *3*, 268. [CrossRef] [PubMed]

40. Allen, K.; Jaeschke, H.; Copple, B.L. Bile acids induce inflammatory genes in hepatocytes: A novel mechanism of inflammation during obstructive cholestasis. *Am. J. Pathol.* **2011**, *178*, 175–186. [CrossRef]

41. Huang, H.; Liu, K.; Ou, H.; Qian, X.; Wan, J. Phgdh serves a protective role in Il-1β induced chondrocyte inflammation and oxidative-stress damage. *Mol. Med. Rep.* **2021**, *23*, 1–10. [CrossRef] [PubMed]

42. Zhou, X.; Zhang, Y.; Wu, X.; Wan, D.; Yin, Y. Effects of Dietary Serine Supplementation on Intestinal Integrity, Inflammation and Oxidative Status in Early-Weaned Piglets. *Cell. Physiol. Biochem.* **2018**, *48*, 993–1002. [CrossRef]

43. Scerri, T.S.; Quaglieri, A.; Cai, C.; Zernant, J.; Matsunami, N.; Baird, L.; Scheppke, L.; Bonelli, R.; Yannuzzi, L.A.; Friedlander, M.; et al. Genome-wide analyses identify common variants associated with macular telangiectasia type 2. *Nat. Genet.* **2017**, *49*, 559–567. [CrossRef]

44. Shen, W.; Lee, S.R.; Mathai, A.E.; Zhang, R.; Du, J.; Yam, M.X.; Pye, V.; Barnett, N.L.; Rayner, C.L.; Zhu, L.; et al. Effect of selectively knocking down key metabolic genes in Müller glia on photoreceptor health. *Glia* **2021**, *69*, 1966–1986. [CrossRef] [PubMed]

45. Shin-ichi, H.; Takehiko, T.; Junko, N.; Masami, S. Dietary and Hormonal Regulation of Serine Synthesis in the Rat. *J. Biochem.* **1975**, *77*, 207–219.

46. Achouri, Y.; Robbi, M.; Van Schaftingen, E. Role of cysteine in the dietary control of the expression of 3-phosphoglycerate dehydrogenase in rat liver. *Biochem. J.* **1999**, *21*, 15–21. [CrossRef]

47. Hill, R.B.; Prosper, J.; Hirschfield, J.S.; Kern, F. Protein starvation and the small intestine. I. The Growth and Morphology of the Small Intestine in Weanling Rats. *Exp. Mol. Pathol.* **1968**, *8*, 66–74. [CrossRef]

Review

Potential Nutrients from Natural and Synthetic Sources Targeting Inflammaging—A Review of Literature, Clinical Data and Patents

Sushruta Koppula [1,†], **Mahbuba Akther** [1,†], **Md Ezazul Haque** [2] and **Spandana Rajendra Kopalli** [3,*]

[1] Department of Integrated Biosciences, College of Biomedical & Health Science, Konkuk University, Chungju 27381, Korea; koppula@kku.ac.kr (S.K.); smritymahbuba@gmail.com (M.A.)

[2] Department of Applied Life Science, Graduate School, BK21 Program, Konkuk University, Chungju 27381, Korea; mdezazulhaque@yahoo.com

[3] Department of Bioscience and Biotechnology, Sejong University, Gwangjin-gu, Seoul 05006, Korea

* Correspondence: spandanak@sejong.ac.kr; Tel.: +82-2-6935-2619

† These authors contributed equally.

Abstract: Inflammaging, the steady development of the inflammatory state over age is an attributable characteristic of aging that potentiates the initiation of pathogenesis in many age-related disorders (ARDs) including neurodegenerative diseases, arthritis, cancer, atherosclerosis, type 2 diabetes, and osteoporosis. Inflammaging is characterized by subclinical chronic, low grade, steady inflammatory states and is considered a crucial underlying cause behind the high mortality and morbidity rate associated with ARDs. Although a coherent set of studies detailed the underlying pathomechanisms of inflammaging, the potential benefits from non-toxic nutrients from natural and synthetic sources in modulating or delaying inflammaging processes was not discussed. In this review, the available literature and recent updates of natural and synthetic nutrients that help in controlling inflammaging process was explored. Also, we discussed the clinical trial reports and patent claims on potential nutrients demonstrating therapeutic benefits in controlling inflammaging and inflammation-associated ARDs.

Keywords: inflammaging; aging related disorders; low grade inflammation; nutrients; natural herbs; pro-inflammatory cytokines

Citation: Koppula, S.; Akther, M.; Haque, M.E.; Kopalli, S.R. Potential Nutrients from Natural and Synthetic Sources Targeting Inflammaging—A Review of Literature, Clinical Data and Patents. *Nutrients* **2021**, *13*, 4058. https://doi.org/10.3390/nu13114058

Academic Editor: Maria Luz Fernandez

Received: 13 October 2021
Accepted: 11 November 2021
Published: 13 November 2021

Publisher's Note: MDPI stays neutral with regard to jurisdictional claims in published maps and institutional affiliations.

Copyright: © 2021 by the authors. Licensee MDPI, Basel, Switzerland. This article is an open access article distributed under the terms and conditions of the Creative Commons Attribution (CC BY) license (https://creativecommons.org/licenses/by/4.0/).

1. Introduction

Aging is a complex physiological and psychological process throughout life where changes occur in many different aspects. Interaction between environmental and genetic factors is the main accelerating force behind aging [1,2]. During the past several decades, research in geroscience has produced several highly interconnected hallmarks for aging progressiveness, namely epigenetics, inflammation, cellular senescence, stem cell exhaustion, mitochondrial dysfunctions, proteostasis, metabolism derangement, and altered intracellular communications [3,4].

As a part of the immune system, inflammation is essential to survive and eliminate the invasion of harmful pathogens, however, this can be detrimental in elderly persons. An array of evidence suggests that an increase in systemic inflammation is a very common phenomenon during aging [5,6]. Inflammaging, first coined by Franceschi et al., is best defined by low-grade, systemic, chronic, asymptomatic, and persistent inflammation [2]. Even though inflammaging is low-grade inflammation, its uncontrolled phenomenon might be a significant risk factor in human aging contributing to their morbidity and mortality [7].

When it comes to elderly, the inflammaging-associated cytokine levels are notably increased, although they remain within the range. The anti-inflammatory response of the body following inflammation is exactly the opposite and controls the inflammatory mechanism. The balance between pro-inflammation and anti-inflammation is essential for the outmost result of inflammation; a similar concept is observed for inflammaging and anti-inflammaging processes. The concept of inflammaging is defined based on the chronic and progressive increase in the pro-inflammatory status and reduction in the ability to respond to different stressors [1].

In general, inflammaging inhibits the ability to fight against infection and wound healing by the impairment in response against new antigens. With progressive aging, a strong correlation between inflammaging and production of proinflammatory cytokines such as tumor necrosis factor-alpha (TNF-α), interleukin (IL)-6, IL-1, and C-reactive protein (CRP) has been established [8–10]. An increase in these proinflammatory cytokines can potentiate the loss of muscle strength, bone metabolism, and nutritional status, which can also lead to age-related disorders (ARDs) such as neurodegenerative diseases, arthritis, cancer, atherosclerosis, type 2 diabetes (T2D), cardiovascular disorders (CVDs), and osteoporosis [11–14].

The inflammaging theory, its pathological changes, and their role in the development of disease, mechanisms, and interventions have been extensively reviewed [15,16]. However, knowledge regarding inflammaging is still incomplete as the cause and risk factors are not completely defined. Further, the measurement of inflammaging also remains vague in clinical settings. Till now, limiting inflammaging by calorie restriction, physical activity along with small-molecule inhibitors is the most reported approach [16–19]. Although implementation of a healthy lifestyle might support in slowing down and delaying the aging process, scientists are endeavoring to understand the possible regulatory mechanisms and discover potential anti-inflammaging molecules by examining several functional foods, dietary nutraceuticals, and pharmacological compounds. In particular, nutritional interventions from natural and synthetic sources in delaying the inflammaging and effective strategies in mitigating the effects of inflammaging and its progress are gaining interest and acquiring new insights [20–22]. In this review, the key inflammatory changes that occur during inflammaging, the available literature, clinical studies, and patents on inhibitory molecules in mitigating or delaying inflammaging from natural and synthetic nutrient compounds, were discussed.

2. Mechanisms of Inflammaging

Earlier studies highlighted various mechanisms that involve the activation of innate immunity together with a rise of proinflammatory mediators and the chronic inflammatory process with aging and oxidation-inflammation theory of aging [23–26]. The primary feature based on these theories highlights an increase in the body's pro-inflammatory status with advancing age. In the following sections, the key inflammaging theories are briefly summarized to elucidate the inflammaging mechanisms with respect to the use of nutrients as anti-inflammaging therapeutic targets.

2.1. Cytokines in Inflammaging

Mounting evidence suggests that inflammaging is associated with elevated levels of pro-inflammatory cytokines such as TNF-α, IL-6, IL-1, interferon-γ (IFN-γ), and IL-18 [8–10]. An increased amount of TNF-α, IL-6, and CRP has been found in elderly patient's serum and strongly correlates with mortality, morbidity, and frailty [27,28]. Studies also indicated that an increase in IL-1, IL-6, TNF-α, and PGE2 in circulation leads to the proinflammatory status in elderly patients and IL-6 can be referred to as one of the important predictive markers for inflammaging [14,29,30]. Notably, this proinflammatory status lead by cytokines creates an inflammatory environment in the tissues and organs, which plays a crucial role in inflammaging [31].

Maintaining a balance between pro- and anti-inflammatory status has been long linked with aging and longevity. Genetic polymorphism in inflammatory cytokines gene might play a significant role to balance and maintain this inflammatory status. A study among an Italian cohort showed an increase in the frequency of -174C single nucleotide polymorphism (SNP) in the IL-6 promoter region in male centenarians while there was an increase in the frequency of -1082G SNP at the 5′ flanking region of the IL-10 gene coding sequence. On the other hand, an increase in the +874A SNP at the IFN-γ gene was found in the female centenarians [32]. Further, genome-wide association study (GWAS) study among Han Chinese centenarians further confirms SNP mapping in IL-6 gene locus (rs2069837) was associated with longevity [33]. Further, a previous study also revealed that polymorphism in the C/G 174 on the IL-6 coding gene is associated with alterations with IL-6 serum concentration and corresponding IL-10 level [34].

2.2. Oxidative Stress in Inflammaging

The free radical concept of aging, wherein accumulation of excessive reactive oxygen species (ROS) produced in our cells is one of the major causative factors of oxidative stress [35]. Moreover, inflammation leads to increased levels of ROS, inducing consistent chronic oxidative stress. Increased ROS formation with weakened oxidative defense has been linked with inflammaging as both ROS and inflammaging exhibit mutual stimulatory roles [36]. Further, oxidative phosphorylation is accelerated among the elderly, which results in the accumulation of oxygen metabolites. An increasing amount of oxygen metabolites can damage the cellular components such as DNA, RNA, lipids, and proteins, thus affecting the cellular homeostasis [37]. Additionally, accumulating oxygen metabolites increases cell membrane porosity and reduces adenosine triphosphate (ATP) levels, which can accelerate the cellular aging process [38]. The functional capacity of the immune cells, individual lifespan, and the redox state are correlated and provide the oxidation-inflammatory theory of aging [24]. This theory suggests that a significant reduction in oxidative stress with the administration of potential antioxidant nutrients possessing anti-inflammatory effects in regular diet might delay the aging process and increase longevity [39].

2.3. Cellular Senescence in Inflammaging

Cellular senescence is best characterized by reduced cell proliferation and cell cycle arrest [40]. Cellular senescence can be caused by shortening of telomere, DNA damage, mitochondrial DNA damage, danger-associated molecular pattern (DAMPs) exposure, epigenetic modifications, point mutation, and also by stress-related signaling pathways. Senescent cells are accumulated in different organs and tissues, which increases exponentially with aging [41]. Even though no prominent marker for senescence exists, a cyclin-dependent kinase inhibitor 2A (P16INK4A) encoding gene CDKN2A was found to possess the strongest association and, therefore, was used as a common biomarker to characterize the cellular senescence [42]. A GWAS study showed that aging-associated diseases such as cardiovascular diseases (CVDs) and T2D are associated with SNPs located near senescence and inflammation [43,44]. Additionally, a common variant rs2811712, which is close to CDKN2A, was found to be associated with poor physical function in elderly [45]. Interestingly, a study suggests that the elimination of p16^{Ink4a}-positive senescent cells decreases aging-associated diseases and increases lifespan [46,47]. DNA damage caused by telomere shortening results in replicative senescence and aging-associated diseases [40,48]. DNA damage response (DDR) increases pro-inflammatory status by activating adjacent cell DDR. The overall DDR in immune cells might accelerate the inflammaging process [49].

2.4. Autophagy in Inflammaging

Autophagy is a common physiological process of recycling and clearing detrimental substances such as misfolded proteins and damaged organelles to maintain cellular homeostasis [39]. However, the imbalance between production and cellular clearance of cell debris and misfolded protein disrupts cellular homeostasis, which may lead to inflammaging [40]. With the progression of age, cellular autophagy is believed to decline [41]. As a result, there is an increase in the accumulation of detrimental substances, production of ROS, and proinflammatory responses, which accelerates the process of aging [42]. Although several pathways are involved in the imbalance of clearing cellular debris, increased production of ROS can result in activating NF-κB signaling and consequent inflammatory responses [50]. A study has linked NLRP3 inflammasome with inflammaging by functional decline [43]. Excessive accumulation of DAMPs can result in the activation of NLRP3 inflammasome, which eventually causes the production of cytokines like IL-1 and IL-18. A study suggests excessive amount of IL-1 and IL-18 are responsible for several aging-associated diseases such as CVD and T2D [16].

3. Anti-Inflammaging Nutrient Compounds from Natural and Synthetic Sources

Increasing evidence suggests that natural and synthetic nutrient interventions have major influence in delaying inflammaging and aid in the prevention of inflammation-associated ARDs. In the following section, we reviewed the selected nutrient compounds that possibly target inflammaging by regulating various inflammation and aging intertwined pathways. A list of selected nutrient compounds reviewed indicating the sources, experimental models, and mechanisms, was shown in Table 1.

3.1. Resveratrol

Resveratrol (trans-3,4,5-trihydroxystilbene, Figure 1A) is a naturally occurring polyphenol abundantly present in many sources such as red variety of grapes, peanuts, blueberries, pines, and rhubarb. In traditional Chinese and Japanese medicine, resveratrol has been used for a long time in the form of extract from *Polygonum cuspidatum* [51]. Pharmacologically, resveratrol has shown strong antioxidant, anti-inflammatory, neuroprotective, anti-microbial, and anti-cancer effects in a number of studies [52–55].

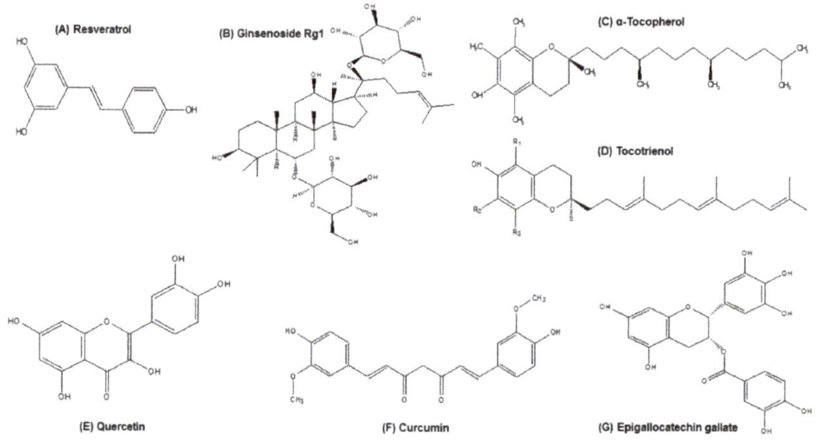

Figure 1. Chemical structures of selected nutrient compounds targeting inflammaging.

Moreover, a strong line of evidence suggests that supplementation with resveratrol can delay the aging process [56]. Liang and group reported that resveratrol can inhibit ROS in vitro in rabbit articular chondrocytes. The authors indicated that pretreatment with resveratrol (100 μM) significantly lowered the sodium nitroprusside-induced ROS. The study also revealed that resveratrol can contribute to a reduction in apoptosis due to its ROS scavenging effects [57].

Previously, resveratrol was reported to work through the activation of silencing information regulator 1 (SIRT1), which can inhibit NF-κB-regulated inflammatory cytokines [58]. Further, another study also claimed that resveratrol mediated an increase in the expression of PPAR-γ, and SIRT1 leads to downregulation of inflammation [59]. Another study suggests that activation of SIRT1 by resveratrol can also downregulate TNF-α-induced IL-1β and IL-6 expression in fibroblast cell line 3T3, and reduces phosphorylation of rapamycin (mTOR) and S6 ribosomal protein (S6RP) [60]. Overall, these data suggest that resveratrol possesses a significant effect against inflammation through SIRT1 pathway, which is also a key signaling pathway in regulating lifespan and thereby the inflammaging process [61].

Interestingly, resveratrol is able to induce a telomerase maintenance factor WRN helicase without affecting cell proliferation, and thus, might contribute to prevent telomerase dysfunction [62]. Further, resveratrol dose-dependently diminished cellular senescence, proving its capability in delaying cellular senescence in endothelial progenitor cells, thus increasing the telomerase activity [63]. In a study by Tung et al., the effect of resveratrol on aged mice was investigated [64]. In their study, young (2 months), adult (12 months), and old (18 months) mice were fed with resveratrol (24 mg/kg/day) and later the progression of proinflammatory markers were investigated in the liver tissues. Data revealed that the proinflammatory cytokines such as IL1β, IL-6, IL-17, and TNF-α increased with age in mice liver. Pretreatment with resveratrol attenuated the increased IL-1β and TNF-α protein and mRNA expression in only old mice. In addition, the authors observed an increase in the amount of ASC (NLRP3 inflammasome component apoptosis-associated speck-like protein containing a CARD), caspase-1, and NALP-3 (NACHT, LRR and PYD domains-containing protein 3) in aging mice, which was also reversed by resveratrol. This study further proves the possible effect of resveratrol in inhibiting inflammaging.

Another study conducted on aged female mice by Jeong et al., further strengthened the links between resveratrol and inflammaging, where pretreatment of 0.1 mg/kg resveratrol in aged female mice reduced IL-1β and TNF-α levels. In addition, treatment of resveratrol also inhibited stroke-induced brain injury and inflammation [65]. Taken together, these studies suggest that resveratrol has multiple targets in attenuating aging-associated chronic inflammatory pathways. Thus, resveratrol can be a useful nutrient compound against inflammaging and inflammaging-associated disorders.

3.2. Ginseng (Panax Ginseng)

Korean red ginseng, scientifically known as Panax ginseng (P. ginseng), has been traditionally used to treat several ailments including immune-related disorders such as inflammatory and autoimmune diseases. It is regarded as a miracle herb found mostly in Korea and China with several active component types of ginsenosides and acidic polysaccharides [66,67]. Recently, the polysaccharides (30 mg/kg) from P. ginseng berries were reported to exert protection against immunosenescent effect in aged mice by reducing the IL-6 and IL-2 cytokine expression. The authors indicated that ginseng-derived polysaccharides might be useful as antiaging agents by regulating the inflammatory immune functions [68]. In another study by Wang and group, the ginsenoside Rg1 from P. ginseng (Figure 1B) attenuated the excessive production of inflammatory cytokines in D-galactase (D-gal)-induced aging mouse. The authors investigated the testicular senescence changes in D-gal-induced aging mice by evaluating the anti-inflammatory parameters including tumor necrosis factor-α, interleukin (IL)-1β, and IL-6 in testicular tissues. Ginsenoside Rg1 (20 mg/kg/day) treated for 2 weeks after D-gal administration reduced the levels of inflammatory cytokines significantly, indicating the anti-inflammaging role of Rg1 on male reproductive function [69].

3.3. Tocotrienol

Vitamin E, a common nutrient found in seeds, cooking oils, nuts, and most foods, is a liposoluble antioxidant that contains two groups, α-tocopherol and tocotrienol (Figure 1C,D). Tocopherol has been studied extensively and is well known for its antioxidant protective role. Tocotrienol is a newer discovery compared to tocopherol and is mostly found in many edible plants such as palm tree, annatto, and achiote tree [70,71]. Several studies with tocotrienol suggest its superiority in the antioxidant activity compared to tocopherol [72]. Although tocopherol and tocotrienol exhibit the same chromanol head, they are structurally different in the hydrophobic tridecyl chain saturation. Tocopherol contains saturated phytol chains whereas tocotrienol contains unsaturated farnesyl isoprenoid chains. Both tocopherol and tocotrienol have four homologs each (alpha, beta, gamma, and delta) and depend on the location and number of methyl groups on the chromanol ring [71].

In a study by Ahn et al., the authors indicated that gamma tocotrienol inhibited IκB-α kinase activation, thus leading to the subsequent inhibition of NF-κB activation [73]. Moreover, Wong et al., investigated the anti-inflammatory role of different homologs of tocotrienols and alpha-tocopherol. The in vivo study suggested that delta and gamma tocotrienol comprises a significant improvement in inflammation, heart, and liver function in rats [74]. Inflammaging has been broadly correlated with IL-6 and CRP. Interestingly in a study by Yam and group, the authors identified that tocotrienol can significantly inhibit IL-6 production in lipopolysaccharide (LPS)-stimulated RAW264.7 macrophages [75]. Also, this study reported the reduction of cyclooxygenase (COX)-2, prostaglandin (PG)-E2 by tocotrienol. In the same line of investigation, a report by Qureshi and colleagues showed inhibition of TNF-α by tocotrienol in Raw264.7 cells. They also demonstrated that in BALB/c mice serum, TNF-α levels were significantly decreased by the treatment of tocotrienol with doses of 1–10 µg/kg [76]. In a study on T2D individuals supplemented with alpha tocopherol for 3 months, a significant reduction in their CRP and IL-6 plasma levels was shown [77]. CRP and IL-6 are two of the most common cytokines found in individuals with inflammaging. Since tocotrienol targets different pathways involved in inflammaging, it might serve as a potential anti-inflammaging candidate.

3.4. Quercetin

Quercetin (2-(3,4-Dihydroxyphenyl)-5,7-dihydroxy-4H-1-benzopyran-4-one, Figure 1E) is a well-studied flavonol containing several health beneficiary properties [78]. Quercetin is abundantly and widely found in various dietary sources such as apples, grapes, tomatoes, onions, berries, and brassica vegetables. Quercetin is also available in natural herbs including *Hypericum perforatum*, *Ginkgo biloba*, and *Sambucus canadensis* [79,80]. Quercetin exhibits a wide range of properties that can play a potential role in immunity and shows overall health benefits through anti-inflammatory, anti-oxidant, anti-viral, and anti-carcinogenic activities [81]. In murine macrophages, a significant reduction of TNF-α production after inducing LPS and in glial cells a reduction in mRNA of TNF-α has been reported [82,83]. Also, in a study conducted in lung A549 cells, quercetin showed a reduction in IL-8 [84]. This flavonol also suppressed the IL-1-stimulated selective release of IL-6 in mast cells significantly [85].

Independent in vitro studies with RPE and THP1 cell line demonstrated inhibition of oxidative stress by quercetin [86,87]. In addition, a recent study conducted by Bao et al. demonstrated that hydrogen peroxide (H_2O_2)-induced oxidative stress is significantly inhibited by the treatment of quercetin in rat pheochromocytoma (PC12) cells. This study also showed a protective effect of quercetin in H_2O_2-induced cell death as quercetin significantly reduced lactate dehydrogenase (LDH) from PC12 cells [88]. Interestingly, recent studies shined a light on the effect of quercetin on aging. A study by Saul et al. demonstrated that quercetin can improve lifespan by 15% in *Caenorhabditis elegans* [89]. As oxidative stress is a leading factor for developing inflammaging and quercetin is known to be widely used as an antioxidant due to its potent ROS and reactive nitrogen species (RNS) scavenging effects, quercetin can be considered as a valuable nutrient in controlling oxidative stress-mediated inflammaging.

3.5. Curcumin

Curcumin (Figure 1F) is the major component of turmeric and is a very commonly used spice and coloring agent in various foods throughout the world. It is a yellow pigment that is naturally found in the rhizome of *Curcuma longa* from the herb family *of Zingiberaceae* [90]. Curcumin has long been used as a traditional medicine in India and other Asian countries for many diseases. Curcumin exhibits anti-oxidant, anti-inflammatory, and anti-carcinogenic properties, which make it an attractive source as a potential nutrient supplement. The anti-inflammatory function of curcumin has been well established in the last few decades. Curcumin can potentially reduce the activity of several inflammation-related transcription factors including NF-κB, AP-1, signal transducer and activator of transcription (STAT), and hypoxia inducible factor-1 (HIF-1). Curcumin inhibits NF-κB activity by inhibiting P65 translocation to the nucleus and suppressing IκB-α degradation. It has also been demonstrated by several authors that the inhibition of NF-κB transcription factor by curcumin leads to reduction in TNF-α, IL-6, and COX-2 [90].

A recent study showed that the anti-inflammatory activity of curcumin through NF-κB depends on its oxidized form. The authors showed inactivation of curcumin electrophiles through the pretreatment of N-acetyl cysteine, which is a precursor of glutathione (GSH). Oxidative stress is involved in several ARDs and the protective effect of curcumin against oxidative stress can favor its use as an anti-inflammaging agent [90]. Additionally, several in vivo studies conducted on mice and rats showed a significant inhibition in oxidative stress. For instance, Sood et al., showed that aluminum-induced oxidative stress has been reduced by free curcumin, as curcumin prevented depletion of GSH [91]. Another study showed that curcumin prevents the increase in malondialdehyde (MDA) and nitrates in the hippocampus of LPS-induced neurobehavioral and neurochemical deficits Swiss albino mice [92].

Curcumin also exhibited its potency against neuroinflammation-mediated aging in in vitro and in vivo models of diseases such as Alzheimer's disease (AD) and Parkinson's disease (PD) [93,94]. Further, Rastogi et al., showed that curcuminoid is capable of inhibiting age-associated mitochondrial impairment in rats [95]. The authors indicated that 100 mg/kg orally-treated curcuminoid inhibited age-associated enzymes such as NADH dehydrogenase, cytochrome c oxidase, Complex I, and total ATP content. The curcuminoid also suppressed neuronal nitric oxide synthase (NOS) in mitochondria. Furthermore, a study on mice showed the ability of curcumin on age-related cognitive dysfunction, where curcumin significantly inhibited oxidative stress [96]. The pleiotropic activity exhibited by curcumin by regulating various inflammatory mechanisms in aging models might open the door for curcumin to be developed as a potential anti-inflammaging agent.

3.6. Epigallocatechin-3-Gallate

Epigallocatechin-3-gallate (EGCG, Figure 1G) is the most abundant catechin found in green tea with immense medicinal benefits [97,98]. Even though green tea contains other catechins, EGCG comprises 50–60% among them all. As a catechin, EGCG contains dihydroxyl or trihydroxyl substitutions on the B ring and the m-5,7-dihydroxyl substitutions on the A ring. Few studies suggest that the B ring is the principal site for the antioxidant property of EGCG, and the polyphenolic structure is believed to be responsible for the ROS quenching [99,100]. EGCG contains several health beneficiaries such as anti-oxidant, anti-inflammatory, and anti-cancer effects. EGCG comprises anti-inflammatory effects through the inhibition of gene expression of TNF-α, COX-2, and iNOS [101]. Studies also showed that EGCG can directly suppress NF-κB and AP-1 in human ECV304 cells [102].

In a study on RAW264.7 cells, EGCG showed its inhibition through TLR4-mediated MyD88 and TRIF signaling pathways [103]. EGCG also exerts antioxidant activity by reducing ROS and suppressing oxidative stress by activating nuclear factor erythroid-2 like factor-2 (Nrf-2) transcription factor [104]. An in vivo study by Nui et al. suggested that 25 mg/kg of EGCG extends lifespan in rats [105]. In a recent study by the same group, 90 male Wistar rats were challenged with a high fat diet and 60 mg/kg EGCG throughout their lifetime. EGCG extended lifespan significantly among high fat diet-fed male rats. Further, EGCG significantly decreased the IL-6, TNF-α level in serum of rats and suppressed the NF-κB expression. Additionally, the authors also determined the level of ROS in the serum of EGCG-induced rats. The ability of EGCG to suppress inflammation and ROS in experimental models might contribute to increased lifespan [106]. In addition, a cohort study in China supported the notion that consumption of green tea increases lifespan [107]. Based on the overall studies, EGCG might be developed as a promising anti-inflammaging nutrient.

3.7. Huperzine A

Huperzine A (HupA; Figure 2A) is a naturally occurring Lycopodium alkaloid isolated from the Chinese popular traditional medicine *Huperzia serrata*. HupA is a potent, reversible, and selective inhibitor of acetylcholinesterase enzyme, and has been used in China since long as a medicament for various disorders such as strains, swellings, contusions, schizophrenia, and fever [108,109]. Further, HupA is believed to have significant effects against cognitive deficits and is an approved drug for treating AD in China [110]. Besides that, several studies have been done to investigate the effect of HupA on multiple directions/pathways. Wang et al., reported that the effect of 0.1 mg/kg HupA intraperitoneal injection decreased neurological deficits in a rat model of transient focal cerebral ischemia model. More importantly, the authors also found that HupA inhibited nuclear translocation of NF-κB and overexpression of several proinflammatory factors [111].

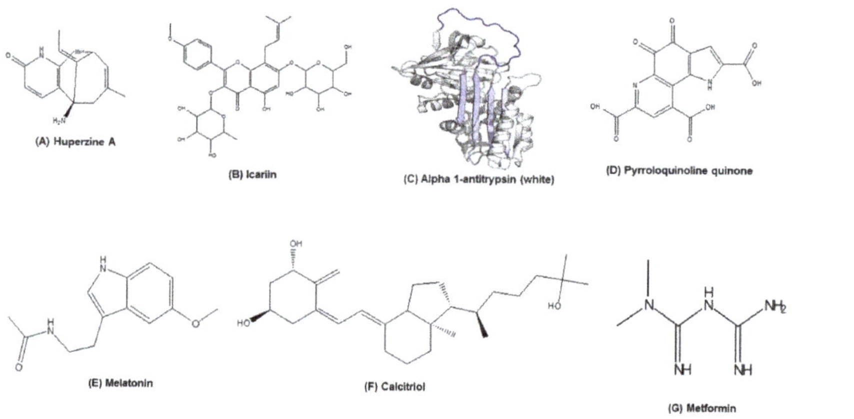

Figure 2. Chemical structures of selected natural and synthetic nutrient compounds targeting inflammaging.

A study conducted by Ruan et al. showed that HupA exhibited hepatoprotective activity based on its anti-inflammaging effects in D-galactose (D-gal)-treated rats [112]. D-gal (300 mg/kg s.c.) administered rats showed increased oxidative damage, hepatic senescence, nuclear factor-kappa B (NF-κB) activation, and inflammatory response. Co-administration with HupA A (0.1 mg/kg s.c.) for 8 weeks significantly attenuated the D-gal-induced changes and also exhibited anti-inflammaging effects by hepatic replicative senescence inhibition in experimental rats. Additionally, the study also demonstrated that the suppression of TNF-α, IL-6, and IL-1β expression level by HupA is via the inhibition of NF-κB pathway. Overall, the results provided a scientific basis on the role of HupA as a potential anti-inflammaging nutrient.

3.8. Icariin

Icariin (ICA; Figure 2B) is a flavonoid and bioactive component of *Herba epimedii*, which is a popular Chinese medicine for treating various diseases [113]. *H. epimedii* is traditionally used for several age-associated diseases such as CVDs, osteoporosis, neurological disorder, and sexual dysfunctions [114]. ICA, the main component of total flavone of *Epimedium* (TFE), was extensively studied and is well known as an aging intervention drug. In vitro studies on human diploid fibroblasts and *C. elegans* showed that ICA extends the lifespan by protecting the length of telomere [115,116]. An in vivo study also confirmed that administration of ICA from 12 months of age increased the health span and mean lifespan in mice. The study showed that the ability of ICA in decreasing oxidative stress and DNA damage might contribute to extended health and life span in mice [117]. Further, ICA was well reported to downregulate the NF-κB pathway, thus contributing to lowering inflammatory pathways. Recently, Chen et al. showed that the downregulation of NF-κB pathway by ICA might be responsible for the upregulation of SIRT6. The study confirmed the inhibition of TNF-α and IL-6 expression by the treatment of ICA (0.02% ICA for 3 months in feed). Additionally, the authors predicted that the effect of ICA in regulating NF-κB pathway through SIRT6 might inhibit cardiac inflammaging, indicating a novel link between the SIRT6 (a regulator of aging) and NF-κB (a regulator of inflammation) and the synergistic effect of ICA in preventing inflammaging [118].

3.9. Blueberry

Blueberry fruit from Ericaceae family contains enormous functional phytoconstituents including polyphenolic and flavonoid components. Pharmacological studies revealed that blueberry possesses anti-diabetic, anti-inflammatory, anti-tumor, and neuroprotective effects [119]. Evidence suggests that blueberry exhibited strong protection against inflammation in pre-clinical and clinical evaluations [120–122]. In a study by Goyarzu et al., the flavonoid-rich diets including the anti-oxidant rich blue berry were well reported to prevent cognitive impairment associated with inflammaging in animal studies [123]. Aged Fischer-344 rats were supplemented with 2% blueberry diet for 4 months, and the cognitive parameters and biochemical parameters were evacuated in comparison with aged control groups. The authors found better performance in cognitive tasks in comparison with aged control. The inflammatory signaling NF-κB expression levels in four brain regions in blueberry-supplemented aged rats showed lower levels when compared with aged control groups. The impaired object recognition memory scores were correlated with high levels of NF-κB in aged control groups, indicating that blueberry-supplemented rat diet may retard brain inflammaging in aged rats.

3.10. Prune (Prunus spinosa L.)

P. spinosa, commonly known as blackthorn from Rosaceae family, is majorly found in Asia, Europe, and the Mediterranean. *P. spinosa* has been used traditionally for centuries as a diuretic, laxative, and antispasmodic, and to prevent inflammatory conditions [124,125]. In a recent report, the wound healing effects of *P. spinosa* fruit extract were evaluated against LPS challenge in young and aged (senescent) human umbilical vein endothelial cells (HUVECs) based on its anti-inflammatory effects. The authors found that *P. spinosa* exerted (40 μg/mL) potent anti-oxidant and anti-inflammaging ability in older cells when compared with younger cells by downregulation of inflammatory markers including IL-1 receptor associated kinase 1 (IRAK-1) and IL-6. Further, the *Prunus* extract (400 μg/mL) increased the life and health span in *C. elegans*. Overall, the study indicated that *P. spinosa* fruit extract might enhance wound healing capacity in senescent conditions, thereby improving the quality of life in aging populations [126].

3.11. Alpha-1 Antitrypsin (AAT)

Alpha-1 antitrypsin (AAT, Figure 2C) is an acute phase glycoprotein produced mainly in liver that protects lungs [127,128]. AAT works as a protease inhibitor on trypsin, chymotrypsin, thrombin, and elastase. AAT is one of the most abundant serine protease inhibitors found in circulating human blood and is able to inhibit proteinase 3 and cathepsin G. However, if human alpha-1 antitrypsin (hAAT) is modified by NO, it can work as a cysteine protease inhibitor [129]. Even though hAAT is abundant in circulation, its expression can be increased by IL-6 and LPS treatment [130]. Further studies by Petrache et al. showed that AAT can inhibit apoptosis in alveolar cell. This study also confirmed that hAAT is able to decrease caspase-3 and oxidative stress at a cellular level [131]. Many studies regarding AAT further proved the role of hAAT in ARDs such as stroke, T1D, and rheumatoid arthritis. A study by Wang et al. showed that AAT is responsible for the suppression of instant blood-mediated inflammatory reaction (IBMR). The authors indicated that significantly lower TNF-α levels, lymphocytic infiltration, and decreased nuclear factor activation were observed in AAT-treated mice compared to control mice. They suggested the potent anti-inflammatory role of AAT is due to the inhibition of JNK phosphorylation [132]. AAT exhibits an anti-inflammatory, immune-regulatory, and cytoprotective role against many diseases. A recent study by Yuan et al. provided direct evidence of hAAT possessing therapeutic potential against inflammaging [133]. In their study, they found that hAAT overexpressing *Drosophila* cell line demonstrated longer lifespan than the control cell line. Interestingly, the identified two aging-associated genes *Relish* and *Diptericin* were significantly lower in the hAAT overexpressing cells. Moreover, RNA sequence analysis showed a significant decrease in the NF-κB-regulated innate immunity genes in hAAT overexpressing Drosophila cell lines.

In another line of evidence, RNA seq analysis also showed a decrease in inflammation-related genes. Experiments with human cell line confirm that hAAT treatment was able to suppresses senescence-associated secretory phenotype (SASP), as it inhibited the IL-6 and IL-8 in X-ray-induced senescence cells. Further, hAAT is an FDA-approved drug with a proven safety profile, making it a potential candidate in inflammaging and aging-associated disorders.

3.12. Pyrroloquinoline Quinone

Pyrroloquinoline Quinone (PPQ; Figure 2D), a quinone nutrient substance widely available in the plant foods and animal tissues, helps in scavenging harmful free radicals, reduces oxidative stress, and suppresses the inflammation markers such as serum CRP and IL-6 [134,135]. A study by Zhang et al. was undertaken to assess the anti-aging potential of PQQ in HepG2 cell cultures. The authors concluded that PQQ (10–30 μM) contributes to damage repair and delays cell senescence [136]. In another study, PQQ delayed inflammaging induced by TNF-α, and abated inflammatory cells as well as inflammatory cytokine infiltration. PQQ also partly delayed the premature ageing phenotype; promoted cell proliferation; decreased the expression of cell cycle arrest protein p16, P19, P21, P27, and P53 expression; and promoted the expression of longevity genes *SIRT1* and *SIRT3* [15,118,136].

In a recent study, PPQ ameliorated D-gal-induced oxidative stress and inflammatory response, resulting in cognitive improvements in mouse. PPQ inhibited the hippocampal MDA and increased the SOD expression, thereby exerting strong antioxidant effects in D-gal-induced mice. Further, PQQ attenuated the increased inflammatory factors (IL-2 and IFN-γ) and the production of prostaglandin E2 (PGE2) in D-gal-induced mice [137]. More recently, the effect of PQQ in cultivated human embryonic lung fibroblasts WI-38 cells with or without TNF-α was studied to establish an inflammaging model in vitro. Data revealed that compared with TNF-α stimulation alone, PQQ (150 nmol/L) showed less SA-β-gal-positive cells, indicating that PQQ attenuated TNF-α-induced inflammaging damage. The authors concluded that PQQ delayed TNF-α-induced cellular senescence and had anti-inflammaging properties [138]. Based on these findings, it is clear that PQQ may influence the generation of pro-inflammatory mediators, including cytokines and prostaglandins, during the aging process, and provides evidence that PPQ might be beneficial in preventing cognitive deficits during the inflammaging process.

3.13. Melatonin

Melatonin (Figure 2E), a naturally occurring hormone from the pineal gland in the brain, has also been widely identified and qualified in various foods from fungi to animals and plants. Melatonin concentration in human serum could significantly increase after the consumption of melatonin-containing food. Melatonin exhibits many bioactivities, such as antioxidant activity anti-inflammatory characteristics, boosts immunity, anticancer activity, cardiovascular protection, anti-diabetic, anti-obese, neuroprotective, and anti-aging activity. Melatonin levels decrease in the course of senescence and are more strongly reduced in ARDs such as coronary heart disease and T2D [139]. The role of melatonin, a highly pleiotropic regulator molecule in the antiaging mechanism associated with inflammaging, was well studied in aged organisms and senescence-accelerated animals [140]. Melatonin behaves under conditions of low-grade inflammation, especially in inflammaging. Melatonin exerts a broad spectrum of effects on physiological functions of relevance to aging, such as metabolic sensing; mitochondrial modulation and presumably also proliferation; antioxidative protection of biomolecules and subcellular structures, in particular, mitochondria; and immunological actions implicated in both the combat against foreign antigens and inflammaging. In different senescence-accelerated prone studies on various organ tissues in male mice and in aged ovariectomized female rat liver, melatonin in the dose range of 1–10 mg/kg/day for 1–3 months downregulated proinflammatory cytokines such as TNF-α, IL-1β, and IL-6, and upregulated the anti-inflammatory IL-10, indicating its role in attenuating inflammaging and ARDs [141–144].

3.14. Calcitriol

Calcitriol (1,25-dihydroxycholecalciferol, Figure 2F) is the inactive form of vitamin D, which is activated by activating enzyme 1 alpha-hydroxylase and converting into 1,25-dihydroxyvitamin D3 (1,25VD3) or vitamin D hormone calcitriol [145]. It is also known as the main circulatory form of vitamin D, which is used as a marker for the evaluation of physiological vitamin D levels. Vitamin D is known for its activity against inflammation and aging-associated diseases. A study on Raw264.7 macrophage cells showed 1,25VD3 is able to downregulate pro-inflammatory gene expression such as COX-2, NF-κB, and AKT [146]. Additionally, a different study showed that CD40 ligand-induced pro-inflammatory cytokines IL-1β and TNF-α are suppressed by the co-treatment with 1,25VD3 [147].

In a double-blinded, placebo-controlled, randomized clinical trial, patients with cystic fibrosis supplemented with cholecalciferol demonstrated that vitamin D suppressed two inflammatory cytokines, IL-6 and TNF-α [148]. Moreover, two different randomized control trials among diabetes patients who were given vitamin D as supplementation and calcium demonstrated a significant decrease in IL-6, IL-1β, TNF-α, and CRP in the serum level of the patients [149,150].

Further, in an in vivo study conducted by Wang et al., a model of *Porphyromonas gingivalis*-infected *db/db* mice was used. The group demonstrated that inflammaging occurred in these diabetic mice, which was measured by increased SASP, increased senescent cells, and periodontal destruction. Interestingly, the induction of 1,25VD3 significantly decreased the serum SASP and periodontal condition. The expression of NF-κB, IL-1β, and STAT-3 was also suppressed significantly in gingival tissue. Additionally, the suppressors of cytokine signaling 3 (SOCS3) level were also decreased with the administration of 1,25VD3 [151]. SOCS3 suppresses Janus kinase (JAK) and signal transducer and activator of transcription (STAT) pathway-mediated inflammatory signaling. Surprisingly, aging increased SOCS-3 expression in the hypothalamus. Thus, 1,25VD3-induced attenuation of SOCS3 might be responsible for the decrease in inflammaging through the NF-κB pathway [152,153].

3.15. BaZiBuShen (BZBS)

In a recent study, a Chinese over-the-counter herbal medicament named BZBS containing various natural herb extracts was investigated for sperm quality and fertility in an aging male mouse model [154]. The prescription contained Semen Cuscutae, Fructus Lycii, Epimedii Folium, Fructus Schisandrae Sphenantherae, Fructus Cnidii, Fructus Rosae Laevigatae, Semen Allii Tuberosi., Radix Morindae Officinalis, Herba Cistanches, Fructus Rubi, Radix Rehmanniae Recens, Radix Cyathulae, Radix Ginseng, Cervi Cornu Pantotrichum, Hippocampus, and Fuctus Toosendan, which are approved by China FDA (No. B20020585). In their study, aged mice were induced by D-galactose (D-gal) and NaNO$_2$ for 65 days and the inflammatory signaling pathways were analyzed in testes tissues. Data revealed that treatment with BZBS (0.7, 1.4, and 2.8 g/kg/day) to D-gal- and NaNO$_2$-induced mice alleviated the increased levels of TNF-α secretion, NF-κB activation, and iNOS expression in aging mice testes tissues. Further, BZBS also modulated the Sirt6 expression in the testes of aging mice. The authors concluded that BZBS rescued the altered testicular morphology and sperm quality in rapidly aging mice induced by D-gal- and

NaNO$_2$, possibly via regulation of inflammatory Sirt6/NF-κB signaling, indicating that the Chinese prescription BZBS can be a promising candidate in ameliorating aging-associated inflammatory testicular damage [154].

3.16. Egg Yolk

Mounting evidence suggests that eggs are rich in biologically effective constituents and shows an immense role in the prevention of chronic infectious diseases. Eggs contain micro and macronutrients including zinc, selenium, retinol, and tocopherols, and low amounts of carbohydrates. Earlier reports revealed that eggs, particularly non-fertilized egg yolk, possess immunomodulatory, anti-inflammatory, and analgesic effects [155]. In a study by Cunill et al., patented egg yolk (PEY) was compared with commercial egg yolk for it anti-inflammaging properties. Mouse RAW 264.7 macrophage cells lines in vitro and LPS (2.5 mg/kg dose, i.p.) administered Wistar rats in vivo were treated with PEY and commercial egg yolk (2000 mg/kg) to two different groups, and the serum levels of cytokines including IL-1β, TNF-α, and MCP-1, were evaluated in both groups. The authors indicated that PEY group showed a significant reduction in concentrations of IL-1 β, TNF-α, and MCP-1 when compared with the commercial egg yolk group, and concluded that the complex biomolecules present in PEY might effectively control the chronic inflammation and aging and help reduce the inflammaging progression [156].

Table 1. Selected nutrients from natural and synthetic sources targeting inflammaging in experimental models.

Nutrient Compounds	Source	Model (In Vitro and In Vivo)	Dose	Mechanism/s	Refs
Resveratrol	Grapes, peanuts	Aged male mice	24 mg/kg/day	↓ IL-1β, TNF-α, COX-2; ↓ ASC, caspase-1 and NALP-3	[64]
		Aged female mice	0.1 mg/kg/10 days	↓ IL-1β and TNF-α ROS scavenger	[65]
APGP	*Panax ginseng* berries	Immunosenescnece old male C57BL/6J mice	5 and 30 mg/kg/daily for 20 days	↓ IL-2, =IL-6	[68]
Ginseng Rg1	*Panax ginseng*	D-gal-induced aging mice	20 mg/kg/28 days	↓ TNF-α, IL-1β, and IL-6.	[69]
Tocotrienol	Palm tree (Palm Vit E)	LPS-induced RAW264.7 cells	10 μg/kg	↓ IL-6, NO, COX-2	[75]
	Rice bran	LPS-stimulated RAW264.7 cells	4, 8, 16 μM	↓ TNF-α	[76]
		LPS-stimulated female mice	2.5, 5.0, and 10.0 μg/kg	↓ TNF-α IL-1β, IL-6, and iNOS	[76]
Tocopherol	*all-rac-α-* tocopherol	T2D patients	1200 IU/day/3 months	↓ CRP and IL-6	[77]
Quercetin	Herbs and fruits	daf-16(mgDf50) mutant strain nematode *Caenorhabditis elegans*	200 μM	↑ lifespan	[89]
Curcumin	Turmeric	Aged Wistar rats	100 mg/kg/3 months	↓ IL-6, TNF-α, mitochondrial impairment and nNOS,.	[95]
		SAMP8 mice	20 and 50 mg/kg per day/25 days	↓ MDA; ↑ p-CaMKII and p-NMDAR1	[96]
EGCG	Green tea	Life time high fat diet fed rats	60 mg/kg/life time	↓ ROS, IL-6, and TNF-α	[106]
HupA	*Huperzia serrata*	D-gal treated rats	0.1 mg/kg/ 8 weeks	↓ TNF-α, IL-6, and IL-1β, ↓ NF-κB	[112]

Table 1. Cont.

Nutrient Compounds	Source	Model (In Vitro and In Vivo)	Dose	Mechanism/s	Refs
ICA	*Herba epimedii*	C57BL/6 aged mice	0.02% for 3 months in feed	↑ Life span; ↓ MDA	[117]
		male BALB/c mice	0.02% ICA for 3 months	↑ SIRT6; ↓ TNF-α, ICAM-1, IL-2, and IL-6 and NF-κB	[118]
Blueberry	*Vaccinium* spp.	Aged Fischer-344 rats	2% in diet for four months	↓ NF-κB and, oxidative stress	[123]
Prune	*Prunus spinosa*	LPS-induced HUVECs	40 µg/mL	↓ IRAK-1, and IL-6	[126]
		Nematode *C. elegans*	400 µg/mL	↑ life span	[126]
PPQ	Plant foods and animal tissues	D-gal-induced mice	100 µg/kg/day/for 6 weeks	↓ IL-2 and IFN-γ	[137]
		TNF-α induced human WI-38 cells	150 nmol/L	↓ TNF-α–induced cellular senescence	[138]
Melatonin	Natural hormone, Foods	Aged ovariectomized female rat	1 mg/kg/day/10 weeks	↑ IL-10; ↓ iNOS, HO-1, IL-6, TNF-α and IL-1β	[141]
		SAMP8 mice	1 mg/kg/day/one month	↓ TNF-α, IL-1β, HO (HO-1 and HO-2), iNOS, MCP1, NFκB1, NFκB2 and NKAP	[142]
		SAMP8 mice	1 and 10 mg/kg day for one month	↓ TNF-α, IL-1β, and IL-6. ↑ IL-10.	[143]
		SAMP8 mice	1 mg/kg/day/one month	↓ TNF-α, IL-1, HO-1, and NFκB; ↑ IL-10.	[144]
Calcitriol	Vitamin D	*Porphyromonas gingivalis*-infected *db/db* mice	5 µg/kg/alternative day for 10 weeks	Regulation of NF-κB, IL-1β, STAT-3	[151]
BZBS	Herbal preparation	D-gal-induced aged mice	0.7, 1.4, and 2.8 g/kg/day for 65 days	Regulation of Sirt6/NF-κB	[154]
PEY	Eggs	LPS-induced RAW 264.7 macrophage cells.	2000 mg/kg	↓ IL-1 β, TNF-α, and MCP-1	[156]

Abbreviations: IL: interleukin; LPS: lipopolysaccharide; TNF-α: tissue necrosis factor-alpha; ROS: reactive oxygen species; NF-κB: nuclear factor-kappa B; CRP: C-reactive protein; Rice NPN: Rice natural peptide network; MCP-1: monocyte chemoattractant protein-1; nNOS: neuronal nitric oxide synthase; IFN-γ: interferon-gamma; IRAK-1: IL-1 receptor associated kinase 1; APGP: acidic-polysaccharide-linked glycopeptide; EGCG: epigallocatechin-3-gallate; HupA: huperzine A; ICA: icariin; PPQ: pyrroloquinoline quinone; BZBS: BaZiBuShen; PEY: patented egg yolk; p-CaMKII: p-calcium/calmodulin-dependent kinase II; p-NMDAR1: p-N-methyl-d-aspartate receptor subunit 1; HO-1: heme oxygenase 1; SAMP8 mice: Senescence-accelerated mice; D-gal: *D*-galactose; ↑: increased; ↓: decreased; =: no change.

4. Clinical Trials and Human Research on Nutrient Compounds for Possible Anti-Inflammaging Effects

Data obtained from clinical trials and research involving human subjects can be ideal in providing strong insights in developing nutrient compounds aimed at controlling inflammaging and regulation of its mechanisms. In the following section we reviewed the available clinical data and studies involving humans on nutrient compounds from natural and synthetic sources targeting inflammaging. A list of selected nutrient compounds with reported clinical trials and human research used in preventing or delaying inflammaging is shown in Table 2.

4.1. Metformin

Metformin (Figure 2G), derived from galegine, a natural product from the plant *Galega officinalis*, has been used in herbal medicine in Europe since medieval times. Metformin, a prescribed drug for T2D, improves glycemic control and shows clear benefits in relation to glucose metabolism and diabetes-related complications. Cohort studies suggested that metformin might work as an anti-aging molecule, as it lengthens the lifespan and has been found to inhibit inflammation [157–159]. Inflammaging is associated with defective autophagy that increases with age. A recent cross sectional study shed light on the mechanism of metformin in diabetes related to age-associated diseases. The study involved young lean subjects (31.2 years.) and BMI-matched older subjects (62 years.) treated with metformin, and the $CD4^+$ T cells from peripheral blood were collected for various assays. The study showed that metformin (1000 mg/day) for 3 months prevented the production of Th17 inflammaging profile. Metformin improved autophagy and mitochondrial function, and thus, reduced inflammaging [160]. Moreover, metformin works in multiple ways against age and ageing-associated diseases through inhibition of ROS production [161]. Studies also showed the effect of metformin against DNA damage and cellular senescence [162,163]. Metformin showed an effect against LPS-induced NF-κB pathway, thus working against inflammation itself [162]. Since metformin is an FDA-approved drug for other conditions including diabetes with well-known safety profile, human trials in assessing the efficacy of metformin for its ability to stimulate autophagy and exhibit anti-inflammaging effects is quite beneficial.

4.2. Zinc

Zinc is the second most prevalent trace element in the human body and is essential for several cellular and metabolic functions and also for the immune system. The adult human body contains 2–3 g of zinc, and it is relatively known as a non-toxic. Many times zinc is given as a supplement to maintain its level in the human body in order to achieve appropriate immune function [164]. The role of zinc in anti-inflammation has been long studied, which shows the ability of zinc to reduce inflammatory cytokines. More interestingly, zinc has been shown to reduce inflammaging-associated cytokines. The conventional inflammatory pathway, NF-κB, is effected by zinc, and in the human monocyte, LPS-induced TNF-α expression is significantly suppressed [165]. A randomized, double-blinded placebo trial of zinc supplementation (45 mg/day for six months) in elderly (56–83 years) showed a significant reduction in CRP, IL-6, and TNF-α levels [166]. In a different study conducted by Jung et al., subjects aged 40 years and older showed a decrease in CRP and IL-6 in the plasma of individuals after zinc supplementation [167]. Studies also suggested that plasma zinc and CRP levels are inversely correlated in elderly patients [168]. Overall, the importance of micronutrients in nutrition such as zinc, especially in geriatric patients, is justified as it plays a significant role in healthy aging and immunosenescence.

4.3. Gotu Kola

Gotu kola (*Centella asiatica* Linn.), is a natural herb extensively used in traditional Ayurvedic system in India and other parts of the world to treat various ailments. In a study by Maramaldi et al., the antiinflammaging capacity of this botanical herb was studied clinically in human explants and volunteers [169]. *C. asiatica* extracts (2 mg/explant) were treated on UV-irradiation-induced human explants maintained alive and obtained from a 58-year-old Caucasian woman, and the expression of the proinflammatory cytokine IL-1α was evaluated. Further, a single-blind, placebo-controlled clinical trial on healthy volunteers (age 40–70 years) also revealed that *C. asiatica* exhibited antiaging and anti-wrinkling properties by improving the skin firmness, wrinkling, elasticity, and collagen density. The authors concluded the antiaging efficacy of *C. asiatica* might be due to its anti-inflammaging effects along with free radical scavenging and antiglycation activities.

4.4. Soy and Whey Proteins

Dairy and soy proteins contain various amino acids and bioactive peptides with immense nutritional values. Earlier studies indicated that soy and whey protein possess anti-oxidant and anti-inflammatory effects including mitigating chronic inflammation during aging [170]. In a crossover designed randomized, acute clinical intervention study, a fat-rich mixed meal was administered with 45 g of whey protein to obese non-diabetic aged subjects (40–68 years). The authors observed an acute suppression of inflammatory markers of low-grade inflammation including monocyte chemotactic protein-1 (MCP-1) expression in the blood samples after 240 min postprandial period [171]. Further, soy protein, which is considered to be a high nutritional protein, is well reported for its potential nutritional intervention for chronic inflammatory conditions in experimental studies [172]. Soy protein was documented to attenuate chronic inflammation through regulation of the NF-κB signaling pathway and cytokine production in *mdx* mice [173]. In a randomized, double-blind, placebo-control, clinical trial involving 131 healthy older women (60 years.), the long term effects (1 yr.) of soy protein (18 g) on the serum lipids and inflammatory markers were evaluated. Data showed that after 1 year of soy protein administration, a significant reduction in IL-6 baseline when compared to control group was observed [174]. Based on the promising nutritional interventions of dietary proteins, future studies involving detailed clinical studies focusing on inflammaging and its associated diseases should be investigated.

4.5. Black Rice

Recently, food-derived bioactive peptide nutrients and optimized diet habits have been emerging as potential sources for the prevention and treatment of several diseases including attenuating age-related inflammation by regulating the balance between pro- and anti-inflammaging [175–177]. In a recent report, rice-derived functional ingredient natural peptide network (rice NPN) significantly reduced TNF-α secretion in human macrophages stimulated by lipopolysaccharide in vitro [178]. Black rice (*Oryza sativa* L. var. japonica), a staple food for Asian populations since ancient times, was well documented for it antioxidant and anti-inflammatory properties due to its rich bioactive functional peptides [179].

A randomized, double blinded, parallel group, placebo-controlled clinical trial with participants (males and females, $n = 30$) aging between 65–75 years was conducted to study the anti-inflammatory effects of rice NPN. Clinical data revealed that 4-week supply of rice NPN (10 g) was found to help improve the physical challenges measured by hand grip test, repeated chair stand test, and short physical performance battery (SPPB) test score in inflammaging populations when compared with placebo group. The inflammation-associated aging-mediated altered glucose, serum LDL, and HDL levels were restored, providing the efficacy data of rice NPN against the inflammaging process both in vitro and in clinical settings.

4.6. Mediterranean Diet

Evidence suggests that the Mediterranean diet (MedDiet) showed beneficial effects by positively influencing the ageing hallmarks and helps in mitigating age-related disease and increasing longevity [177]. In a study by Martucci and group, the nutritious supplements provided by (MedDiet) rich in whole-grain cereals, vegetable, fruits, legumes, fish, olive oils, and nuts aided in neurohormetic and neuroprotective effects. MedDiet also regulates the balance between pro and anti-inflammaging conditions. The authors indicated that hormetic interventions by both nutritional and physical activity, control the inflammaging processes by decreasing the senescent cells accumulation [176]. Further, data from the United Kingdom Arm of the NU-AGE randomized controlled trial revealed that 1-year consumption of MedDiet along with Vitamin D by elderly subjects (65–79 years) showed MedDiet-dependent changes in T cell degranulation, cytokine production, and co-receptor expression. The elderly placebo group exhibited increased signs of IL-12 cytokine production, which might contribute to inflammaging. However, although not significant, the MedDiet supplemented group showed a declining tendency in IL-12 expression, indicating beneficial effects in elderly subjects. The authors suggested that prolonged MedDiet intervention studies are necessary to confirm the anti-inflammaging effects of MedDiet [180].

Table 2. List of selected nutrient compounds with clinical trials and human research targeting inflammaging.

Nutrient Compounds	Source	Study	Model (In Vitro and In Vivo)	Dose	Mechanism/s	Refs
Metformin	*Galega officinalis*	Cross sectional study	Human subjects	1000 mg/day/3 months	↑ autophagy and mitochondrial function	[160]
Zinc	Nutrient trace element	Randomized, double blinded placebo trial	Aged human subjects (56–83 years)	45 mg/d for 6 months)	↓ CRP, IL-6 and TNF-α levels	[166]
Gotu Kola	*Centella asiatica*	Single-blind, placebo-controlled clinical trial	Aged human subjects	2 mg/explant/ 5 days	↓ IL-1α	[169]
Whey protein	Dietary protein	Crossover designed randomized, acute clinical intervention study	Obese non-diabetic human subjects	45 g/12-week	↓ MCP-1 expression	[171]
Soy protein	Dietary protein	Randomized, double-blind, placebo-control, clinical trial	Healthy older women (>70 years.)	18 g/day/1 year	↓ IL-6 baseline	[174]
Rice NPN	Black rice	Randomized, double-blind, placebo-control, clinical trial	LPS induced macrophages. Aged subjects (65–75 years)	10 g dose/12 weeks	↓ TNF-α. Restored glucose, LDL and HDL levels	[178]
MedDiet	Mediterranean diet	Randomized controlled trial	Elderly subject (65–79 years)	MedDiet for 1 year	↓ IL-12 expression	[180]

Abbreviations: IL: Interleukin; LPS: Lipopolysaccharide; TNF-α: Tissue necrosis factor-alpha; HDL: high density lipoprotein; LDL: Low density lipoprotein; CRP: C-Reactive protein; Rice NPN: Rice natural peptide network; MCP-1: Monocyte chemoattractant protein-1, ↑: increased, ↓: decreased.

5. Patents Claims on Nutrient Compounds for Possible Anti-Inflammaging Effects

Till date, patents with direct mechanistic evidence on the role of nutrient natural compounds targeting inflammaging has not been published. However, we identified a few patent claims on nutrient compounds and natural product mixtures for their anti-aging effects based on anti-inflammatory properties. A list of selected patented nutrient compounds claimed in targeting inflammaging is shown in Table 3.

5.1. Cyclodextrins

In a recent patent by Asdera LLC, USA, the inventers claimed that cyclodextrin (α and β; Figure 3A,B) might be useful in the treatment and prevention of malignancies, in neurodegeneration (AD and PD), and other aspects of aging including the T2D and atherosclerosis [181]. Cyclodextrin is recognized as a safe food additive and its derivative 2-hydroxypropyl-p-cyclodextrin (HP-β-CD) has been reported to improve autophagy, which is a common factor in the etiology of ARDs [182]. In NSG mice infected with human MDA-MB-231 breast cancer cells, cyclodextrin (800 mg/kg) reduced the plasma cytokine levels such as IL-1β, IL-18, IL-6, and IL-8, significantly confirming that cyclodextrins reduce inflammation. It is well known that during inflammaging conditions there is an upregulation of inflammatory responses and downregulation of autophagy [183]. The inventors claim that cyclodextrins might be essential during early life in reducing inflammation, thereby improving autophagy. The authors concluded that cyclodextrins reduce chronic inflammation associated with aging.

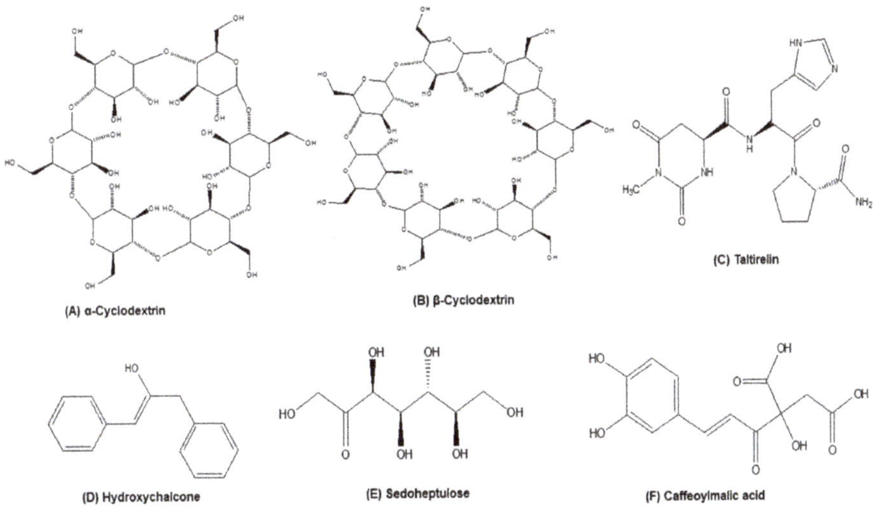

Figure 3. Chemical structures of selected patent agents targeting inflammaging.

5.2. Taltirelin

In a patent by Eolas Research LTD, Great Britain, the inventors claimed that a compound named taltirelin (N-[[(4S)-Hexahydro-l-methyl-2,6-dioxo-4-pyrimidinyl] carbonyl]-L-histidyl-L-prolinamide; Figure 3C) showed immense potential in the treatment of ARDs associated with cellular senescence, inflammaging, and autophagy [184]. It is well documented that the senescent cells develop SASP involving enhanced secretion of pro-inflammatory mediators contributing to inflammaging. Taltirelin, which is a thyrotropin-releasing hormone analogue, was well documented for its effects on improving neurological functions. The inventors indicated that taltirelin (10 mg/kg/day) treated to old fibroblast cells increased the expression of forkhead box protein (FOX) O1 and FOXO3 genes, which are regarded to be actively involved in various mechanisms including inflammaging, autophagy, and apoptosis. Taltirelin also increased the expression of SIRT1, which is known to be involved in several ARDs. Further, the EFNB1 gene is well documented to be involved in the adipose inflammatory response and in obesity. Taltirelin significantly increased the Enhrins (EFNB1) in old fibroblast cells compared with the control cells. Nrf2 is known to be involved in immunity and inflammation. During aging, there is a decrease in the Nrf2 activity. In this patent, the inventors claim that taltirelin significantly increased the Nrf2 activity in old fibroblasts compared with untreated controls.

5.3. Chalcones

Chalcones and their derivatives found in plants are well known to be used as medicaments and nutraceutical agents. In a patent application by III.XTII B.V, Netherlands, the inventors claimed that hydroxychalcone (33 µM; Figure 3D) exhibited a curative effect by reducing the blood IL-1β/Il-18 levels, inhibited NLRP3 inflammasome-mediated IL-18 expression in macrophages or dendritic cells, and inhibited caspase-1 in THP-1 macrophages or dendritic cells stimulated by LPS/ATP. In aging assay, hydroxychalcone treatment (50 µM) to N2, bristol (wild-type) strain nematodes increased the length of survival time of nematodes. The inventors concluded that hydroxychalcone possibly treats low grade inflammation and shows beneficial effects in targeting inflammaging and other ARDs [185].

5.4. Sedoheptulose

Sedoheptulose (Figure 3E), a monosaccharide with seven carbon atoms and a ketone functional group, is found abundantly in nature. It can be produced both naturally from fruits and vegetables, or synthetically. Sedoheptulose was pharmacologically reported to protect diabetic nephropathy by alleviating inflammatory response. In a patent applied by Medizinische Universität Wien 1090 Vienna, Austria, the inventors generated a transgenic mice overexpressing sedoheptulose kinase (CARKL) and tested various parameters including inflammatory markers, redox regulation, and physical activity by standard assays. The inventors claimed that sedoheptulose in the dose range of up to 150 mg/mL, decreased the LPS (7 mg/kg)-stimulated increase in inflammatory cytokines in endotoxemia in vivo murine model. The inventors indicated that sedoheptulose-administered animals showed suppressed inflammatory responses by reducing the inflammatory cytokine levels especially TNF-α or IL-6, or other inflammation markers when compared to wild type subjects. The inventors concluded that sedoheptulose might mitigate chronic low grade inflammation, thereby positively delaying inflammaging and controlling other ARDs [186].

5.5. Herbal Mixtures

In a patent by PharmacoGentetics Limited, Shatin, China, the inventors claimed that the natural herb mixtures including Radix *Bupleurum chinense*, Rhizoma *Corydalis yanhusuo*, Caulis *Polygonum multiflorum*, and Flos *Albizia julibrissin* delayed the aging process in D-gal-exposed mice. The inventors suggested that treatments helping in alleviating memory functions, attenuating oxidative stress, and reducing pro-inflammatory mediators might be crucial in preventing the accelerated aging seen in several ARDs. The inventors claimed that the mixture (120 mg/kg) treated for 2 months along with D-gal (150 mg/kg) significantly ameliorated the increased MDA (oxidative stress marker) and pro-inflammatory cytokines including TNF-α and IL-6 in the brain tissues of mice. The inventors concluded that the anti-inflammaging capacities of the claimed composition might be useful in preventing or delaying the inflammation-mediated aging-related disorders such as AD and PD (US 10,722,547 B2).

5.6. Nutrient Cosmeceutical Preparations

In another patent by LVMH Recherche GIE, France, the inventors claim that the aqueous extract of rose fruits (Evarant or Rose jardin de Granville variety) exhibited skin aging protective functions by ameliorating UV radiation-induced inflammaging of the skin. The inventors showed that rose fruit extract at 65 µg/mL concentration reduced the proinflammatory mediators such as IL-1α, IL-1β, IL-8, TNF-α, and NF-κB in keratinocyte cultures or co-cultures of sensory neurons and human keratinocytes induced by light stress (UV) chronic inflammation. The inventors indicated that the preventive effects of rose fruit extract on skin aging are linked with controlling inflammaging processes [187]. The same group also claimed that rose wood extract from similar species also exhibited potential anti-aging capacity on skin by preventing inflammaging [188].

In a study by a South Korean group of inventors from Seoul National University and Hyundai Bioland Co LTD, a cosmetic and functional food composition containing caffeoylmalic acid (Figure 3F) isolated from *Canavalia gladiata* reduced skin photo-aging. The inventors indicated that caffeoylmalic acid (20 µM) inhibited the UV radiation-induced increased expression of COX-2 in human keratinocyte HaCaT cells lines, one of the inflammaging targets [189]. Further, another South Korean company "COSMAX Inc" claimed that the cosmetic composition containing an extracts mixture of *Juglans nigra*, *Sophora japonica*, and *Pinus densiflora* exhibited potent anti-aging and anti-inflammatory effects by inhibiting the ROS and inflammatory mediators, thereby preventing skin aging mediated by UV-induced inflammation. The inventors concluded that equal portions of the three extracts significantly inhibited UV-induced IL-1β mRNA expression in human HS68 fibroblast cells, showing their efficacy for anti-aging and anti-inflammatory activities [190].

Table 3. List of selected nutrient compounds with patent claims targeting inflammaging.

Compound Name	Source	Model (In Vitro and In Vivo)	Dose	Mechanism/s	Refs
Cyclodextrin	Naturally occurring food additive	NSG mice infected with human MDA-MB-231 breast cancer cells	800 mg/kg once weekly for 6 weeks	↓ IL-1β, IL-18, IL-6, and IL-8. ↓ autophagy	[181]
Taltirelin	Thyrotropin-releasing hormone analogue	Old fibroblast cells	10 mg/kg/day	↑ Nrf2 activity	[184]
Chalcones	Plant derived nutrient	THP-1 macrophages	33 μM (in vitro)	↓ IL-1β/Il-18 levels	[185]
		N2, bristol (wild-type) strain nematodes	50 μM (in vivo)	↓ NLRP3 and ↓ caspase-1	
Sedoheptulose	Naturally from fruits and vegetables	LPS stimulated endotoxemia in vivo murine model and sedoheptulose kinase overexpressing mice	Up to 150 mg/mL	↓ TNF-α or IL-6	[186]

Abbreviations: LPS: Lipopolysaccharide; IL: Interleukin; TNF-α: Tissue necrosis factor-alpha; Nrf2: Nuclear factor erythroid-2-related factor 2; NLRP3: NOD-like receptor pyrin domain-containing protein 3, ↑: increased, ↓: decreased.

6. Conclusions and Future Perspectives

Inflammaging is highly correspondent with ARDs. Although the manifestation of ARDs is not solely dependent on inflammaging, the pathomechanisms observed in ARDs can be correlated with inflammation. As aging is irreversible, the strategy to delay the aging process without favoring diseases could be the foremost. Thus, the idea of using natural nutrient compounds or nutraceuticals to reduce inflammaging can be very effective in controlling various geriatric syndromes and increasing the lifespan of aging populations to live healthier and longer. Over the past few decades, the literature has been suggesting that nutrients and food supplements from natural and synthetic sources hold promising agents in controlling the inflammatory conditions associated with aging and ARDs. In this review, we discussed several nutrient compounds that have proven to be promising in cellular and animal experimental models targeting inflammaging and increasing life span. Further, reports from clinical trials, human research, and patents claimed so far on selected nutrients with possible therapeutics on inflammaging and inflammaging-associated disorders was also discussed. A schematic diagram with selected nutrient compounds targeting inflammaging at various mechanistic pathways was shown (Figure 4).

Very few studies including the rice NPN, zinc supplementation, calcitriol, and Med-Diet, have been investigated in clinical settings for their possible influence as potential anti-inflammaging agents. Further, the patents applied in recent years were focused primarily on anti-aging and anti-inflammatory effects in stimulated cells and animal tissues. A few other patents targeting inflammaging were mainly made from the cosmeceutical perspective, naming them as anti-aging or anti-photo aging agents by either preventing or delaying skin aging. Although the reported clinical studies and patent claims suggested to improve the over health conditions and life span by delaying the inflammaging processes, none of the agents were approved to be introduced in to the drug or nutraceutical market claiming therapeutic anti-inflammaging effects in aged populations associated with ARDs. Though targeting inflammaging by nutrient compounds is a prospective strategy to reduce the occurrence of ARDs and increase the lifespan of individuals, the direct link between inflammaging and ARDs has yet to be elucidated. To date, compounds targeting inflammation that can inhibit ARDs are believed to be the best choices against inflammaging. However, elucidating the specific pathomechanisms of respective diseases in relation to inflammaging could bring a phenomenal improvement. Moreover, the crucial question of whether there is a causal relationship between inflammaging and diseases needs to be resolved by performing extensive integrated biological and clinical research. Whether early modulation of inflammaging prevents or delays the onset of ARDs should be tested thoroughly by implementing stringent inflammaging experimental models. Since the verification of detailed inflammaging mechanisms in humans is still unclear, elaborated, well-controlled, and larger dose–response randomized clinical trials on the use of potential nutrient compounds in elderly populations and patients with ARDs are still necessary to identify threshold effects targeting inflammaging.

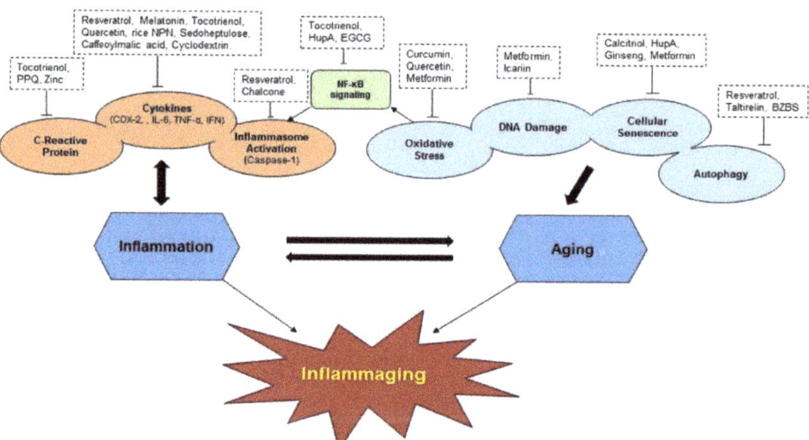

Figure 4. Schematic diagram on selected nutrient compounds and possible sites of action on inflammaging pathways. Inflammaging is caused by interactions at multiple levels of inflammatory and aging mechanisms. Aging-mediated activation involving oxidative stress, DNA damage, cellular senescence, autophagy, and parallel activation of inflammatory responses including increased CRP levels, pro-inflammatory cytokine levels (IL-6, IL-1, IL-1β, TNF-α, IFN, etc.), and inflammasome activation (caspase-1) are some of the key players known to involved in inflammaging and its propagation. Possible involvement of selected reviewed nutrient compounds was shown at various signaling steps. COX-2: Cyclooxygenase, IL: Interleukin, TNF-α: Tissue necrosis factor-alpha, IFN: Interferon, NF-κB: Necrosis Factor-Kappa B, PPQ: Pyrroloquinoline quinone, Rice NPN: Rice natural peptide network, HupA: Huperzine A, EGCG: Epigallocatechin-3-gallate, BZBS: BaZiBuShen.

Author Contributions: S.K., M.A. and S.R.K. conceptualized and designed the manuscript; S.K. and M.A. wrote the manuscript; S.K., M.A., M.E.H. and S.R.K. revised the manuscript; S.K., M.A., M.E.H. and S.R.K. participated in drafting the article and revising it critically. S.K. and S.R.K. involved in funding acquisition. All authors have read and agreed to the published version of the manuscript.

Funding: This research received no external funding.

Acknowledgments: This work was supported by Konkuk University, Republic of Korea and Sejong University, Republic of Korea.

Conflicts of Interest: The authors declare no conflict of interest.

References

1. Nicita-Mauro, V.; Lo Balbo, C.; Mento, A.; Nicita-Mauro, C.; Maltese, G.; Basile, G. Smoking, aging and the centenarians. *Exp. Gerontol.* **2008**, *43*, 95–101. [CrossRef]
2. Franceschi, C.; Bonafe, M.; Valensin, S.; Olivieri, F.; De Luca, M.; Ottaviani, E.; De Benedicts, G. Inflamm-aging: An evolutionary perspective on immunosenescence. *Ann. N. Y. Acad. Sci.* **2000**, *908*, 244–254. [CrossRef] [PubMed]
3. Reay, J.L.; Kennedy, D.O.; Scholey, A.B. Effects of Panax ginseng, consumed with and without glucose, on blood glucose levels and cognitive performance during sustained 'mentally demanding' tasks. *J. Psychopharmacol.* **2006**, *20*, 771–781. [CrossRef] [PubMed]
4. López-Otín, C.; Blasco, M.A.; Partridge, L.; Serrano, M.; Kroemer, G. The hallmarks of aging. *Cell* **2013**, *153*, 1194–1217. [CrossRef] [PubMed]

5. Ferrucci, L.; Corsi, A.; Lauretani, F.; Bandinelli, S.; Bartali, B.; Taub, D.D.; Guralnik, J.M.; Longo, D.L. The origins of age-related proinflammatory state. *Blood* **2005**, *105*, 2294–2299. [CrossRef]

6. Dinarello, C.A. Interleukin 1 and interleukin 18 as mediators of inflammation and the aging process. *Am. J. Clin. Nutr.* **2006**, *83*, 447S–455S. [CrossRef]

7. Kennedy, B.K.; Berger, S.L.; Brunet, A.; Campisi, J.; Cuervo, A.M.; Epel, E.S.; Franceschi, C.; Lithgow, G.J.; Morimoto, R.I.; Pessin, J.E.; et al. Geroscience: Linking aging to chronic disease. *Cell* **2014**, *159*, 709–713. [CrossRef]

8. Fülöp, T.; Larbi, A.; Pawelec, G. Human T cell aging and the impact of persistent viral infections. *Front. Immunol.* **2013**, *4*, 271. [CrossRef]

9. Pawelec, G. Immunosenenescence: Role of cytomegalovirus. *Exp. Gerontol.* **2014**, *54*, 1–5. [CrossRef]

10. Puzianowska-Kuźnicka, M.; Owczarz, M.; Wieczorowska-Tobis, K.; Nadrowski, P.; Chudek, J.; Slusarczyk, P.; Skalska, A.; Jonas, M.; Franek, E.; Mossakowska, M. Interleukin-6 and C-reactive protein, successful aging, and mortality: The PolSenior study. *Immun. Ageing* **2016**, *13*, 21. [CrossRef]

11. Akiyama, H.; Barger, S.; Barnum, S.; Bradt, B.; Bauer, J.; Cole, G.M.; Cooper, N.R.; Eikelenboom, P.; Emmerling, M.; Fiebich, B.L.; et al. Inflammation and Alzheimer's disease. *Neurobiol. Aging* **2000**, *21*, 383–421. [CrossRef]

12. Yeh, S.-S.; Schuster, M.W. Geriatric cachexia: The role of cytokines. *Am. J. Clin. Nutr.* **1999**, *70*, 183–197. [CrossRef] [PubMed]

13. Manolagas, S.C.; Jilka, R.L. Bone marrow, cytokines, and bone remodeling—Emerging insights into the pathophysiology of osteoporosis. *New Engl. J. Med.* **1995**, *332*, 305–311. [CrossRef] [PubMed]

14. Cesari, M.; Penninx, B.W.J.H.; Pahor, M.; Lauretani, F.; Corsi, A.M.; Williams, G.R.; Guralnik, J.M.; Ferrucci, L. Inflammatory markers and physical performance in older persons: The InCHIANTI study. *J. Gerontol. Ser. A Biol. Sci. Med. Sci.* **2004**, *59*, M242–M248. [CrossRef] [PubMed]

15. Xia, S.; Zhang, X.; Zheng, S.; Khanabdali, R.; Kalionis, B.; Wu, J.; Wan, W.; Tai, X. An update on inflamm-aging: Mechanisms, prevention, and treatment. *J. Immunol. Res.* **2016**, *2016*, 1–12. [CrossRef]

16. Ferrucci, L.; Fabbri, E. Inflammageing: Chronic inflammation in ageing, cardiovascular disease, and frailty. *Nat. Rev. Cardiol.* **2018**, *15*, 505–522. [CrossRef]

17. Gabandé-Rodríguez, E.; Gómez de las Heras, M.M.; Mittelbrunn, M. Control of inflammation by calorie restriction mimetics: On the crossroad of autophagy and mitochondria. *Cells* **2019**, *9*, 82. [CrossRef]

18. Nilsson, M.I.; Bourgeois, J.M.; Nederveen, J.P.; Leite, M.R.; Hettinga, B.P.; Bujak, A.L.; May, L.; Lin, E.; Crozier, M.; Rusiecki, D.R.; et al. Lifelong aerobic exercise protects against inflammaging and cancer. *PLoS ONE* **2019**, *14*, e0210863. [CrossRef]

19. Wiley, C.D.; Schaum, N.; Alimirah, F.; Lopez-Dominguez, J.A.; Orjalo, A.V.; Scott, G.; Desprez, P.-Y.; Benz, C.; Davalos, A.R.; Campisi, J. Small-molecule MDM2 antagonists attenuate the senescence-associated secretory phenotype. *Sci. Rep.* **2018**, *8*, 2410. [CrossRef]

20. Vel Szic, K.S.; Declerck, K.; Vidaković, M.; Vanden Berghe, W. From inflammaging to healthy aging by dietary lifestyle choices: Is epigenetics the key to personalized nutrition? *Clin. Epigenetics* **2015**, *7*, 33. [CrossRef]

21. Losappio, V.; Infante, B.; Leo, S.; Troise, D.; Calvaruso, M.; Vitale, P.; Renzi, S.; Stallone, G.; Castellano, G. Nutrition-based management of inflammaging in CKD and renal replacement therapies. *Nutrients* **2021**, *13*, 267. [CrossRef]

22. Calder, P.C.; Bosco, N.; Bourdet-Sicard, R.; Capuron, L.; Delzenne, N.; Doré, J.; Franceschi, C.; Lehtinen, M.J.; Recker, T.; Salvioli, S.; et al. Health relevance of the modification of low grade inflammation in ageing (inflammageing) and the role of nutrition. *Ageing Res. Rev.* **2017**, *40*, 95–119. [CrossRef]

23. Giunta, S. Is inflammaging an auto[innate]immunity subclinical syndrome? *Immun. Ageing* **2006**, *3*, 12. [CrossRef]

24. Fuente, M.; Miquel, J. An update of the oxidation-inflammation theory of aging: The involvement of the immune system in oxi-inflamm-aging. *Curr. Pharm. Des.* **2009**, *15*, 3003–3026. [CrossRef]

25. Salminen, A.; Huuskonen, J.; Ojala, J.; Kauppinen, A.; Kaarniranta, K.; Suuronen, T. Activation of innate immunity system during aging: NF-kB signaling is the molecular culprit of inflamm-aging. *Ageing Res. Rev.* **2008**, *7*, 83–105. [CrossRef]

26. Mishto, M.; Santoro, A.; Bellavista, E.; Bonafé, M.; Monti, D.; Franceschi, C. Immunoproteasomes and immunosenescence. *Ageing Res. Rev.* **2003**, *2*, 419–432. [CrossRef]

27. Maggio, M.; Guralnik, J.M.; Longo, D.L.; Ferrucci, L. Interleukin-6 in aging and chronic disease: A magnificent pathway. *J. Gerontol. Ser. A Biol. Sci. Med. Sci.* **2006**, *61*, 575–584. [CrossRef]

28. Fried, L.P.; Tangen, C.M.; Walston, J.; Newman, A.B.; Hirsch, C.; Gottdiener, J.; Seeman, T.; Tracy, R.; Kop, W.J.; Burke, G.; et al. Frailty in older adults: Evidence for a phenotype. *J. Gerontol. Ser. A Biol. Sci. Med. Sci.* **2001**, *56*, M146–M157. [CrossRef] [PubMed]

29. De Martinis, M.; Franceschi, C.; Monti, D.; Ginaldi, L. Inflamm-ageing and lifelong antigenic load as major determinants of ageing rate and longevity. *FEBS Lett.* **2005**, *579*, 2035–2039. [CrossRef] [PubMed]

30. Bruunsgaard, H.; Andersen-Ranberg, K.; Hjelmborg, J.; Pedersen, B.K.; Jeune, B. Elevated levels of tumor necrosis factor alpha and mortality in centenarians. *Am. J. Med.* **2003**, *115*, 278–283. [CrossRef]

31. Di Bona, D.; Vasto, S.; Capurso, C.; Christiansen, L.; Deiana, L.; Franceschi, C.; Hurme, M.; Mocchegiani, E.; Rea, M.; Lio, D.; et al. Effect of interleukin-6 polymorphisms on human longevity: A systematic review and meta-analysis. *Ageing Res. Rev.* **2009**, *8*, 36–42. [CrossRef] [PubMed]

32. Pes, G.M.; Lio, D.; Carru, C.; Deiana, L.; Baggio, G.; Franceschi, C.; Ferrucci, L.; Oliveri, F.; Scola, L.; Crivello, A.; et al. Association between longevity and cytokine gene polymorphisms. A study in Sardinian centenarians. *Aging Clin. Exp. Res.* **2004**, *16*, 244–248. [CrossRef]

33. Zeng, Y.; Nie, C.; Min, J.; Liu, X.; Li, M.; Chen, H.; Xu, H.; Wang, M.; Ni, T.; Li, Y.; et al. Novel loci and pathways significantly associated with longevity. *Sci. Rep.* **2016**, *6*, 21243. [CrossRef] [PubMed]

34. Rea, I.M.; Ross, O.A.; Armstrong, M.; McNerlan, S.; Alexander, D.H.; Curran, M.D.; Middleton, D. Interleukin-6-gene C/G 174 polymorphism in nonagenarian and octogenarian subjects in the BELFAST study. Reciprocal effects on IL-6, soluble IL-6 receptor and for IL-10 in serum and monocyte supernatants. *Mech. Ageing Dev.* **2003**, *124*, 555–561. [CrossRef]

35. Cannizzo, E.S.; Clement, C.C.; Sahu, R.; Follo, C.; Santambrogio, L. Oxidative stress, inflamm-aging and immunosenescence. *J. Proteom.* **2011**, *74*, 2313–2323. [CrossRef] [PubMed]

36. Zuo, L.; Prather, E.R.; Stetskiv, M.; Garrison, D.E.; Meade, J.R.; Peace, T.I.; Zhou, T. Inflammaging and oxidative stress in human diseases: From molecular mechanisms to novel treatments. *Int. J. Mol. Sci.* **2019**, *20*, 4472. [CrossRef]

37. Cheeseman, K.H.; Slater, T.F. An introduction to free radical biochemistry. *Br. Med. Bull.* **1993**, *49*, 481–493. [CrossRef]

38. Dorn, G.W. Molecular mechanisms that differentiate apoptosis from programmed necrosis. *Toxicol. Pathol.* **2013**, *41*, 227–234. [CrossRef]

39. Marchal, J.; Pifferi, F.; Aujard, F. Resveratrol in mammals: Effects on aging biomarkers, age-related diseases, and life span. *Ann. N. Y. Acad. Sci.* **2013**, *1290*, 67–73. [CrossRef]

40. Campisi, J.; d'Adda di Fagagna, F. Cellular senescence: When bad things happen to good cells. *Nat. Rev. Mol. Cell Biol.* **2007**, *8*, 729–740. [CrossRef]

41. Herbig, U.; Ferreira, M.; Condel, L.; Carey, D.; Sedivy, J.M. Cellular senescence in aging primates. *Science* **2006**, *311*, 1257. [CrossRef] [PubMed]

42. De Jesus, B.B.; Blasco, M.A. Assessing cell and organ senescence biomarkers. *Circ. Res.* **2012**, *111*, 97–109. [CrossRef] [PubMed]

43. Jeck, W.R.; Siebold, A.P.; Sharpless, N.E. Review: A meta-analysis of GWAS and age-associated diseases. *Aging Cell* **2012**, *11*, 727–731. [CrossRef] [PubMed]

44. Johnson, S.C.; Dong, X.; Vijg, J.; Suh, Y. Genetic evidence for common pathways in human age-related diseases. *Aging Cell* **2015**, *14*, 809–817. [CrossRef] [PubMed]

45. Melzer, D.; Frayling, T.M.; Murray, A.; Hurst, A.J.; Harries, L.W.; Song, H.; Khaw, K.; Luben, R.; Surtees, P.G.; Bandinelli, S.S.; et al. A common variant of the p16INK4a genetic region is associated with physical function in older people. *Mech. Ageing Dev.* **2007**, *128*, 370–377. [CrossRef] [PubMed]

46. Baker, D.J.; Childs, B.G.; Durik, M.; Wijers, M.E.; Sieben, C.J.; Zhong, J.; Saltness, R.A.; Jeganathan, K.B.; Verzosa, G.C.; Pezeshki, A.; et al. Naturally occurring p16Ink4a-positive cells shorten healthy lifespan. *Nature* **2016**, *530*, 184–189. [CrossRef] [PubMed]

47. Baker, D.J.; Wijshake, T.; Tchkonia, T.; LeBrasseur, N.K.; Childs, B.G.; van de Sluis, B.; Kirkland, J.L.; van Deursen, J.M. Clearance of p16Ink4a-positive senescent cells delays ageing-associated disorders. *Nature* **2011**, *479*, 232–236. [CrossRef]

48. Hewitt, G.; Jurk, D.; Marques, F.D.M.; Correia-Melo, C.; Hardy, T.; Gackowska, A.; Anderson, R.; Taschuk, M.; Mann, J.; Passos, J.F. Telomeres are favoured targets of a persistent DNA damage response in ageing and stress-induced senescence. *Nat. Commun.* **2012**, *3*, 708. [CrossRef]

49. Olivieri, F.; Albertini, M.C.; Orciani, M.; Ceka, A.; Cricca, M.; Procopio, A.D.; Bonafè, M. DNA damage response (DDR) and senescence: Shuttled inflamma-miRNAs on the stage of inflamm-aging. *Oncotarget* **2015**, *6*, 35509–35521. [CrossRef]

50. Sies, H.; Berndt, C.; Jones, D.P. Oxidative stress. *Annu. Rev. Biochem.* **2017**, *86*, 715–748. [CrossRef]

51. Cucciolla, V.; Borriello, A.; Oliva, A.; Galletti, P.; Zappia, V.; Ragione, F. Della Resveratrol: From basic science to the clinic. *Cell Cycle* **2007**, *6*, 2495–2510. [CrossRef]

52. Sawda, C.; Moussa, C.; Turner, R.S. Resveratrol for Alzheimer's disease. *Ann. N. Y. Acad. Sci.* **2017**, *1403*, 142–149. [CrossRef] [PubMed]

53. De Paula, D.T.S.; de Carvalho, G.S.G.; Almeida, A.C.; Lourenço, M.; da Silva, A.D.; Coimbra, E.S. Synthesis, cytotoxicity and antileishmanial activity of Aza-stilbene derivatives. *Mediterr. J. Chem.* **2013**, *2*, 493–502. [CrossRef]

54. Riba, A.; Deres, L.; Sumegi, B.; Toth, K.; Szabados, E.; Halmosi, R. Cardioprotective effect of resveratrol in a postinfarction heart failure model. *Oxidative Med. Cell. Longev.* **2017**, *2017*, 1–10. [CrossRef] [PubMed]

55. Varoni, E.M.; Lo Faro, A.F.; Sharifi-Rad, J.; Iriti, M. Anticancer molecular mechanisms of resveratrol. *Front. Nutr.* **2016**, *3*. [CrossRef] [PubMed]

56. Sadowska-Bartosz, I.; Bartosz, G. Effect of antioxidants supplementation on aging and longevity. *BioMed Res. Int.* **2014**, *2014*, 1–17. [CrossRef]

57. Liang, Q.; Wang, X.; Chen, T. Resveratrol protects rabbit articular chondrocyte against sodium nitroprusside-induced apoptosis via scavenging ROS. *Apoptosis* **2014**, *19*, 1354–1363. [CrossRef]

58. Finkel, T.; Deng, C.-X.; Mostoslavsky, R. Recent progress in the biology and physiology of sirtuins. *Nature* **2009**, *460*, 587–591. [CrossRef]

59. Said, R.S.; El-Demerdash, E.; Nada, A.S.; Kamal, M.M. Resveratrol inhibits inflammatory signaling implicated in ionizing radiation-induced premature ovarian failure through antagonistic crosstalk between silencing information regulator 1 (SIRT1) and poly(ADP-ribose) polymerase 1 (PARP-1). *Biochem. Pharmacol.* **2016**, *103*, 140–150. [CrossRef]

60. Zhu, X.; Liu, Q.; Wang, M.; Liang, M.; Yang, X.; Xu, X.; Zou, H.; Qiu, J. Activation of Sirt1 by resveratrol inhibits TNF-α induced inflammation in fibroblasts. *PLoS ONE* **2011**, *6*, e27081. [CrossRef]

61. Haigis, M.C.; Guarente, L.P. Mammalian sirtuins–emerging roles in physiology, aging, and calorie restriction. *Genes Dev.* **2006**, *20*, 2913–2921. [CrossRef]

62. Uchiumi, F.; Watanabe, T.; Hasegawa, S.; Hoshi, T.; Higami, Y.; Tanuma, S. The effect of resveratrol on the werner syndrome RecQ helicase gene and telomerase activity. *Curr. Aging Sci.* **2011**, *4*, 1–7. [CrossRef] [PubMed]

63. Wang, X.-B.; Zhu, L.; Huang, J.; Yin, Y.-G.; Kong, X.-Q.; Rong, Q.-F.; Shi, A.-W.; Cao, K.-J. Resveratrol-induced augmentation of telomerase activity delays senescence of endothelial progenitor cells. *Chin. Med. J.* **2011**, *124*, 4310–4315. [PubMed]

64. Tung, B.T.; Rodríguez-Bies, E.; Talero, E.; Gamero-Estévez, E.; Motilva, V.; Navas, P.; López-Lluch, G. Anti-inflammatory effect of resveratrol in old mice liver. *Exp. Gerontol.* **2015**, *64*, 1–7. [CrossRef] [PubMed]

65. Jeong, S.I.; Shin, J.A.; Cho, S.; Kim, H.W.; Lee, J.Y.; Kang, J.L.; Park, E.-M. Resveratrol attenuates peripheral and brain inflammation and reduces ischemic brain injury in aged female mice. *Neurobiol. Aging* **2016**, *44*, 74–84. [CrossRef] [PubMed]

66. Yayeh, T.; Jung, K.-H.; Jeong, H.-Y.; Park, J.-H.; Song, Y.-B.; Kwak, Y.-S.; Kang, H.-S.; Cho, J.-Y.; Oh, J.-W.; Kim, S.-K.; et al. Korean red ginseng saponin fraction downregulates proinflammatory mediators in LPS stimulated RAW264.7 cells and protects mice against endotoxic shock. *J. Ginseng Res.* **2012**, *36*, 263–269. [CrossRef] [PubMed]

67. Nabavi, S.F.; Sureda, A.; Habtemariam, S.; Nabavi, S.M. Ginsenoside Rd and ischemic stroke; a short review of literatures. *J. Ginseng Res.* **2015**, *39*, 299–303. [CrossRef]

68. Kim, M.; Yi, Y.-S.; Kim, J.; Han, S.Y.; Kim, S.H.; Seo, D.B.; Cho, J.Y.; Shin, S.S. Effect of polysaccharides from a Korean ginseng berry on the immunosenescence of aged mice. *J. Ginseng Res.* **2018**, *42*, 447–454. [CrossRef]

69. Wang, Z.; Chen, L.; Qiu, Z.; Chen, X.; Liu, Y.; Li, J.; Wang, L.; Wang, Y. Ginsenoside Rg1 ameliorates testicular senescence changes in D-gal-induced aging mice via anti-inflammatory and antioxidative mechanisms. *Mol. Med. Rep.* **2018**, *17*, 6269–6276. [CrossRef]

70. Qureshi, A.A.; Burger, W.C.; Peterson, D.M.; Elson, C.E. The structure of an inhibitor of cholesterol biosynthesis isolated from barley. *J. Biol. Chem.* **1986**, *261*, 10544–10550. [CrossRef]

71. Sen, C.K.; Khanna, S.; Roy, S. Tocotrienols: Vitamin E beyond tocopherols. *Life Sci.* **2006**, *78*, 2088–2098. [CrossRef] [PubMed]

72. Suzuki, Y.J.; Tsuchiya, M.; Wassall, S.R.; Choo, Y.M.; Govil, G.; Kagan, V.E.; Packer, L. Structural and dynamic membrane properties of. Alpha.-tocopherol and. Alpha.-tocotrienol: Implication to the molecular mechanism of their antioxidant potency. *Biochemistry* **1993**, *32*, 10692–10699. [CrossRef] [PubMed]

73. Ahn, K.S.; Sethi, G.; Krishnan, K.; Aggarwal, B.B. γ-Tocotrienol inhibits nuclear factor-κB signaling pathway through inhibition of receptor-interacting protein and TAK1 leading to suppression of antiapoptotic gene products and potentiation of apoptosis. *J. Biol. Chem.* **2007**, *282*, 809–820. [CrossRef]

74. Wong, W.-Y.; Ward, L.C.; Fong, C.W.; Yap, W.N.; Brown, L. Anti-inflammatory γ- and δ-tocotrienols improve cardiovascular, liver and metabolic function in diet-induced obese rats. *Eur. J. Nutr.* **2017**, *56*, 133–150. [CrossRef]

75. Yam, M.-L.; Abdul Hafid, S.R.; Cheng, H.-M.; Nesaretnam, K. Tocotrienols suppress proinflammatory markers and cyclooxygenase-2 expression in RAW264.7 macrophages. *Lipids* **2009**, *44*, 787–797. [CrossRef]

76. Qureshi, A.A.; Reis, J.C.; Papasian, C.J.; Morrison, D.C.; Qureshi, N. Tocotrienols inhibit lipopolysaccharide-induced pro-inflammatory cytokines in macrophages of female mice. *Lipids Health Dis.* **2010**, *9*, 143. [CrossRef]

77. Devaraj, S.; Jialal, I. Alpha tocopherol supplementation decreases serum C-reactive protein and monocyte interleukin-6 levels in normal volunteers and type 2 diabetic patients. *Free. Radic. Biol. Med.* **2000**, *29*, 790–792. [CrossRef]

78. Boots, A.W.; Haenen, G.R.M.M.; Bast, A. Health effects of quercetin: From antioxidant to nutraceutical. *Eur. J. Pharmacol.* **2008**, *585*, 325–337. [CrossRef]

79. Häkkinen, S.H.; Kärenlampi, S.O.; Heinonen, I.M.; Mykkänen, H.M.; Törrönen, A.R. Content of the flavonols quercetin, myricetin, and kaempferol in 25 edible berries. *J. Agric. Food Chem.* **1999**, *47*, 2274–2279. [CrossRef]

80. Wiczkowski, W.; Romaszko, J.; Bucinski, A.; Szawara-Nowak, D.; Honke, J.; Zielinski, H.; Piskula, M.K. Quercetin from shallots (Allium cepa L. var. aggregatum) is more bioavailable than its glucosides. *J. Nutr.* **2008**, *138*, 885–888. [CrossRef] [PubMed]

81. Aguirre, L.; Arias, N.; Macarulla, M.T.; Gracia, A.; Portillo, M.P. Beneficial effects of quercetin on obesity and diabetes. *Open Nutraceuticals J.* **2011**, *4*, 189–198. [CrossRef]

82. Manjeet, K.R.; Ghosh, B. Quercetin inhibits LPS-induced nitric oxide and tumor necrosis factor-α production in murine macrophages. *Int. J. Immunopharmacol.* **1999**, *21*, 435–443. [CrossRef]

83. Bureau, G.; Longpré, F.; Martinoli, M.-G. Resveratrol and quercetin, two natural polyphenols, reduce apoptotic neuronal cell death induced by neuroinflammation. *J. Neurosci. Res.* **2008**, *86*, 403–410. [CrossRef] [PubMed]

84. Geraets, L.; Moonen, H.J.J.; Brauers, K.; Wouters, E.F.M.; Bast, A.; Hageman, G.J. Dietary flavones and flavonoles are inhibitors of Poly(ADP-ribose)polymerase-1 in pulmonary epithelial cells. *J. Nutr.* **2007**, *137*, 2190–2195. [CrossRef]

85. Kandere-Grzybowska, K.; Kempuraj, D.; Cao, J.; Cetrulo, C.L.; Theoharides, T.C. Regulation of IL-1-induced selective IL-6 release from human mast cells and inhibition by quercetin. *Br. J. Pharmacol.* **2006**, *148*, 208–215. [CrossRef]

86. Kook, D.; Wolf, A.H.; Yu, A.L.; Neubauer, A.S.; Priglinger, S.G.; Kampik, A.; Welge-Lüssen, U.C. the protective effect of quercetin against oxidative stress in the human RPE in vitro. *Investig. Opthalmology Vis. Sci.* **2008**, *49*, 1712. [CrossRef]

87. Hatahet, T.; Morille, M.; Shamseddin, A.; Aubert-Pouëssel, A.; Devoisselle, J.M.; Bégu, S. Dermal quercetin lipid nanocapsules: Influence of the formulation on antioxidant activity and cellular protection against hydrogen peroxide. *Int. J. Pharm.* **2017**, *518*, 167–176. [CrossRef] [PubMed]

88. Bao, D.; Wang, J.; Pang, X.; Liu, H. Protective effect of quercetin against oxidative stress-induced cytotoxicity in rat pheochromo-cytoma (PC-12) cells. *Molecules* **2017**, *22*, 1122. [CrossRef] [PubMed]

89. Saul, N.; Pietsch, K.; Menzel, R.; Stürzenbaum, S.R.; Steinberg, C.E.W. Quercetin-mediated longevity in *Caenorhabditis elegans*: Is DAF-16 involved? *Mech. Ageing Dev.* **2008**, *129*, 611–613. [CrossRef]

90. Zhou, H.; Beevers, S.C.; Huang, S. The targets of curcumin. *Curr. Drug Targets* **2011**, *12*, 332–347. [CrossRef]

91. Sood, P.K.; Nahar, U.; Nehru, B. Curcumin attenuates aluminum-induced oxidative stress and mitochondrial dysfunction in rat brain. *Neurotox. Res.* **2011**, *20*, 351–361. [CrossRef] [PubMed]

92. Jangra, A.; Kwatra, M.; Singh, T.; Pant, R.; Kushwah, P.; Sharma, Y.; Saroha, B.; Datusalia, A.K.; Bezbaruah, B.K. Piperine augments the protective effect of curcumin against lipopolysaccharide-induced neurobehavioral and neurochemical deficits in mice. *Inflammation* **2016**, *39*, 1025–1038. [CrossRef]

93. Giunta, B.; Fernandez, F.; Nikolic, W.V.; Obregon, D.; Rrapo, E.; Town, T.; Tan, J. Inflammaging as a prodrome to Alzheimer's disease. *J. Neuroinflammation* **2008**, *5*, 51. [CrossRef] [PubMed]

94. Calabrese, V.; Santoro, A.; Monti, D.; Crupi, R.; Di Paola, R.; Latteri, S.; Cuzzocrea, S.; Zappia, M.; Giordano, J.; Calabrese, E.J.; et al. Aging and Parkinson's disease: Inflammaging, neuroinflammation and biological remodeling as key factors in pathogenesis. *Free. Radic. Biol. Med.* **2018**, *115*, 80–91. [CrossRef] [PubMed]

95. Rastogi, M.; Ojha, R.P.; Sagar, C.; Agrawal, A.; Dubey, G.P. Protective effect of curcuminoids on age-related mitochondrial impairment in female Wistar rat brain. *Biogerontology* **2014**, *15*, 21–31. [CrossRef]

96. Sun, C.Y.; Qi, S.S.; Zhou, P.; Cui, H.R.; Chen, S.X.; Dai, K.Y.; Tang, M.L. Neurobiological and pharmacological validity of curcumin in ameliorating memory performance of senescence-accelerated mice. *Pharmacol. Biochem. Behav.* **2013**, *105*, 76–82. [CrossRef] [PubMed]

97. Singh, B.N.; Shankar, S.; Srivastava, R.K. Green tea catechin, epigallocatechin-3-gallate (EGCG): Mechanisms, perspectives and clinical applications. *Biochem. Pharmacol.* **2011**, *82*, 1807–1821. [CrossRef]

98. Yang, C.S.; Wang, X.; Lu, G.; Picinich, S.C. Cancer prevention by tea: Animal studies, molecular mechanisms and human relevance. *Nat. Rev. Cancer* **2009**, *9*, 429–439. [CrossRef]

99. Balentine, D.A.; Wiseman, S.A.; Bouwens, L.C.M. The chemistry of tea flavonoids. *Crit. Rev. Food Sci. Nutr.* **1997**, *37*, 693–704. [CrossRef]

100. Valcic, S.; Burr, J.A.; Timmermann, B.N.; Liebler, D.C. Antioxidant chemistry of green tea catechins. New oxidation products of (−)-epigallocatechin gallate and (−)-epigallocatechin from their reactions with peroxyl radicals. *Chem. Res. Toxicol.* **2000**, *13*, 801–810. [CrossRef]

101. Zhong, Y.; Chiou, Y.-S.; Pan, M.-H.; Shahidi, F. Anti-inflammatory activity of lipophilic epigallocatechin gallate (EGCG) derivatives in LPS-stimulated murine macrophages. *Food Chem.* **2012**, *134*, 742–748. [CrossRef] [PubMed]

102. Khoi, P.N.; Park, J.S.; Kim, J.H.; Xia, Y.; Kim, N.H.; Kim, K.K.; Jung, Y.D. (−)-Epigallocatechin-3-gallate blocks nicotine-induced matrix metalloproteinase-9 expression and invasiveness via suppression of NF-κB and AP-1 in endothelial cells. *Int. J. Oncol.* **2013**, *43*, 868–876. [CrossRef] [PubMed]

103. Youn, H.S.; Lee, J.Y.; Saitoh, S.I.; Miyake, K.; Kang, K.W.; Choi, Y.J.; Hwang, D.H. Suppression of MyD88- and TRIF-dependent signaling pathways of toll-like receptor by (−)-epigallocatechin-3-gallate, a polyphenol component of green tea. *Biochem. Pharmacol.* **2006**, *72*, 850–859. [CrossRef]

104. Smith, R.; Tran, K.; Smith, C.; McDonald, M.; Shejwalkar, P.; Hara, K. The role of the Nrf2/ARE antioxidant system in preventing cardiovascular diseases. *Diseases* **2016**, *4*, 34. [CrossRef] [PubMed]

105. Niu, Y.; Na, L.; Feng, R.; Gong, L.; Zhao, Y.; Li, Q.; Li, Y.; Sun, C. The phytochemical, EGCG, extends lifespan by reducing liver and kidney function damage and improving age-associated inflammation and oxidative stress in healthy rats. *Aging Cell* **2013**, *12*, 1041–1049. [CrossRef] [PubMed]

106. Yuan, H.; Li, Y.; Ling, F.; Guan, Y.; Zhang, D.; Zhu, Q.; Liu, J.; Wu, Y.; Niu, Y. The phytochemical epigallocatechin gallate prolongs the lifespan by improving lipid metabolism, reducing inflammation and oxidative stress in high-fat diet-fed obese rats. *Aging Cell* **2020**, *19*. [CrossRef]

107. Li, X.; Yu, C.; Guo, Y.; Bian, Z.; Si, J.; Yang, L.; Chen, Y.; Ren, X.; Jiang, G.; Chen, J.; et al. Tea consumption and risk of ischaemic heart disease. *Heart* **2017**, *103*, 783–789. [CrossRef]

108. Ma, X.; Tan, C.; Zhu, D.; Gang, D.R.; Xiao, P. Huperzine A from Huperzia species—An ethnopharmacolgical review. *J. Ethnopharmacol.* **2007**, *113*, 15–34. [CrossRef]

109. Skolnick, A.A. Old Chinese herbal medicine used for fever yields possible new Alzheimer disease therapy. *JAMA* **1997**, *277*, 776. [CrossRef]

110. Erdogan Orhan, I.; Orhan, G.; Gurkas, E. An overview on natural cholinesterase inhibitors-a multi-targeted drug class-and their mass production. *Mini Rev. Med. Chem.* **2011**, *11*, 836–842. [CrossRef]

111. Wang, Z.-F.; Wang, J.; Zhang, H.-Y.; Tang, X.-C. Huperzine A exhibits anti-inflammatory and neuroprotective effects in a rat model of transient focal cerebral ischemia. *J. Neurochem.* **2008**, *106*, 1594–1603. [CrossRef] [PubMed]

112. Ruan, Q.; Liu, F.; Gao, Z.; Kong, D.; Hu, X.; Shi, D.; Bao, Z.; Yu, Z. The anti-inflamm-aging and hepatoprotective effects of huperzine A in d-galactose-treated rats. *Mech. Ageing Dev.* **2013**, *134*, 89–97. [CrossRef]

113. Sze, S.C.W.; Tong, Y.; Ng, T.B.; Cheng, C.L.Y.; Cheung, H.P. Herba Epimedii: Anti-oxidative properties and its medical implications. *Molecules* **2010**, *15*, 7861–7870. [CrossRef] [PubMed]

114. Pei, L.-K.; Guo, B.-L.; Sun, S.-Q.; Huang, W.-H. Study on the identification of some species of Herba Epimedii with FTIR. *Guang Pu Xue Yu Guang Pu Fen Xi Guang Pu* **2008**, *28*, 55–60. [PubMed]

115. Hu, Z.-W.; Shen, Z.-Y.; Huang, J.-H. Experimental study on effect of epimedium flavonoids in protecting telomere length of senescence cells HU. *Chin. J. Integr. Tradit. West. Med.* **2004**, *24*, 1094–1097.

116. Cai, W.-J.; Zhang, X.-M.; Huang, J.-H. Effect of epimedium flavonoids in retarding aging of C. elegans. *Chin. J. Integr. Tradit. West. Med.* **2008**, *28*, 522–525.

117. Zhang, S.-Q.; Cai, W.-J.; Huang, J.-H.; Wu, B.; Xia, S.-J.; Chen, X.-L.; Zhang, X.-M.; Shen, Z.-Y. Icariin, a natural flavonol glycoside, extends healthspan in mice. *Exp. Gerontol.* **2015**, *69*, 226–235. [CrossRef]

118. Chen, Y.; Sun, T.; Wu, J.; Kalionis, B.; Zhang, C.; Yuan, D.; Huang, J.; Cai, W.; Fang, H.; Xia, S. Icariin intervenes in cardiac inflammaging through upregulation of SIRT6 enzyme activity and inhibition of the NF-Kappa B pathway. *BioMed Res. Int.* **2015**, *2015*, 1–12. [CrossRef]

119. Patel, S. Blueberry as functional food and dietary supplement: The natural way to ensure holistic health. *Mediterr. J. Nutr. Metab.* **2014**, *7*, 133–143. [CrossRef]

120. Wu, L.-H.; Xu, Z.-L.; Dong, D.; He, S.-A.; Yu, H. Protective effect of anthocyanins extract from blueberry on TNBS-induced IBD model of mice. *Evid. -Based Complementary Altern. Med.* **2011**, *2011*, 1–8. [CrossRef]

121. McAnulty, L.S.; Nieman, D.C.; Dumke, C.L.; Shooter, L.A.; Henson, D.A.; Utter, A.C.; Milne, G.; McAnulty, S.R. Effect of blueberry ingestion on natural killer cell counts, oxidative stress, and inflammation prior to and after 2.5 h of running. *Appl. Physiol. Nutr. Metab.* **2011**, *36*, 976–984. [CrossRef]

122. Johnson, M.H.; de Mejia, E.G.; Fan, J.; Lila, M.A.; Yousef, G.G. Anthocyanins and proanthocyanidins from blueberry-blackberry fermented beverages inhibit markers of inflammation in macrophages and carbohydrate-utilizing enzymes in vitro. *Mol. Nutr. Food Res.* **2013**, *57*, 1182–1197. [CrossRef]

123. Goyarzu, P.; Malin, D.H.; Lau, F.C.; Taglialatela, G.; Moon, W.D.; Jennings, R.; Moy, E.; Moy, D.; Lippold, S.; Shukitt-Hale, B.; et al. Blueberry supplemented diet: Effects on object recognition memory and nuclear factor-kappa B levels in aged rats. *Nutr. Neurosci.* **2004**, *7*, 75–83. [CrossRef]

124. Nour, V.; Trandafir, I.; Cosmulescu, S. Bioactive compounds, antioxidant activity and nutritional quality of different culinary aromatic herbs. *Not. Bot. Horti Agrobot. Cluj-Napoca* **2017**, *45*, 179–184. [CrossRef]

125. Fraternale, D.; Giamperi, L.; Bucchini, A.; Sestili, P.; Paolillo, M.; Ricci, D. Prunus spinosa fresh fruit juice: Antioxidant activity in cell-free and cellular systems. *Nat. Prod. Commun.* **2009**, *4*, 1665–1670. [CrossRef]

126. Coppari, S.; Colomba, M.; Fraternale, D.; Brinkmann, V.; Romeo, M.; Rocchi, M.B.L.; Di Giacomo, B.; Mari, M.; Guidi, L.; Ramakrishna, S.; et al. Antioxidant and anti-inflammaging ability of prune (*Prunus Spinosa* L.) extract result in improved wound healing efficacy. *Antioxidants* **2021**, *10*, 374. [CrossRef]

127. Hersh, C.P.; Campbell, E.J.; Scott, L.R.; Raby, B.A. Alpha-1 antitrypsin deficiency as an incidental finding in clinical genetic testing. *Am. J. Respir. Crit. Care Med.* **2019**, *199*, 246–248. [CrossRef] [PubMed]

128. Gramegna, A.; Aliberti, S.; Confalonieri, M.; Corsico, A.; Richeldi, L.; Vancheri, C.; Blasi, F. Alpha-1 antitrypsin deficiency as a common treatable mechanism in chronic respiratory disorders and for conditions different from pulmonary emphysema? A commentary on the new European Respiratory Society statement. *Multidiscip. Respir. Med.* **2018**, *13*, 39. [CrossRef] [PubMed]

129. Miyamoto, Y.; Akaike, T.; Alam, M.S.; Inoue, K.; Hamamoto, T.; Ikebe, N.; Yoshitake, J.; Okamoto, T.; Maeda, H. Novel functions of human α1-protease inhibitor after S-Nitrosylation: Inhibition of cysteine protease and antibacterial activity. *Biochem. Biophys. Res. Commun.* **2000**, *267*, 918–923. [CrossRef] [PubMed]

130. Perlmutter, D.H.; Punsal, P.I. Distinct and additive effects of elastase and endotoxin on expression of alpha 1 proteinase inhibitor in mononuclear phagocytes. *J. Biol. Chem.* **1988**, *263*, 16499–16503. [CrossRef]

131. Petrache, I.; Fijalkowska, I.; Zhen, L.; Medler, T.R.; Brown, E.; Cruz, P.; Choe, K.-H.; Taraseviciene-Stewart, L.; Scerbavicius, R.; Shapiro, L.; et al. A novel antiapoptotic role for α 1 -antitrypsin in the prevention of pulmonary emphysema. *Am. J. Respir. Crit. Care Med.* **2006**, *173*, 1222–1228. [CrossRef]

132. Wang, J.; Sun, Z.; Gou, W.; Adams, D.B.; Cui, W.; Morgan, K.A.; Strange, C.; Wang, H. α-1 Antitrypsin enhances islet engraftment by suppression of instant blood-mediated inflammatory reaction. *Diabetes* **2017**, *66*, 970–980. [CrossRef] [PubMed]

133. Yuan, Y.; DiCiaccio, B.; Li, Y.; Elshikha, A.S.; Titov, D.; Brenner, B.; Seifer, L.; Pan, H.; Karic, N.; Akbar, M.A.; et al. Anti-inflammaging effects of human alpha-1 antitrypsin. *Aging Cell* **2018**, *17*, e12694. [CrossRef] [PubMed]

134. Sasakura, H.; Moribe, H.; Nakano, M.; Ikemoto, K.; Takeuchi, K.; Mori, I. Lifespan extension by peroxidase/dual oxidase-mediated ROS signaling through pyrroloquinoline quinone in C. elegans. *J. Cell Sci.* **2017**, *130*, 2631–2643. [CrossRef] [PubMed]

135. Harris, C.B.; Chowanadisai, W.; Mishchuk, D.O.; Satre, M.A.; Slupsky, C.M.; Rucker, R.B. Dietary pyrroloquinoline quinone (PQQ) alters indicators of inflammation and mitochondrial-related metabolism in human subjects. *J. Nutr. Biochem.* **2013**, *24*, 2076–2084. [CrossRef]

136. Zhang, J.; Meruvu, S.; Bedi, Y.S.; Chau, J.; Arguelles, A.; Rucker, R.; Choudhury, M. Pyrroloquinoline quinone increases the expression and activity of Sirt1 and -3 genes in HepG2 cells. *Nutr. Res.* **2015**, *35*, 844–849. [CrossRef]

137. Zhou, X.; Yao, Z.; Peng, Y.; Mao, S.; Xu, D.; Qin, X.; Zhang, R. PQQ ameliorates D-galactose induced cognitive impairments by reducing glutamate neurotoxicity via the GSK-3β/Akt signaling pathway in mouse. *Sci. Rep.* **2018**, *8*, 8894. [CrossRef]

138. Hao, J.; Ni, X.; Giunta, S.; Wu, J.; Shuang, X.; Xu, K.; Li, R.; Zhang, W.; Xia, S. Pyrroloquinoline quinone delays inflammaging induced by TNF-α through the p16/p21 and Jagged1 signalling pathways. *Clin. Exp. Pharmacol. Physiol.* **2020**, *47*, 102–110. [CrossRef] [PubMed]

139. Cardinali, D.P.; Hardeland, R. Inflammaging, metabolic syndrome and melatonin: A call for treatment studies. *Neuroendocrinology* **2017**, *104*, 382–397. [CrossRef] [PubMed]

140. Hardeland, R. Melatonin and the theories of aging: A critical appraisal of melatonin's role in antiaging mechanisms. *J. Pineal Res.* **2013**, *55*, 325–356. [CrossRef]
141. Kireev, R.A.; Tresguerres, A.C.F.; Garcia, C.; Ariznavarreta, C.; Vara, E.; Tresguerres, J.A.F. Melatonin is able to prevent the liver of old castrated female rats from oxidative and pro-inflammatory damage. *J. Pineal Res.* **2008**, *45*, 394–402. [CrossRef] [PubMed]
142. Cuesta, S.; Kireev, R.; Forman, K.; García, C.; Escames, G.; Ariznavarreta, C.; Vara, E.; Tresguerres, J.A.F. Melatonin improves inflammation processes in liver of senescence-accelerated prone male mice (SAMP8). *Exp. Gerontol.* **2010**, *45*, 950–956. [CrossRef] [PubMed]
143. Cuesta, S.; Kireev, R.; García, C.; Forman, K.; Escames, G.; Vara, E.; Tresguerres, J.A.F. Beneficial effect of melatonin treatment on inflammation, apoptosis and oxidative stress on pancreas of a senescence accelerated mice model. *Mech. Ageing Dev.* **2011**, *132*, 573–582. [CrossRef] [PubMed]
144. Forman, K.; Vara, E.; Garcia, C.; Kireev, R.; Cuesta, S.; Escames, G.; Tresguerres, J.A.F. Effect of a combined treatment with growth hormone and melatonin in the cardiological aging on male SAMP8 mice. *J. Gerontol. Ser. A Biol. Sci. Med Sci.* **2011**, *66A*, 823–834. [CrossRef] [PubMed]
145. Zittermann, A. Vitamin D in preventive medicine: Are we ignoring the evidence? *Br. J. Nutr.* **2003**, *89*, 552–572. [CrossRef]
146. Wang, Q.; He, Y.; Shen, Y.; Zhang, Q.; Chen, D.; Zuo, C.; Qin, J.; Wang, H.; Wang, J.; Yu, Y. Vitamin D inhibits COX-2 expression and inflammatory response by targeting thioesterase superfamily member 4. *J. Biol. Chem.* **2014**, *289*, 11681–11694. [CrossRef]
147. Almerighi, C.; Sinistro, A.; Cavazza, A.; Ciaprini, C.; Rocchi, G.; Bergamini, A. 1α,25-Dihydroxyvitamin D3 inhibits CD40L-induced pro-inflammatory and immunomodulatory activity in Human Monocytes. *Cytokine* **2009**, *45*, 190–197. [CrossRef]
148. Grossmann, R.E.; Zughaier, S.M.; Liu, S.; Lyles, R.H.; Tangpricha, V. Impact of vitamin D supplementation on markers of inflammation in adults with cystic fibrosis hospitalized for a pulmonary exacerbation. *Eur. J. Clin. Nutr.* **2012**, *66*, 1072–1074. [CrossRef]
149. Tabesh, M.; Azadbakht, L.; Faghihimani, E.; Tabesh, M.; Esmaillzadeh, A. Calcium-vitamin D cosupplementation influences circulating inflammatory biomarkers and adipocytokines in vitamin D-insufficient diabetics: A randomized controlled clinical trial. *J. Clin. Endocrinol. Metab.* **2014**, *99*, E2485–E2493. [CrossRef]
150. Neyestani, T.R.; Nikooyeh, B.; Alavi-Majd, H.; Shariatzadeh, N.; Kalayi, A.; Tayebinejad, N.; Heravifard, S.; Salekzamani, S.; Zahedirad, M. Improvement of vitamin D status via daily intake of fortified yogurt drink either with or without extra calcium ameliorates systemic inflammatory biomarkers, including adipokines, in the subjects with type 2 diabetes. *J. Clin. Endocrinol. Metab.* **2012**, *97*, 2005–2011. [CrossRef]
151. Wang, Q.; Zhou, X.; Zhang, P.; Zhao, P.; Nie, L.; Ji, N.; Ding, Y.; Wang, Q. 25-Hydroxyvitamin D3 positively regulates periodontal inflammaging via SOCS3/STAT signaling in diabetic mice. *Steroids* **2020**, *156*, 108570. [CrossRef]
152. Yasukawa, H.; Nagata, T.; Oba, T.; Imaizumi, T. SOCS3: A novel therapeutic target for cardioprotection. *JAK-STAT* **2012**, *1*, 234–240. [CrossRef] [PubMed]
153. Peralta, S.; Carrascosa, J.; Gallardo, N.; Ros, M.; Arribas, C. Ageing increases SOCS-3 expression in rat hypothalamus: Effects of food restriction. *Biochem. Biophys. Res. Commun.* **2002**, *296*, 425–428. [CrossRef]
154. Li, L.; Chen, B.; An, T.; Zhang, H.; Xia, B.; Li, R.; Zhu, R.; Tian, Y.; Wang, L.; Zhao, D.; et al. BaZiBuShen alleviates altered testicular morphology and spermatogenesis and modulates Sirt6/P53 and Sirt6/NF-κB pathways in aging mice induced by D-galactose and NaNO$_2$. *J. Ethnopharmacol.* **2021**, *271*, 113810. [CrossRef] [PubMed]
155. Mahmoudi, M.; Ebrahimzadeh, M.A.; Pourmorad, F.; Rezaie, N.; Mahmoudi, M.A. Anti-inflammatory and analgesic effects of egg yolk: A comparison between organic and machine made. *Eur. Rev. Med Pharmacol. Sci.* **2013**, *17*, 472–476.
156. Cunill, J.; Babot, C.; Santos, L.; Serrano, J.C.E.; Jové, M.; Martin-Garí, M.; Portero-Otín, M. In vivo anti-inflammatory effects and related mechanisms of processed egg yolk, a potential anti-inflammaging dietary supplement. *Nutrients* **2020**, *12*, 2699. [CrossRef]
157. Cameron, A.R.; Morrison, V.L.; Levin, D.; Mohan, M.; Forteath, C.; Beall, C.; McNeilly, A.D.; Balfour, D.J.K.; Savinko, T.; Wong, A.K.F.; et al. Anti-inflammatory effects of metformin irrespective of diabetes status. *Circ. Res.* **2016**, *119*, 652–665. [CrossRef]
158. Malínská, H.; Oliyarnyk, O.; Škop, V.; Šilhavý, J.; Landa, V.; Zídek, V.; Mlejnek, P.; Šimáková, M.; Strnad, H.; Kazdová, L.; et al. Effects of metformin on tissue oxidative and dicarbonyl stress in transgenic spontaneously hypertensive rats expressing human C-reactive protein. *PLoS ONE* **2016**, *11*, e0150924. [CrossRef]
159. Barzilai, N.; Crandall, J.P.; Kritchevsky, S.B.; Espeland, M.A. Metformin as a tool to target aging. *Cell Metab.* **2016**, *23*, 1060–1065. [CrossRef]
160. Bharath, L.P.; Agrawal, M.; McCambridge, G.; Nicholas, D.A.; Hasturk, H.; Liu, J.; Jiang, K.; Liu, R.; Guo, Z.; Deeney, J.; et al. Metformin enhances autophagy and normalizes mitochondrial function to alleviate aging-associated inflammation. *Cell Metab.* **2020**, *32*, 44–55. [CrossRef]
161. Batandier, C.; Guigas, B.; Detaille, D.; El-Mir, M.; Fontaine, E.; Rigoulet, M.; Leverve, X.M. The ROS production induced by a reverse-electron flux at respiratory-chain complex 1 is hampered by metformin. *J. Bioenerg. Biomembr.* **2006**, *38*, 33–42. [CrossRef] [PubMed]
162. Moiseeva, O.; Deschênes-Simard, X.; St-Germain, E.; Igelmann, S.; Huot, G.; Cadar, A.E.; Bourdeau, V.; Pollak, M.N.; Ferbeyre, G. Metformin inhibits the senescence-associated secretory phenotype by interfering with IKK/NF-κ B activation. *Aging Cell* **2013**, *12*, 489–498. [CrossRef] [PubMed]

163. Algire, C.; Moiseeva, O.; Deschênes-Simard, X.; Amrein, L.; Petruccelli, L.; Birman, E.; Viollet, B.; Ferbeyre, G.; Pollak, M.N. Metformin reduces endogenous reactive oxygen species and associated DNA damage. *Cancer Prev. Res.* **2012**, *5*, 536–543. [CrossRef] [PubMed]
164. Maret, W.; Sandstead, H.H. Zinc requirements and the risks and benefits of zinc supplementation. *J. Trace Elem. Med. Biol.* **2006**, *20*, 3–18. [CrossRef]
165. Von Bülow, V.; Dubben, S.; Engelhardt, G.; Hebel, S.; Plümäkers, B.; Heine, H.; Rink, L.; Haase, H. Zinc-dependent suppression of TNF-α production is mediated by protein kinase a-induced inhibition of Raf-1, IκB Kinase β, and NF-κB. *J. Immunol.* **2007**, *179*, 4180–4186. [CrossRef]
166. Bao, B.; Prasad, A.S.; Beck, F.W.; Fitzgerald, J.T.; Snell, D.; Bao, G.W.; Singh, T.; Cardozo, L.J. Zinc decreases C-reactive protein, lipid peroxidation, and inflammatory cytokines in elderly subjects: A potential implication of zinc as an atheroprotective agent. *Am. J. Clin. Nutr.* **2010**, *91*, 1634–1641. [CrossRef]
167. Jung, S.; Kim, M.K.; Choi, B.Y. The relationship between zinc status and inflammatory marker levels in rural korean adults aged 40 and older. *PLoS ONE* **2015**, *10*, e0130016. [CrossRef]
168. De Paula, R.C.S.; Aneni, E.C.; Costa, A.P.R.; Figueiredo, V.N.; Moura, F.A.; Freitas, W.M.; Quaglia, L.A.; Santos, S.N.; Soares, A.A.; Nadruz, W.; et al. Low zinc levels is associated with increased inflammatory activity but not with atherosclerosis, arteriosclerosis or endothelial dysfunction among the very elderly. *BBA Clin.* **2014**, *2*, 1–6. [CrossRef]
169. Maramaldi, G.; Togni, S.; Franceschi, F.; Lati, E. Anti-inflammaging and antiglycation activity of a novel botanical ingredient from African biodiversity (Centevita™). *Clin. Cosmet. Investig. Dermatol.* **2014**, *7*, 1. [CrossRef]
170. Draganidis, D.; Karagounis, L.G.; Athanailidis, I.; Chatzinikolaou, A.; Jamurtas, A.Z.; Fatouros, I.G. Inflammaging and skeletal muscle: Can protein intake make a difference? *J. Nutr.* **2016**, *146*, 1940–1952. [CrossRef]
171. Holmer-Jensen, J.; Karhu, T.; Mortensen, L.S.; Pedersen, S.B.; Herzig, K.-H.; Hermansen, K. Differential effects of dietary protein sources on postprandial low-grade inflammation after a single high fat meal in obese non-diabetic subjects. *Nutr. J.* **2011**, *10*, 115. [CrossRef] [PubMed]
172. Burris, R.L.; Ng, H.-P.; Nagarajan, S. Soy protein inhibits inflammation-induced VCAM-1 and inflammatory cytokine induction by inhibiting the NF-κB and AKT signaling pathway in apolipoprotein E–deficient mice. *Eur. J. Nutr.* **2014**, *53*, 135–148. [CrossRef] [PubMed]
173. Messina, S.; Bitto, A.; Aguennouz, M.; Vita, G.L.; Polito, F.; Irrera, N.; Altavilla, D.; Marini, H.; Migliorato, A.; Squadrito, F.; et al. The soy isoflavone genistein blunts nuclear factor kappa-B, MAPKs and TNF-α activation and ameliorates muscle function and morphology in mdx mice. *Neuromuscul. Disord.* **2011**, *21*, 579–589. [CrossRef]
174. Mangano, K.M.; Hutchins-Wiese, H.L.; Kenny, A.M.; Walsh, S.J.; Abourizk, R.H.; Bruno, R.S.; Lipcius, R.; Fall, P.; Kleppinger, A.; Kenyon-Pesce, L.; et al. Soy proteins and isoflavones reduce interleukin-6 but not serum lipids in older women: A randomized controlled trial. *Nutr. Res.* **2013**, *33*, 1026–1033. [CrossRef]
175. Moughan, P.J.; Rutherfurd, S.M.; Montoya, C.A.; Dave, L.A. Food-derived bioactive peptides—A new paradigm. *Nutr. Res. Rev.* **2014**, *27*, 16–20. [CrossRef] [PubMed]
176. Martucci, M.; Ostan, R.; Biondi, F.; Bellavista, E.; Fabbri, C.; Bertarelli, C.; Salvioli, S.; Capri, M.; Franceschi, C.; Santoro, A. Mediterranean diet and inflammaging within the hormesis paradigm. *Nutr. Rev.* **2017**, *75*, 442–455. [CrossRef]
177. Shannon, O.M.; Ashor, A.W.; Scialo, F.; Saretzki, G.; Martin-Ruiz, C.; Lara, J.; Matu, J.; Griffiths, A.; Robinson, N.; Lillà, L.; et al. Mediterranean diet and the hallmarks of ageing. *Eur. J. Clin. Nutr.* **2021**, *75*, 1176–1192. [CrossRef]
178. Kennedy, K.; Keogh, B.; Lopez, C.; Adelfio, A.; Molloy, B.; Kerr, A.; Wall, A.M.; Jalowicki, G.; Holton, T.A.; Khaldi, N. An artificial intelligence characterised functional ingredient, derived from rice, inhibits TNF-α and significantly improves physical strength in an inflammaging population. *Foods* **2020**, *9*, 1147. [CrossRef]
179. Palungwachira, P.; Tancharoen, S.; Phruksaniyom, C.; Klungsaeng, S.; Srichan, R.; Kikuchi, K.; Nararatwanchai, T. Antioxidant and anti-inflammatory properties of anthocyanins extracted from Oryza sativa L. in primary dermal fibroblasts. *Oxidative Med. Cell. Longev.* **2019**, *2019*, 1–18. [CrossRef]
180. Maijo, M.; Ivory, K.; Clements, S.J.; Dainty, J.R.; Jennings, A.; Gillings, R.; Fairweather-Tait, S.; Gulisano, M.; Santoro, A.; Franceschi, C.; et al. One-year consumption of a mediterranean-like dietary pattern with vitamin D3 supplements induced small scale but extensive changes of immune cell phenotype, co-receptor expression and innate immune responses in healthy elderly subjects: Results from the United Kingdom arm of the NU-AGE trial. *Front. Physiol.* **2018**, *9*, 997. [CrossRef]
181. Wittkowski, K.M. Use of Cyclodextrins in Diseases and Disorders Involving Phospholipid Dysregulation. U.S. Patent Application No. 16/647,476, 28 September 2018.
182. Song, W.; Wang, F.; Lotfi, P.; Sardiello, M.; Segatori, L. 2-hydroxypropyl-β-cyclodextrin promotes transcription factor EB-mediated activation of autophagy. *J. Biol. Chem.* **2014**, *289*, 10211–10222. [CrossRef] [PubMed]
183. Plaza-Zabala, A.; Sierra-Torre, V.; Sierra, A. Autophagy and microglia: Novel partners in neurodegeneration and aging. *Int. J. Mol. Sci.* **2017**, *18*, 598. [CrossRef] [PubMed]
184. Forbes, I.; Forbes, W. Taltirelin Use. International Patent Application No. PCT/GB2020/051315, 29 May 2020.
185. Verlinden, S.F.F. Chalcones and Derivatives for Use in Medicaments and Nutraceuticals. U.S. Patent Application No. 17/261,129, 24 July 2019.
186. Arvand, H.; Oswald, W.; Csorsz, N.; Rodrig, M. Use of Sedoheptulose for Prevention or Treatment of Inflammation. U.S. Patent Application No. 15/622,235, 26 May 2020.

187. Franchi, J.; Pecher, V. Aqueous Extract of Rose Fruits as a Skin Neuro-Protective Agent. International Patent Application No. FR1873710A, 20 December 2020.
188. Franchi, J.; Pecher, V.; Juan, M. Rosewood Extract. International Patent Application No. EP19816335.4A, 10 December 2019.
189. Lee, S.K.; Park, H.J.; Moon, D.H.; Park, S.Y.; Kim, G.Y.; Lee, K.H.; Shin, S.S. Cosmetic and Functional Food Composition Comprising Caffeoylmalic Acid Having Whitening and Anti-Photoaging Activities. International Patent Application No. KR1020200089496A, 20 July 2020.
190. Jeong, S.G.; Kyung, S.Y.; Yoon, S.K.; Yeo, H.J.; Kang, S.H.; Park, M.S. Cosmetic Composition for Anti-Aging and Anti-Inflammatory Activities Comprising Extracts Mixture of Juglans Nigra, Sophora Japonica, and Pinus Densiflora. International Patent Application No. KR1020180105141A, 25 May 2020.

Article

Identification and Functional Evaluation of Polyphenols That Induce Regulatory T Cells

Tsukasa Fujiki [1], Ryosuke Shinozaki [2], Miyako Udono [3] and Yoshinori Katakura [2,3,*]

[1] Faculty of Pharmaceutical Sciences, Nagasaki International University, 2825-7 Huis Ten Bosch, Sasebo, Nagasaki 859-3298, Japan; fujiki@niu.ac.jp

[2] Graduate School of Bioresources and Bioenvironmental Sciences, Kyushu University, 744 Motooka, Nishi-ku, Fukuoka 819-0395, Japan; shinozaki@s.kyushu-u.ac.jp

[3] Faculty of Agriculture, Kyushu University, 744 Motooka, Nishi-ku, Fukuoka 819-0395, Japan; mudono@grt.kyushu-u.ac.jp

* Correspondence: katakura@grt.kyushu-u.ac.jp; Tel./Fax: +81-92-802-4727

Abstract: Regulatory T cells (Tregs) and $CD4^+/CD25^+$ T cells play an important role in the suppression of excessive immune responses, homeostasis of immune function, and oral tolerance. In this study, we screened for food-derived polyphenols that induce Tregs in response to retinaldehyde dehydrogenase (*RALDH2*) activation using macrophage-like THP-1 cells. THP-1 cells were transfected with an EGFP reporter vector whose expression is regulated under the control of mouse *Raldh2* promoter and named THP-1 (Raldh2p-EGFP) cells. The THP-1 (Raldh2p-EGFP) cells were treated with 33 polyphenols after inducing their differentiation into macrophage-like cells using phorbol 12-myristate 13-acetate. Of the 33 polyphenols, five (kaempferol, quercetin, morin, luteolin and fisetin) activated *Raldh2* promoter activity, and both quercetin and luteolin activated the endogenous *Raldh2* mRNA expression and enzymatic activity. Furthermore, these two polyphenols increased transforming growth factor beta 1 and forkhead box P3 mRNA expression, suggesting that they have Treg-inducing ability. Finally, we verified that these polyphenols could induce Tregs in vivo and consequently induce IgA production. Oral administration of quercetin and luteolin increased IgA production in feces of mice. Therefore, quercetin and luteolin can induce Tregs via *RALDH2* activation and consequently increase IgA production, suggesting that they can enhance intestinal barrier function.

Keywords: regulatory T cells; retinaldehyde dehydrogenase; IgA; quercetin; luteolin

Citation: Fujiki, T.; Shinozaki, R.; Udono, M.; Katakura, Y. Identification and Functional Evaluation of Polyphenols That Induce Regulatory T Cells. *Nutrients* **2022**, *14*, 2862. https://doi.org/10.3390/nu14142862

Academic Editor: Maria Annunziata Carluccio

Received: 23 June 2022
Accepted: 11 July 2022
Published: 13 July 2022

Publisher's Note: MDPI stays neutral with regard to jurisdictional claims in published maps and institutional affiliations.

Copyright: © 2022 by the authors. Licensee MDPI, Basel, Switzerland. This article is an open access article distributed under the terms and conditions of the Creative Commons Attribution (CC BY) license (https://creativecommons.org/licenses/by/4.0/).

1. Introduction

Regulatory T cells (Tregs) and $CD4^+/CD25^+$ T cells play an important role in suppressing excessive immune responses, maintaining homeostasis of immune function [1], and further regulating oral immune tolerance [2]. There are two types of Tregs: naturally occurring Tregs, which are directly differentiated from undifferentiated cells in the thymus, and induced Tregs (iTregs), which are differentiated from naive $CD4^+$ T cells upon antigen stimulation in peripheral tissues, such as the intestinal tract. iTregs are considered important in the regulation of antigen-specific immune responses in the periphery. Retinoic acid (RA) has been found to be involved in the induction of Treg and Th17 cell differentiation and in the regulation of immune cell differentiation and function [3]. Additionally, RA promotes the differentiation of forkhead box P3 $FOXP3^+$ iTregs and inhibits the differentiation of Th17 cells in a transforming growth factor beta (TGF)-β-dependent manner [3].

After Vitamin A is converted from retinyl ester into retinol in the liver, it is released into the bloodstream, where it binds to retinol-binding proteins and circulates in the body. Retinaldehyde dehydrogenase (RALDH) catalyzes the conversion of retinol into RA. RALDH1 to RALDH3 exist as isoforms of RALDH; dendritic cells in intestine-related tissues mainly express *RALDH2* [4]. Intestinal dendritic cells and mucosal intrinsic layer macrophages produce RA in a RALDH2-dependent manner, and activation of the *RALDH2* gene in these cells plays an important role in Treg induction [5,6]. At present, no studies have evaluated Treg induction via *RALDH2* activation mediated by food components. Therefore, we aimed to identify polyphenols that activate *RALDH2* expression and further evaluate their function in vivo. We found that quercetin and luteolin can induce Tregs via *RALDH2* activation and consequently increase IgA production, suggesting that they can enhance intestinal barrier function.

2. Materials and Methods

2.1. Cell Culture and Reagents

THP-1 cells of human acute monocytic leukemia were cultured in RPMI 1640 medium (Nissui Pharmaceutical, Tokyo, Japan) supplemented with 10% fetal bovine serum (FBS; Life Technologies, Gaithersburg, MD, USA) at 37 °C and 5% CO_2. All polyphenols were purchased from Fujifilm Wako Pure Chemical (Osaka, Japan). All polyphenols were dissolved in dimethyl sulfoxide (DMSO) at the concentration of 10 mM. These polyphenol stocks were diluted 1000-fold and added to the cells. DMSO was used as a control.

2.2. Establishment of a Reporter System to Screen for Polyphenols That Activate the Raldh2 Promoter

Polymerase chain reaction (PCR) was performed using the primers 5′-ATTAATAACTG ACTTACCAGCTCGT-3′ and 5′-GCTAGCGGCGATCTCGCTGGAAGTCA-3′ and mouse genomic DNA to clone the mouse *Raldh2* promoter. After replacing the CMV promoter of the EGFP-C3 vector with the *Raldh2* promoter, the vector was transfected into THP-1 cells. Transfected cells were selected using 800 μg/mL of G418 (Fujifilm Wako Pure Chemical, Osaka, Japan) to establish a stable cell line of THP-1 (Raldh2p-EGFP cells).

2.3. Screening of Polyphenols That Activate Raldh2 Promoter via IN Cell Analyzer 2200

Effects of the polyphenols on *Raldh2* promoter activity in differentiated THP-1 cells were evaluated by monitoring changes in enhanced green fluorescent protein (EGFP) fluorescence derived from THP-1 (Raldh2p–EGFP) cells using the IN Cell Analyzer 2200 (Cytiva, Tokyo, Japan). THP-1 (Raldh2p–EGFP) cells were seeded in 96-well blackplates (Greiner Bio-one, Tokyo, Japan) at a density of 6×10^5 cells/mL, treated with 100 ng/mL phorbol 12-myristate 13-acetate (PMA), and cultured for 48 h. After culturing, polyphenols were directly added to the cells at the final concentration of 10 μM and cells were further cultured for 24 h. Cells were then fixed with 4% formaldehyde for 15 min at room temperature. After washing the cells with phosphate-buffered saline (PBS), the cells were stained with 1 μg/mL Hoechst 33,342 solution (Dojindo, Kumamoto, Japan) for 20 min. The relative EGFP fluorescence intensity per cell was measured using IN Cell Analyzer 2200.

2.4. Quantitative Reverse Transcription-PCR (RT-qPCR)

THP-1 cells were seeded in 60 mm dishes at a density of 6×10^5 cells/mL, induced to differentiate via addition of 100 ng/mL PMA, and then subsequently cultured for 48 h at 37 °C. Cells were then cultured in the presence of 10 μM of polyphenol for 48 h. RNA was isolated using High Pure RNA Isolation kit (Roche Diagnostics GmbH, Mannheim, Germany), and cDNA was prepared using ReverTra Ace qPCR RT Master Mix (Toyobo, Osaka, Japan) according to the manufacturer's instructions. RT-qPCR was performed using Thunderbird SYBR qPCR mix (Toyobo) and Thermal Cycler Dice Real Time System TP-800 (TaKaRa Bio, Shiga, Japan). The samples were analyzed in triplicate, and gene expression levels were normalized to the corresponding β-actin levels. The PCR primer sequences used were as follows: human β-actin: forward primer 5′-TGGCACCCAGCACAATGAA-3′ and reverse primer 5′-CTAAGTCATAGTCCGCCTAGAAGC-3′; *Raldh2*, forward primer 5′-GCAATGCAAGCTGGGACTGT-3′ and reverse primer 5′-CCCGCAAGCCAAATTCTCCC-3′; *TGFB1*, forward primer 5′-AACCGGCCTTTCCTGCTTCT-3′ and reverse primer 5′-ACGCAGCAGTTCTTCTCCGT-3′; *FOXP3*, forward primer 5′-AGTGGCCCGGATGTGAGAAG-3′ and reverse primer 5′-ACATTGTGCCCTGCCCTTCT-3′.

2.5. Flow Cytometry

THP-1 cells were differentiated using 100 ng/mL PMA and then treated with 10 μM quercetin or 10 μM luteolin for 24 h. The ALDEFLUOR reagent system (Stemcell Technologies, Cambridge, MA, USA) was used to monitor cellular aldehyde dehydrogenase activity using a CytoFlex flow cytometer (Beckman Coulter, Miami, FL, USA).

2.6. Preparation of Human Peripheral Blood Mononuclear Cells (PBMCs)

PBMCs were isolated from collected human peripheral blood using Leucosep (Greiner Bio-one). Cells were washed with PBS, seeded into 5 mL dishes at a cell density of 1.0×10^6 cells/mL, and cultured in RPMI 1640 medium containing 10% FBS for 24 h. On the next day, cells were seeded into 24-well plates at a cell density of 1.0×10^6 cells/mL and cultured in the presence of 10 μM polyphenol for 24 h. RNA preparation, cDNA synthesis and qRT-PCR were performed according to the methods described in Section 2.4.

2.7. Animal Experiments

Seven-week-old male BALB/c mice (Japan SLC Co., Ltd., Shizuoka, Japan) were assigned to six groups ($n = 6$) and orally administered with luteolin and quercetin at 0.2 and 2 mg/kg body weight, respectively. Mice were fed food and water ad libitum, and oral administration was performed once a day at 10:00 AM. Mice were housed individually for 7 d in a 12 h:12 h light/dark cycle at 23 °C and 60% humidity. Feces were collected daily before oral administration. The collected feces were weighed, suspended in PBS containing protease inhibitor cocktail, and centrifuged, and the supernatant was collected and stored at −85 °C. All animal experiments were conducted in accordance with the "Guidelines for the Handling and Use of Animals" of Nagasaki International University.

2.8. Measurement of Fecal IgA Content via Enzyme-Linked Immunosorbent Assay (ELISA)

The amount of IgA secreted into the intestinal tract of mice was measured via ELISA. Mouse fecal samples were dissolved in PBS using cOmplete Mini protease inhibitor cocktail (Roche Diagnostics GmbH). A microtiter plate (Nunc, Naperville, IL, USA) was coated with anti-mouse IgA antibody (eBioscience, Burlingame, CA, USA) diluted in 0.1 M sodium carbonate buffer (pH 9.6) and incubated at 37 °C for 2 h. The plate was washed three times with PBS containing 0.05% Tween 20 (PBS-T). The supernatant of the mouse fecal solution was serially diluted and added to the plate, which was incubated overnight at 4 °C. After washing three times with PBS-T, diluted horseradish peroxidase-conjugated goat anti-mouse IgA antibody (eBioscience) was added and the plate was incubated for 2 h at 37 °C. After washing five times with PBS-T, the 3,3′,5,5′-tetramethylbenzidine substrate (eBioscience) was added and the plate was incubated at room temperature for 15 min. The absorbance at 450 nm was measured using an ELISA plate reader.

3. Results

3.1. Screening for Polyphenols That Activate the Raldh2 Promoter

We screened for polyphenols that activate the *Raldh2* promoter in differentiated THP-1 (Raldh2p-GFP) cells. Changes in fluorescence intensity derived from THP-1 (Raldh2p-GFP) cells after adding polyphenols were monitored. Among the 33 polyphenols tested, five polyphenols including kaempferol, quercetin, morin, luteolin and fisetin were found to activate the *Raldh2* promoter (Figure 1A).

Next, the effects of these polyphenols on the expression of endogenous *RALDH2* in differentiated THP-1 cells were evaluated via RT-qPCR. Among the five polyphenols, quercetin and luteolin significantly increased the expression of endogenous *RALDH2* mRNA in THP-1 cells (Figure 1B).

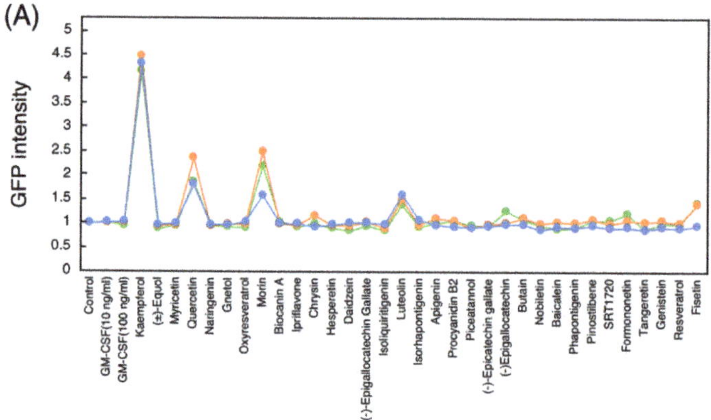

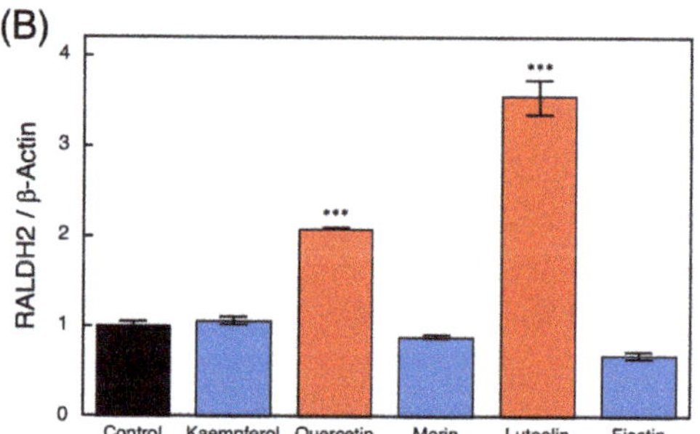

Figure 1. Screening for polyphenols that activate the *Raldh2* promoter in THP-1 cells. (**A**) THP-1 (Raldh2p-GFP) cells differentiated using phorbol 12-myristate 13-acetate were treated with 33 polyphenols (10 µM) and incubated for 24 h. Changes in EGFP fluorescence were monitored using IN Cell Analyzer 2200 (green, orange and blue show three independent experiments). (**B**) The effect of quercetin and luteolin on the expression of endogenous *RALDH2* mRNA in differentiated THP-1 cells was assessed via quantitative reverse transcription-polymerase chain reaction. Dimethyl sulfoxide (DMSO) was used as a control. Two-sided Student's *t*-test was used to test for significant differences compared to the results of the controls. Significance was defined as *** $p < 0.001$. Raldh2: retinaldehyde dehydrogenase 2; EGFP: enhanced green fluorescent protein.

3.2. Effects of the Raldh2-Affecting Polyphenols on THP-1 Cells

Next, we tested the effects of *RALDH2*-affecting polyphenols on THP-1 cells. First, changes in RALDH2 enzyme activity upon polyphenol treatment were evaluated using the ALDEFLUOR reagent system which can detect enzymatic activity of aldehyde dehydrogenase including RALDH2. Differentiated THP-1 cells were treated with 10 µM quercetin or luteolin and incubated for 24 h. We quantified the change in RALDH2 activity upon quercetin and luteolin treatment using the change in median fluorescence as an indicator of RALDH2 activity. The results showed that treatment with quercetin and luteolin increased the median fluorescence intensity by 25.8% and 15.4%, respectively, compared to the control, indicating that the respective polyphenols enhanced RALDH2 activity (Figure 2A,B).

Next, we determined the effects of the two polyphenols on the expression of TGF-β in THP-1 cells. TGF represents an important cytokine involved in Treg induction. Quercetin and luteolin treatment significantly affected *TGFB1* mRNA levels (Figure 2B). These results suggest that quercetin and luteolin may activate macrophages by affecting *RALDH2* expression and inducing TGF expression, thereby creating an environment in which Tregs can be further induced.

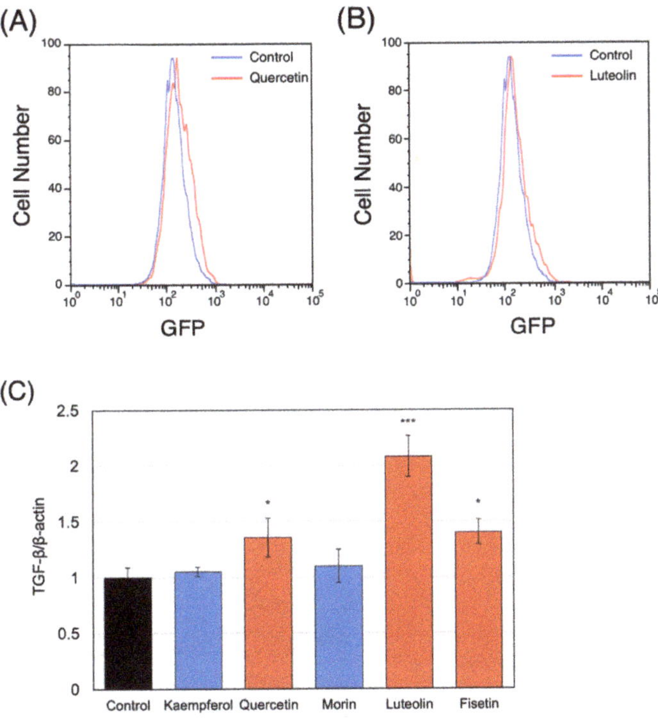

Figure 2. Quercetin and luteolin activated the endogenous RALDH enzymatic activity in THP-1 cells. Enzymatic activity of RALDH2 in differentiated THP-1 cells treated with (**A**) quercetin and (**B**) luteolin were measured using the ALDEFLUOR staining kit. (**C**) Quercetin and luteolin treatment increased the *TGFB1* mRNA expression level in the differentiated THP-1 cells. Two-sided Student's *t*-test was used to test for significant differences compared to results of the controls. Significance was defined as * $p < 0.05$, *** $p < 0.001$. RALDH2: retinaldehyde dehydrogenase 2; TGFB1: transforming growth factor beta 1.

3.3. Effect of RALDH2 Promoter-Activating Polyphenols on Treg Induction

We investigated whether macrophage-activating polyphenols can induce Tregs using human PBMCs. Potential Treg induction in PBMCs via polyphenol treatment was determined based on FOXP3 expression, a known marker of Treg cells, as an indicator. PBMCs were incubated with quercetin or luteolin for 24 h, and endogenous *FOXP3* expression level was determined via qRT-PCR. Quercetin and luteolin treatment significantly induced endogenous *FOXP3* in PBMCs (Figure 3), indicating that these polyphenols could induce Tregs in vivo.

3.4. Effect of Orally Administered Polyphenols on Treg Function In Vivo

Several studies have shown that Tregs contribute to IgA production in vivo. Therefore, we tested whether polyphenols, which could activate macrophages and induce Tregs in PBMCs, can enhance *Raldh2* expression and induce IgA production as a result of Treg induction in vivo. We tested the effect of polyphenols on IgA production by measuring the amount of IgA in the feces of mice orally administered with polyphenols. Compared to the control, quercetin and luteolin treatment significantly induced IgA production in vivo. In particular, quercetin treatment strongly induced IgA production even at low concentrations (Figure 4). Therefore, quercetin and luteolin can induce *Raldh2* expression in macrophages and induce Tregs, consequently inducing IgA production in vivo. However, as shown in

Figure 4, dose-dependent results could not be obtained with quercetin in particular. This may be due to differences in the in vivo environment, differences in local concentration-dependent availability, or polymerization, etc. We aim to verify this point in future studies.

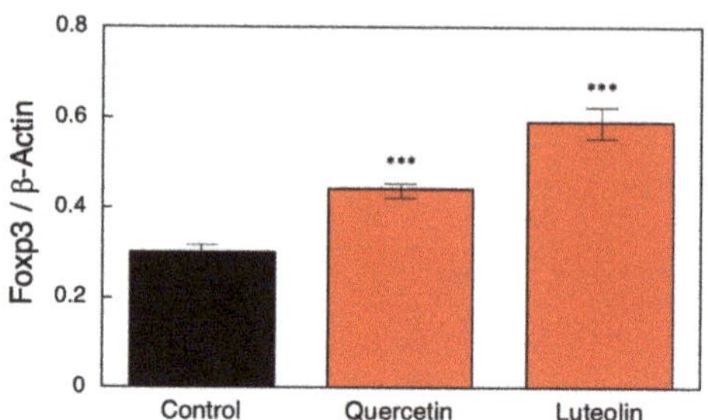

Figure 3. Quercetin and luteolin treatment affected *FOXP3* mRNA expression in PBMCs. *FOXP3* mRNA levels were determined via quantitative polymerase chain reaction in triplicate. Two-sided Student's *t*-test was used to test for significant differences compared to results of the controls. Significance was defined as *** $p < 0.001$. FOXP3: forkhead box P3; PBMCs: peripheral blood mononuclear cells.

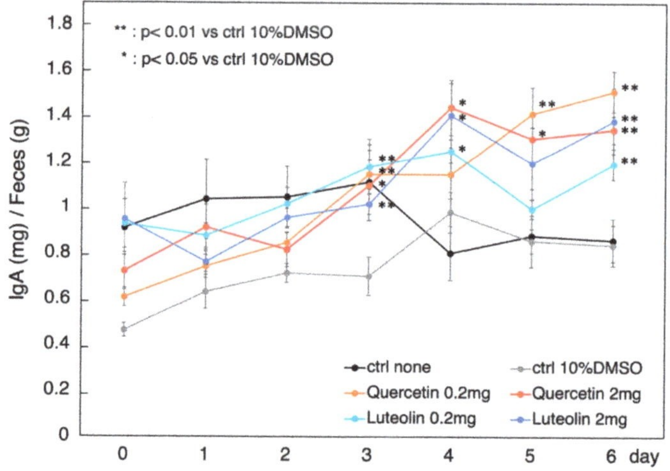

Figure 4. Effect of oral administration of *Raldh2*-activating polyphenols on IgA production in the intestines of BALB/c mice. Quercetin and luteolin were administered orally to BALB/c mice ($n = 6$), and total IgA was measured via enzyme-linked immunosorbent assays using mouse fecal samples. IgA expression is expressed as the mean ± standard error of mean mg/g fecal weight using a two-sided Student's *t*-test. Significant differences were tested in comparison to results of the controls. Significance was defined as * $p < 0.05$, ** $p < 0.01$. *Raldh2*: retinaldehyde dehydrogenase.

4. Discussion

Previous studies have shown that RA induces Treg differentiation and inhibits Th17 differentiation [1,3]. Therefore, this study focused on RALDH, which is known to be involved in RA synthesis, to identify polyphenols that activate the *RALDH2* gene and to clarify its function. THP-1 cells induced to differentiate into macrophage-like cells via PMA treatment were used as the cell line for tracking changes in *RALDH2* expression.

Screening revealed that five polyphenols (kaempferol, quercetin, morin, luteolin and fisetin) increased *Raldh2* promoter expression, two of which (quercetin and luteolin) enhanced endogenous *RALDH2* expression in differentiated THP-1 cells. The bioactivities of quercetin and luteolin have been reported, including inhibition of cholesterol absorption in the intestinal tract, strengthening of the intestinal barrier mediated by quercetin [7,8], and the antidepressant effect of luteolin [9]. Furthermore, these polyphenols have been reported to increase the number of Tregs, increase the production of Treg-related cytokines, and reduce arthritis via anti-inflammatory effects in a mouse model of rheumatoid arthritis [10]. In vitro, luteolin has been reported to exhibit anti-inflammatory effects in mouse models of enteritis and dextran sulfate sodium-induced colitis [11].

The function of quercetin and luteolin was evaluated in this study. RA is known to play an important role in IgA production by inducing B cell homing to the intrinsic layer of the small intestine and expression of $\alpha 4 \beta 7$ and C-C chemokine receptor 9 [12]. The two polyphenols identified in this study also enhanced *RALDH2* expression and induced Tregs, suggesting that oral administration of quercetin and luteolin to mice may increase IgA production in the intestine and suppress inflammation [13]. The IgA content in the feces of mice treated with these two polyphenols was increased, indicating that the polyphenols induced Tregs and enhanced IgA production in the intestinal tract as a result of *RALDH2* induction. Although other polyphenols, such as isoliquiritigenin and naringenin, exhibit Treg-inducing activity [1], we demonstrated that quercetin and luteolin induce Tregs and consequently induce intestinal IgA production as well as *RALDH2* enhancement in this study. These two polyphenols are thought to be responsible for the enhancement of barrier function and defense against infections via the enhancement of IgA production in the intestinal tract. The detailed molecular mechanisms of the enhancement of *RALDH2* expression by quercetin and luteolin should be clarified in the future. Furthermore, any additional functions of these polyphenols mediated via Treg induction should be elucidated.

Author Contributions: Conceptualization, Y.K., R.S. and T.F.; methodology, T.F., R.S. and M.U.; formal analysis, T.F. and R.S.; investigation, T.F. and R.S.; resources, T.F. and Y.K.; data curation, T.F., R.S. and Y.K.; writing—original draft preparation, T.F. and Y.K.; writing—review and editing, T.F. and Y.K.; visualization, T.F.; supervision, Y.K.; project administration, T.F. and Y.K. All authors have read and agreed to the published version of the manuscript.

Funding: This research received no external funding.

Institutional Review Board Statement: The study protocol was conducted in an appropriate manner in consideration of animal welfare and bioethics based on the Nagasaki International University Guidelines for Animal Experiments. The animal experiments conducted in this experiment have been approved by the Nagasaki International University Research Ethics Committee and the Animal Experiment Committee (Approval No. 156).

Informed Consent Statement: Not applicable.

Data Availability Statement: The data that support the findings of this study are available from the corresponding author, Y.K., upon reasonable request.

Conflicts of Interest: The authors declare no conflict of interest.

References

1. Guo, A.; He, D.; Xu, H.B.; Geng, C.A.; Zhao, J. Promotion of regulatory T cell induction by immunomodulatory herbal medicine licorice and its two constituents. *Sci. Rep.* **2015**, *5*, 14046. [CrossRef] [PubMed]
2. Pabst, O.; Mowat, A.M. Oral tolerance to food protein. *Mucosal Immun.* **2012**, *5*, 232–239. [CrossRef] [PubMed]
3. Mucida, D.; Park, Y.; Kim, G.; Turovskaya, O.; Scott, I.; Kronenberg, M.; Cheroutre, H. Reciprocal TH17 and regulatory T cell differentiation mediated by retinoic acid. *Science* **2007**, *317*, 256–260. [CrossRef] [PubMed]
4. Iwata, M.; Hirakiyama, A.; Eshima, Y.; Kagechika, H.; Kato, C.; Song, S.Y. Retinoic acid imprints gut-homing specificity on T cells. *Immunity* **2004**, *21*, 527–538. [CrossRef] [PubMed]
5. Iwata, M. Retinoic acid production by intestinal dendritic cells and its role in T-cell trafficking. *Semin. Immunol.* **2009**, *21*, 8–13. [CrossRef] [PubMed]
6. Denning, T.L.; Wang, Y.C.; Patel, S.R.; Williams, I.R.; Pulendran, B. Lamina propria macrophages and dendritic cells differentially induce regulatory and interleukin 17-producing T cell responses. *Nat. Immunol.* **2007**, *8*, 1086–1094. [CrossRef]

7. Nekohashi, M.; Ogawa, M.; Ogihara, T.; Nakazawa, K.; Kato, H.; Misaka, T.; Abe, K.; Kobayashi, S. Luteolin and quercetin affect the cholesterol absorption mediated by epithelial cholesterol transporter Niemann–Pick c1-like 1 in caco-2 cells and rats. *PLoS ONE* **2014**, *9*, e97901. [CrossRef] [PubMed]

8. Suzuki, T.; Hara, H. Quercetin enhances intestinal barrier function through the assembly of zonula [corrected] occludens-2, occludin, and claudin-1 and the expression of claudin-4 in Caco-2 cells. *J. Nutr.* **2009**, *139*, 965–974. [CrossRef] [PubMed]

9. De la Peña, J.B.; Kim, C.A.; Lee, H.L.; Yoon, S.Y.; Kim, H.J.; Hong, E.Y.; Kim, G.H.; Ryu, J.H.; Lee, Y.S.; Kim, K.M.; et al. Luteolin mediates the antidepressant-like effects of *Cirsium japonicum* in mice, possibly through modulation of the GABAA receptor. *Arch. Pharm. Res.* **2014**, *37*, 263–269. [CrossRef] [PubMed]

10. Yang, Y.; Zhang, X.; Xu, M.; Wu, X.; Zhao, F.; Zhao, C. Quercetin attenuates collagen-induced arthritis by restoration of Th17/Treg balance and activation of Heme Oxygenase 1-mediated anti-inflammatory effect. *Int. Immunopharmacol.* **2018**, *54*, 153–162. [CrossRef] [PubMed]

11. Nishitani, Y.; Yamamoto, K.; Yoshida, M.; Azuma, T.; Kanazawa, K.; Hashimoto, T.; Mizuno, M. Intestinal anti-inflammatory activity of luteolin: Role of the aglycone in NF-κB inactivation in macrophages co-cultured with intestinal epithelial cells. *Biofactors* **2013**, *39*, 522–533. [CrossRef] [PubMed]

12. Mora, J.R.; Iwata, M.; Eksteen, B.; Song, S.Y.; Junt, T.; Senman, B.; Otipoby, K.L.; Yokota, A.; Takeuchi, H.; Ricciardi-Castagnoli, P.; et al. Generation of gut-homing IgA-secreting B cells by intestinal dendritic cells. *Science* **2006**, *314*, 1157–1160. [CrossRef] [PubMed]

13. Orihara, K.; Narita, M.; Tobe, T.; Akasawa, A.; Ohya, Y.; Matsumoto, K.; Saito, H. Circulating Foxp^{3+}CD^{4+} cell numbers in atopic patients and healthy control subjects. *J. Allergy Clin. Immunol.* **2007**, *120*, 960–962. [CrossRef] [PubMed]

MDPI
St. Alban-Anlage 66
4052 Basel
Switzerland
Tel. +41 61 683 77 34
Fax +41 61 302 89 18
www.mdpi.com

Nutrients Editorial Office
E-mail: nutrients@mdpi.com
www.mdpi.com/journal/nutrients

www.ingramcontent.com/pod-product-compliance
Lightning Source LLC
LaVergne TN
LVHW070958170726
843515LV00005B/1020